BIOCHEMISTRY RESEARCH TRENDS

HANDBOOK OF RESEARCH ON NANOMATERIALS, NANOCHEMISTRY AND SMART MATERIALS

BIOCHEMISTRY RESEARCH TRENDS

Additional books in this series can be found on Nova's website under the Series tab.

Additional E-books in this series can be found on Nova's website under the E-book tab.

NANOTECHNOLOGY SCIENCE AND TECHNOLOGY

Additional books in this series can be found on Nova's website under the Series tab.

Additional E-books in this series can be found on Nova's website under the E-book tab.

HANDBOOK OF RESEARCH ON NANOMATERIALS, NANOCHEMISTRY AND SMART MATERIALS

A. K. HAGHI
AND
G. E. ZAIKOV
EDITORS

NOVA BIOMEDICAL

New York

NOTICE TO THE READER

The Publisher has taken reasonable care in the preparation of this book, but makes no expressed or implied warranty of any kind and assumes no responsibility for any errors or omissions. No liability is assumed for incidental or consequential damages in connection with or arising out of information contained in this book. The Publisher shall not be liable for any special, consequential, or exemplary damages resulting, in whole or in part, from the readers' use of, or reliance upon, this material. Any parts of this book based on government reports are so indicated and copyright is claimed for those parts to the extent applicable to compilations of such works.

Independent verification should be sought for any data, advice or recommendations contained in this book. In addition, no responsibility is assumed by the publisher for any injury and/or damage to persons or property arising from any methods, products, instructions, ideas or otherwise contained in this publication.

This publication is designed to provide accurate and authoritative information with regard to the subject matter covered herein. It is sold with the clear understanding that the Publisher is not engaged in rendering legal or any other professional services. If legal or any other expert assistance is required, the services of a competent person should be sought. FROM A DECLARATION OF PARTICIPANTS JOINTLY ADOPTED BY A COMMITTEE OF THE AMERICAN BAR ASSOCIATION AND A COMMITTEE OF PUBLISHERS.

Additional color graphics may be available in the e-book version of this book.

Library of Congress Cataloging-in-Publication Data

ISBN: 978-1-61942-525-5

Library of Congress Control Number: 2012947877

Published by Nova Science Publishers, Inc. † New York

Contents

Preface

The collection of topics in this book aims to reflect the diversity of recent advances in nanomaterials, nanochemistry and smart materials with a broad perspective that may be useful for scientists as well as for graduate students and engineers. The book offers scope for academics, researchers, and engineering professionals to present their research and development works that have potential for applications in several disciplines of engineering and science. Contributions range from new methods to novel applications of existing methods to gain understanding of the material and/or structural behavior of new and advanced systems. This book presents leading-edge research from around the world in these dynamic fields.

Chapter I - Clothing is a person's second skin, since it covers great parts of the body and has a large surface area in contact with the environment. Therefore, clothing is proper interface between environment and human body and could act as an ideal tool to enhance personal protection. Over the years, growing concern regarding health and safety of persons in various sectors, such as industries, hospitals, research institutions, battlefields and other hazardous conditions, has led to intensive research and development in field of personal protective clothing. Nowadays, there are different types of protective clothing. The simplest and most preliminary of this equipment is made from rubber or plastic that is completely impervious to hazardous substances, air and water vapor. Another approach to protective clothing is laminating activated carbon into multilayer fabric in order to absorb toxic vapors from environment and prevent penetration to the skin. The use of activated carbon is considered only a short-term solution because it loses its effectiveness upon exposure to sweat and moisture. The use of semi-permeable membranes as a constituent of the protective material is another approach. In this way, reactive chemical decontaminants encapsulates in microparticles or fills in microporous hollow fibers. The microparticle or fiber walls are permeable to toxic vapors but impermeable to decontaminants, so that the toxic agents diffuse selectively into them and neutralize. All of these equipments could trap such toxic pollutions but usually are impervious to air and water vapor and thus retain body heat. In other words, a negative relationship always exists between thermal comfort and protection performance for currently available protective clothing. For example, nonwoven fabrics with high air permeability exhibit low barrier performance, whereas microporous materials, laminated fabrics and tightly constructed wovens offer higher level of protection but lower air permeability. Thus there still exists a very real demand for improved protective clothing that can offer acceptable levels of impermeability to highly toxic pollutions of low molecular.

Chapter II - Recently, the words "nanobiocomposites" or "biopolymer nanocomposites" are most frequently observed in environmentally friendly research studies. The synthetic polymers have been widely used in a various application of nanocomposites. However, they become a major source of waste after use due to their poor biodegradability. On the other hand, most of the synthetic polymers are without biocompatibility in *vivo* and *vitro* environments. Hence, scientists were interested in biopolymers as biodegradable materials. Later, several groups of natural biopolymers such as polysaccharide, proteins, and nucleic acids were used in various applications. Nevertheless, the use of these materials has been limited due to relatively poor mechanical properties. Therefore, researcher efforts have been made to improve the properties of biopolymers as a matrix by using of reinforcement.

Chitosan (CHT) is a polysaccharide biopolymer that has been widely used as a matrix in nanobiocomposites. Chitosan represents high biocompatibility and biodegradability properties, although these biopolymers have an essential requirement to additional material with high mechanical properties. Following discovery of carbon nanotube, results of characterization represented unique electrical and mechanical properties. Thereby, many research studies have focused on improving the physical properties of biopolymer nanocomposites by using of the fundamental behavior of carbon nanotubes.

It is the aim of this review to summarize recent advances in the production of carbon nanotubes/chitosan nanocomposites by several methods. Specifically, the authors will discuss the authors' recent work in preparing CNTs/CHT nanofiber composites by using electrospinning method.

Chapter III - Nanotechnology has become in recent years a topic of great interest to scientists and engineers and is now established as prioritized research area in many countries. The reduction of the size to the nano-meter range brings an array of new possibilities in terms of material properties, in particular with respect to achievable surface-to-volume ratios. Electrospinning of nanofibers is a novel process for producing superfine fibers by forcing a solution through a spinnerette with an electric field. An emerging technology of manufacturing of thin natural fibers is based on the principle of electrospinning process. In conventional fiber spinning, the mechanical force is applied to the end of a jet, whereas in the electrospinning process, the electric body force act on element of charged fluid. Electrospinning has emerged as a specialized processing technique for the formation of sub-micron fibers (typically between 100 nm and 1 μ m in diameter), with high specific surface areas. Due to their high specific surface area, high porosity, and small pore size, the unique fibers have been suggested for wide range of applications. Electrospinning of nanofibers offers unique capabilities for producing novel natural nanofibers and fabrics with controllable pore structure.

Some 4-9% of cotton fiber is lost at textile mill in so-called opening and cleaning, which involves mechanically separating compressed clumps of fibers for removal of trapped debris. Another 1% is lost in drawing and roving-pulling lengths of fiber into longer and longer segments, which are then twisted together for strength. An average of 20% is lost during combing and yarn production. Typically, waste cotton is used in relatively low-value products such as cotton balls, yarn, and cotton batting. A new process for electrospinning waste cotton using a less harmful solvent has been developed.

Chapter IV - Natural polymers have potential pharmaceutical applications because of their low toxicity, biocompatibility, and excellent biodegradability. In recent years, biodegradable polymeric systems have gained importance for design of surgical devices, artificial organs, drug delivery systems with different routes of administration, carriers of immobilized enzymes and cells, biosensors, ocular inserts, and materials for orthopedic applications. These polymers are classified as either synthetic (polyesters, polyamides, polyanhydrides) or natural (polyamino acids, polysaccharides). Polysaccharide-based polymers represent a major class of biomaterials, which includes agarose, alginate, carageenan, dextran, and chitosan. Chitosan [_(1,4)2- amino-2-d-glucose] is a cationic biopolymer produced by alkaline N-deacetylation of chitin, which is the main component of the shells of crab, shrimp, and krill. Chitosan is a functional linear polymer derived from chitin, the most abundant natural polysaccharide on the earth after cellulose, and it is not digested in the upper GI tract by human digestive enzymes. Chitosan is a copolymer consisting of 2-amino-2-deoxyd- glucose and 2-acetamido-2-deoxy-d-glucose units linked with β-(1-4) bonds. It should be susceptible to glycosidic hydrolysis by microbial enzymes in the colon because it possesses glycosidic linkages similar to those of other enzymatically depolymerized polysaccharides. Among diverse approaches that are possible for modifying polysaccharides, grafting of synthetic polymer is a convenient method for adding new properties to a polysaccharide with minimum loss of its initial properties. Graft copolymerization of vinyl monomers onto polysaccharides using free radical initiation has attracted the interest of many scientists. Up to now, considerable works have been devoted to the grafting of vinyl monomers onto the substrates, especially starch and cellulose. Existence of polar functionally groups as carboxylic acid are needed not only for bio-adhesive properties but also for pH-sensitive properties of polymer, because the increase of MAA content in the hydrogels provides more hydrogen bonds at low pH and more electrostatic repulsion at high pH. It is as a part of the authors' research program on chitosan modification to prepare materials with pH-sensitive properties for use as drug delivery.

In this study, the authors hypothesized that the absorption of paclitaxel could be enhanced by administration with chitosan-polymethacrylic acid nanoparticles because of their greater permeability properties.

Chapter V - The use of nanoparticles in developing materials has been introduced in the last few years. It has been observed that the extremely fine size of nanoparticles strongly affects their physical and chemical properties and enables them to fabricate new materials with novelty function. Thus integrating nanoparticles with existing cement-based building materials may enable us to fabricate new products with some outstanding properties. Among the nanoparticles, nano silica has been used to improve the properties of cement-based materials. Some efforts on excellent mechanical properties and microstructure of cement composites with nano-SiO_2 have been reported. Gengig Li showed that incorporating nanoparticles to high-volume fly ash concrete can significantly increase the initial pozzolanic activity of fly ash. He also concluded that nano-SiO_2 can enhance the short-term and long-term strength of high-volume high-strength concrete. K. Lin reported that nano-SiO_2 particles could potentially improve the negative influences caused by enhanced the compressive and flexural strength of cement concrete and mortar. As the rate of pozzolanic reaction is proportional to the amount of surface available for reaction and owing to the high specific surface of nanoparticles, they possess high pozzolanic activity. Nano-SiO_2 effectively consumes calcium hydroxide crystals (CH), which array in the interfacial transition zone between hardened cement paste and aggregates and produce hydrated calcium silicate (CSH), which enhances the strength of cement paste. In addition, due to nano-scale size of particles, nano-SiO$_2$ can fill the ultrafine pores in cement matrix. This physical effect leads to reduction in porosity of transition zone in the fresh concrete. This mechanism strengthens the bond between the matrix and the aggregates and improves the cement paste properties. Furthermore, it has been found that when the ultrafine particles of nano-SiO$_2$ uniformly disperse in the paste, thanks to high activity, generate a large number of nucleation sites for the precipitation of the hydration products, which accelerates cement hydration.

Chapter VI - During the past two decades, researchers from different disciplines have advanced the emerging field of nanotechnology at a phenomenal pace. In particular, the discovery of carbon nanotubes (CNTs) in 1991, by Iijima has speeded the development of nanotechnology because of superior mechanical, electronic and other physical and chemical properties of CNTs over other materials known to man. A single-walled carbon nanotube (SWCNT) is best described as a rolled-up tubular shell of graphene sheet, while a multiwalled nanotube (MWCNT) is a rolled-up stack of graphene sheets into concentric shells of carbons with adjacent shells separation of 0.34 nm. Mechanical properties of carbon nanotubes have shown to exceed those of other existing materials known to man, with high elastic modulus of greater than 1 TPa comparable to that of diamonds and strengths many times higher than the strongest steel at a fraction of the weight. Carbon nanotubes can be either metallic or semiconducting depending upon their structure, characterized by chirality, and also have very high thermal conductivity.

Generally, the theoretical modeling and analysis of nanostructured materials can be classified into two main categories. One category is atomic modeling, which includes techniques such as the classical molecular dynamics (MD), tight-binding molecular dynamics, and the ab initio method. In this respect, a comprehensive review on these methods can be found. The other category is continuum modeling, which is increasingly being viewed as an alternative way of modeling materials on the atomistic scale. Although the MD simulations have generated abundant results for understanding the behavior of nanostructured materials, they are limited by the size and time scales of such atomic systems. Moreover, performing controlled experiments at the nanoscale is very difficult and prohibitively expensive. Consequently, continuum mechanics as a computationally efficient technique has still been the dominant tool for modeling large-scale systems at the nanometer scale. In the classical (local) continuum models, CNTs are taken as linear elastic thin shells. This continuum approximation is appropriate when the radius of CNTs is considerably larger than the interlayer spacing, as remarked by Peng et al.

Chapter VII - Natural polymers have potential pharmaceutical applications because of their low toxicity, biocompatibility, and excellent biodegradability. In recent years, biodegradable polymeric systems have gained importance for design of surgical devices, artificial organs, drug delivery systems with different routs of administration, carriers of immobilized enzymes and cells, biosensors, ocular inserts, and materials for orthopedic applications. These polymers are classified as either synthetic (polyesters, polyamides, polyanhydrides) or natural (polyamino acids, polysaccharides). Polysaccharide-based polymers represent a major class of biomaterials, which includes agarose, alginate, carageenan, dextran, and chitosan. Chitosan[_(1,4)2-amino-2-d-glucose] is a cationic biopolymer produced by alkaline N-deacetylation of chitin, which is the main component of the shells of crab, shrimp, and krill. Chitosan is a functional linear polymer derived from chitin, the most abundant natural polysaccharide on the earth after cellulose, and it is not digested in the upper GI tract by human digestive enzymes. Chitosan is a copolymer consisting of 2-amino-2-deoxy-d-glucose and 2-acetamido-2-deoxy-d-glucose units linked with β-(1-4) bonds. It should be susceptible to glycosidic hydrolysis by microbial enzymes in the colon because it possesses glycosidic linkages similar to those of other enzymatically depolymerized polysaccharides. Among diverse approaches that are possible for modifying polysaccharides, grafting of synthetic polymer is a convenient method for adding new properties to a polysaccharide with minimum loss of its initial properties. Graft copolymerization of vinyl monomers onto polysaccharides using free radical initiators has attracted the interest of many scientists. Up to now, considerable works have been devoted to the grafting of vinyl monomers onto the substrates, especially Starch and cellulose. Existence of polar functionally groups as carboxylic acid are needed not only for bio-adhesive properties but also for pH-sensitive properties of polymer. The increase of MAA content in the hydrogels provides more hydrogen bonds at low pH and more electrostatic repulsion at high pH. A part of the authors' research program is chitosan modification to prepare materials with pH-sensitive properties for uses as drug delivery. The free radical graft copolymerization polymethacrylic acid onto chitosan was carried out at 70 °C, bis-acrylamide as a cross-linking agent and persulfate as an initiator. Polymer bonded drug usually contain one solid drug bonded together in a matrix of a solid polymeric binder. They can be produced by polymerizing a monomer such as methacrylic acid (MAA), mixed with a particulate drug, by means of a chemical polymerization catalyst, such as AIBN or by means of high-energy radiation, such as X-ray or γ-rays. The modified hydrogel and satranidazole as a model drug were converted to nanoparticles by freeze-drying method. The equilibrium swelling studies and in vitro release profiles were carried out in enzyme-free simulated gastric and intestinal fluids [SGF (pH 1) and SIF (pH 7.4), respectively)]. The influences of different factors, such as content of MAA in the feed monomer and swelling, were studied.

Chapter VIII - To investigate the properties of nanostructures, and in particular carbon nanotubes (CNTs), a strong tool is computational experiments. In this regard, computational nanomechanics have an indispensable role in the field of current nanotechnology and nanosciences. Different methods are applied in the literature in order to model carbon nanotubes. Primarily, Yakobson et al. implemented molecular dynamics (MD) simulations to investigate the buckling behavior of these nanostructures under different loading conditions. They also tried to extract a fitted continuum model based on thin-shell theories. Indeed, MD simulations played a stronger tool in modeling carbon nanotubes, in comparison with continuum models, due to the fact that they can describe the atomistic level of these structures. As a result, an extensive study on carbon nanotubes has been conducted based on MD simulations, in recent years. However, the important deficiencies of these simulations are their time expense and their limitations in the size scale of the structures under investigation. Subsequently, with the current computational computer abilities, MD simulations cannot be performed on very large systems. On the other hand, continuum models can overcome these problems, but as mentioned before, in some cases, they cannot appropriately model the real behavior of the system (e.g., properties of carbon nanotubes containing defects). However, in the last two decades, continuum models have been numerously applied to various problems of nanostructured materials.

Chapter IX - Over the recent decades, scientists became interested for creation of polymer nanofibers due to their promising potential in many engineering and medical applications. According to various outstanding properties such as very small fiber diameters, large surface area per mass ratio, high porosity along with small pore sizes and flexibility, electrospun nanofiber mats have found numerous applications in diverse areas. For example, in biomedical field, nanofibers play a substantial role in tissue engineering, drug delivery, and wound dressing. Electrospinning is a sophisticated and efficient method by which fibers produced with diameters in nanometer scale are entitled as nanofibers. In electrospinning process, a strong electric field applies on a droplet of polymer solution (or melt) held by its surface tension at the tip of a syringe needle (or a capillary tube). As a result, the pendent drop will become highly electrified, and the induced charges distributes over its surface. Increasing the intensity of electric field, the surface of the liquid drop will be distorted to a conical shape known as the Taylor cone. Once the electric field strength exceeds a threshold value, the repulsive electric force dominates the surface tension of the liquid and a stable jet emerges from the cone tip. The charged jet then accelerates toward the target and rapidly thins and dries because of elongation and solvent evaporation. As the jet diameter decreases, the surface charge density increases and the resulting high repulsive forces split the jet to smaller jets. This phenomenon may take place several times, leading to many small jets. Ultimately, solidification carry out and fibers deposits on the surface of the collector as a randomly oriented nonwoven mat. Figure 1 shows a schematic illustration of electrospinning setup.

Chapter X - With potential applications ranging from protective clothing and filtration technology to reinforcement of composite materials and tissue engineering, nanofibers offer remarkable opportunity in the development of multifunctional material systems. The emergence of various applications for nanofibers is stimulated from their outstanding properties such as very small diameters, huge surface area-per-mass ratio and high porosity along with small pore size. Moreover, the high degree of orientation and flexibility beside superior mechanical properties are extremely important for diverse applications. In this study, aligned and molecularly oriented PAN nanofibers were prepared using a novel technique comprised of two needles with opposite voltage and a rotating drum for applying take-up mechanism. The electrospinning process was optimized for increasing of productivity and improving the mechanical properties through controlling internal structure of the generated fibers.

Chapter XI - Because of the high conductivity of copper, electro-less copper plating is currently used to manufacture conductive fabrics with high shielding effectiveness (SE). It can be performed at any step of the textile production, such as yarn, stock, fabric or clothing.

Electro-less copper plating as a non-electrolytic method of deposition from solution on fabrics has been studied by some researchers. The early reported copper electro-less deposition method uses a catalytic redox reaction between metal ions and dissolved reduction agent of formaldehyde at high temperature and alkaline medium. Despite technique advantages, such as low cost, excellent conductivity, easy formation of a continuous and uniform coating, experimental safety risks appear through formation of hazardous gaseous product during plating process, especially for industrial scale.

Further research has been conducted to substitute formaldehyde with other reducing agents coupled with oxidation accelerator such as sodium hypophosphite and nickel sulphate. Incorporation of Nickel and Phosphorus particles provide good potential for creation of fabrics with a metallic appearance and good handling characteristics. These properties are practically viable if plating process followed by finishing process in optimized pH and in presence of ferrocyanide. Revealing the performance of electro-less plating of Cu-Ni-P alloy on cotton fabrics is an essential research area in textile finishing processing and for technological design.

The main aim of this chapter is to explore the possibility of applying electro-less plating of Cu-Ni-P alloy onto cotton fabric to obtain highest level of conductivity, washing and abrasion fastness, room condition durability and EMI shielding effectiveness. The fabrication and properties of Cu-Ni-P alloy plated cotton fabric are investigated in accordance with standard testing methods.

Chapter XII - At the present time, development of new methods and perfection of already known and approved methods of metal nanoparticles obtaining are in the center of attention. According to one of modern classifications, methods of nanoparticles' obtainment can be subdivided conditionally on physical and chemical. In case of physical method using, nanoparticles are formed of separate atoms as a result of metals sublimation and their subsequent condensation on various substrates or are formed owing to crushing of the big metal particles by means of corresponding devices (colloidal mills, ultrasonic generators, etc.). On the data of nanoparticles, obtaining chemical methods process of chemical reduction in a solution of metal complex ions in the conditions favouring to the subsequent formation small metal clusters or aggregates is important.

Depending on the reducing agent nature, chemical methods subdivide into classical, using chemical reducers (hydrazine, alkaline metal tetrahydrideborates, hydrogen, etc.), X-ray and electrochemical in which a reducer is solvated electron generated accordingly by ionising radiation in a solution or electrochemically (an external current source, contact exchange) on electrode surfaces.

Cobalt nanoparticles find a wide application at composite materials and obtaining alloys. At the same time, they can be claimed in areas where their magnetic properties are important.

Studying of possibility and conditions of cobalt nanoparticles synthesis in water solutions by an electrochemical method is the purpose of the work.

Chapter XIII - In the practice of the breakage of various kinds of starting materials, there are two main mechanical-technological lines that have become widely used, for they embrace coarseness of the starting material, requirements towards the end product, its physical-mechanical properties, etc. They are:

Breakage during one or several intakes in one apparatus with the simultaneous classification of the mass flow or without it;

Multistep breakage in several sequentially installed apparatus; here, the breakage process is accompanied by separation at the sizing screens.

The main criterion for choosing a mechanical-technological line for size-reduction is its efficiency, which means the yield of fractions with predetermined sizes for a certain period, energy inputs for receiving one unit of product, the number of personnel, laboriousness of maintenance and repair of the equipment, its reliability, and occupied production space.

Chapter XIV - The development of new polymer composites with predetermined properties is the urgent problem of modern material science. Adhesion of polymer binder to the filler material (fabrics, fibers, disperse particles) is one of the important factors influenced by the strength properties of polymer composites. Thus, the prediction of polymer composite strength properties based on the polymer adhesion characteristics determination is the main task to develop the scientific principles of new polymer binder design. The work of adhesion

(W_a) depends on the surface energy of components (σ_1 and σ_2, respectively) and the energy of polymer-filler interface (σ_{12}): $W_a = \sigma_1 + \sigma_2 - \sigma_{12}$. So, the development of express techniques permitted to determine these energetic characteristics of interface, to calculate the work of adhesion and to optimize the choice of polymer binder composition for the particular type of filler, is the actual problem. The contact angle measurements permit solving this problem.

Chapter XV - At the present stage, only active innovation should provide operational effectiveness of agro-based industries. The same is true of crops protected ground. Implementation of the latest advances in science technology and best practices can lead to improvement in innovative crop production, such as the developments in the fields of nanotechnology and nano-materials.

The main purpose of the use of nanotechnology and nanomaterials in hydroponic crop production is to ensure high yield and product quality, coupled with rational use of energy and resources.

In the context of a protected ground, where the main microclimatic parameters are regulated, the problems of root environment meeting the requirements of plants and allowing regulation of their mineral nutrition come to the forefront. In selecting priority areas for implementation of these technologies, it is also necessary to take into account the social value of output products, as food must not only meet physiological needs of human beings in the life-essential substances but also carry out environmental-preventive and therapeutic targets.

Chapter XVI - Among using now polymers the polyethylene and poly(vinyl chloride) (PVC) are the very widespread materials. That is connected with their good dielectric performances, high chemical stability, strength, etc. These materials fall into vinyl type polymers with repeating monomeric unit $[-CH_2-CH_2-]_n$, where in the case of PVC, one of hydrogen atoms is replaced by atom of chlorine. However, in many cases, application of these materials is essentially restricted because of low values of surface energy, which causes their poor adhesion performances and surface wettability. Noted properties largely depend on chemical composition and structure of surface layer and can be regulated by two basic expedients: removal of feeble boundary polymer layers and forming of new active centers changing their surface energy.

One of methods of formation of new functional structures on the polymeric materials surface is gas-phase modification of solid matrixes, based on the molecular layering principles (ML). The essence of this method is that on a solid surface, the chemical reactions are carried out in requirements of maximal removal from equilibrium between reagents brought from the outside and functional groups of matrix. Molecular layering has appeared rather effective for regulating macroscopic properties of phenolphormaldehyde, polyamide, epoxy and other polymeric materials. According to IR-spectroscopy results, during chemical modification, grafting of element containing structures to the superficial reactive centers of polymers (olefinic bonds, oxygen-containing groups, hydrogen at tertiary atom of carbon, etc.) takes place. That results in changing of thermal-oxidative, electret, diffusive and other properties of materials.

The purpose of the present work was the examination of effect of chemical composition of poly(vinyl chloride) and high-pressure polyethylene (HPPE) films, modified by vapors of volatile halogenides (PCl_3, $TiCl_4$, $VOCl_3$ and $Si(CH_3)_2Cl_2$) on energy performances, wettability and topography of polymers surface.

Chapter XVII - Cotton is one of the major textile fibers, and it has a unique combination of properties, including softness, durability, high strength, good dyeability and biodegradability, and for many centuries, it has found use in textile production. Reactive dyes with vinylsulphone groups are widely used to dyc cotton fibers, because of the simplicity of application, the great choice of commercial products and their cheapness. Even though a long tradition has given a solid and an in-depth knowledge of cotton textile fibers and of dyeing processes, the new research borders are moving to a development of the inherent textile materials properties: for this purpose, chemical finishing procedures are widely used. Textile materials can be treated with different functional finishes, such as water and oil repellent, durable press, soil-release, flame retardant, antistatic, and antimicrobial. Water-repellent finishing on fabrics is mostly imparted by the incorporation of low surface energy compounds, accompanied by the increase of the contact angle of liquids on its surface. The most recent approaches to improved repellency are based on the use of nanoparticles, such as highly branched 3D surface functional macromolecules called dendrimers, whose effect mechanism depends on being in a position to build-up crystal structures in nano-range, which produce wash-permanent, water-repellent and highly abrasion-resistant effects. When combined with fluoropolymers, dendrimers force them to co-crystallize, leading to a self-organization of the whole system and to an enrichment of the fluoro polymers on the most outer layer of the textile. Functionality and properties of dendrimers can be changed by filling their cavities or modifying the core and chain-ends. Conformational flexibility of branches is capable of placing dendrimers hydrophilic interior in contact with aqueous sub-phase and extending their chains into the air above the air–water interface. Particle size of these repellent finishes plays a vital role because, when the inorganic particle size is reduced, the surface area is increased; this leads to good interaction with the matrix polymer, and a highest performance is achieved. Alteration of surface properties by textile finishing applications and creation of a smoother reflection surface by a reduced superficial particle size could also give a color change. In this study, the effect of particle sizes on surface roughness and color assessment after finishing of cotton fabrics was evaluated by surface reflectance, absorbance of finishes and color coordinates measurement. For this purpose, three types of commercially available dendrimer water repellent (DWR), fluorocarbon included dendrimer water-oil repellent (DWOR) and fluorocarbon water-oil repellent (FWOR) reagents were impregnated to dyed cotton fabrics and polymerized under optimum conditions. The dyeing of the fabric samples was carried out by three different colors commercial dyes. The reflectance and color coordinates of treated and untreated samples were measured by reflectance spectrophotometer according to the CIELAB, a CIE defined color space, which supports the accepted theory of color perception based on three separate color receptors in the eye (red, green and blue) and is currently one of the most popular color spaces. Finally, water and oil repellency performances, treated-substrate characterization and fabrics mechanical properties were extensively investigated to estimate if finishing agent's application gives rise to other changes, besides color alterations.

Chapter XVIII - Today, as well as it was a lot of centuries back, prevention and extinguishing of fires is one of the global problems standing in front of humanity. In the world, annually, thousands of the new reagents appear directed on the decision of this problem. But as the authors can see, the problems such as, for example, extinguishing of forest fires or creation of low inflammable polymeric materials is far from the solution.

The combustion of natural and polymer materials, like the combustion process of any other fuel material, is a combination of complex physical and chemical processes, which include the transformation of initial products. This whole conversion process may be divided in stages, with specific physical and chemical processes occurring in each of these stages. In contrast to the combustion of gases, the combustion process of condensed substances has a multi-phase character. Each stage of the initial transformation of a substance correlates with a corresponding value (combustion wave) with specific physical and chemical properties (state of aggregation, temperature range, concentration of the reacting substances, kinetic parameters of the reaction, etc.).

Char-forming materials often swell and intumesce during their degradation (combustion), and the flame-retardant approach is to promote the formation of such intumescent char. The study of new polymer flame retardants has been directed at finding ways to increase the fire resistance.

Chapter XIX - At present time, great attention is given to the study of properties of polymeric nanocomposites produced on the basis of well-known thermoplastics (PP, PE, PS, PMMA, polycarbonates, polyamides) and carbon nanotubes (CN). CNs are considered to have the wide set of important properties like thermal stability, reduced combustibility, electroconductivity, etc. Thermoplastic polymer nanocomposites are generally produced with the use of melting technique.

Development of synthetic methods and the thermal characteristics study of PP/multiwalled carbon nanotube (MWCNT) nanocomposites were taken as an objective in this paper.

A number of papers pointed at synthesis and research of thermal properties of nanocomposites (atactic polypropylene (aPP)/MWCNT) were reported. It is remarkable that PP/MWCNT composites with minor level of nanocarbon content (1-5% by weight) were determined to obtain an increase in thermal and thermal-oxidative stability in the majority of these publications.

Thermal stability of aPP and aPP/MWCNT nanocomposites with the various concentrations of MWCNT was studied in the paper. It was shown that thermal degradation processes are similar for aPP and aPP/MWCNT nanocomposites and initial degradation temperatures are the same. However, the maximum mass loss rate temperature of PP/MWCNT nanocomposites with 1 and 5% wt of MWCNT raised by 40° - 70°C as compared with pristine aPP.

Kashiwagi et al. published the results of study of thermal and combustion properties of PP/MWCNT nanocomposites. A significant decrease of maximum heat release rate was detected during combustion research with use of cone calorimeter. A formation of char network structure during the combustion process was considered to be the main reason for combustibility decrease. The carbonization influence upon combustibility of polymeric nanocomposites was widely presented in literature. Notably, Kashiwagi et al. were the first to hypothesize that abnormal dependence of maximum heat release rate upon MWCNT concentration is closely related with thermal conductivity growth of PP/MWCNT nanocomposites during high-temperature pyrolysis and combustion.

Chapter XX - At present, nanocomposites polymer/organoclay studies attained very big widespread interest. However, the majority of works fulfilled on this theme has mainly an applied character, and theoretical aspects of polymers reinforcement by organoclays are studied much less. So, the authors developed for this purpose multiscale micromechanical model, on the basis of which representation about organoclay "effective particle" was assumed. The indicated "effective particle" includes in itself both nanofiller platelets and adjoining to them (or located between them), polymeric matrix layers. Despite the complexity of model, it essentially lacks number. First, the above-indicated complexity results in necessity of parameters large number usage, a part of which is difficult enough and sometimes impossible to determine. Secondly, this model is based on micromechanical models application, which, in essence, exhausted their resources. Thirdly, at the entire of its complexity, the model does not take into account such basis for polymer nanocomposites factors as a real level of interfacial adhesion nanofiller-polymeric matrix and polymer chain flexibility for nanocomposite matrix.

The other model, used for the same purpose, is based on the following percolation relationship application.

Chapter XXI - Quantum chemical calculation of a molecules of 2-methylsulfanil-4-difluoromethoxypyrimidine, 2-ethylsulfanil-4-difluoromethoxypyrimidine, 2-isopropyl sulfanil-4-difluoromethoxypyrimidine, 2-isobutylsulfanil-4-difluoromethoxypyrimidine, 2-methylsulfanil-4-difluoromethoxy-6-methylpyrimidine, 2-ethylsulfanil-4-difluoro methoxy-6-methylpyrimidine, 2-isobutylsulfanil-4-difluoromethoxy-6-methylpyrimidine is executed by methods MNDO and AB INITIO in base 6-311G** with optimization of geometry on all parameters for the first time. The optimized geometrical and electronic structure of these compounds is received. Acid force of these compounds is theoretically appreciated (MNDO: $26 \leq pKa \leq 27$, AB INITIO: $28 \leq pKa \leq 32$). It is established that 2-methylsulfanil-4-difluoromethoxypyrimidine, 2-ethylsulfanil-4-difluoromethoxy pyrimidine, 2-isopropyl sulfanil-4-difluoromethoxypyrimidine, 2-isobutylsulfanil-4-difluoromethoxypyrimidine, 2-methylsulfanil-4-difluoromethoxy-6-methylpyrimidine, 2-ethylsulfanil-4-difluoromethoxy-6-methylpyrimidine, 2-isobutylsulfanil-4-difluoromethoxy-6-methylpyrimidine relate to a class of weak H-acids ($pKa > 14$).

Chapter XXII - It is demonstrated that for two-particle interactions, the principle of adding reciprocals of energy characteristics of subsystems is fulfilled for the processes flowing along the potential gradient, and the principle of their algebraic addition − for the processes against the potential gradient.

Chapter XXIII - The notion of spatial-energy parameter (P-parameter) is introduced based on the modified Lagrangian equation for relative motion of two interacting material points and is a complex characteristic of important atomic values responsible for interatomic interactions and having the direct connection with electron density inside an atom. Wave properties of P-parameter are found; its wave equation having a formal analogy with the equation of Ψ-function is given.

With the help of P-parameter technique, numerous calculations of exchange structural interactions are done; the applicability of the model for the evaluation of the intensity of fundamental interactions is demonstrated; initial theses of quark screw model are given.

Chapter XXIV - The aggregation of the initial nanofiller powder particles in more or less large particles aggregates always occurs in the course of technological process of making particulate-filled polymer composites in general and elastomeric nanocomposites in particular. The aggregation process tells on composites (nanocomposites) macroscopic properties. For nanocomposites, nanofiller aggregation process gains special significance, since its intensity can be the one that nanofiller particles aggregates size exceeds 100 nm − the value, which assumes (although and conditionally enough) as an upper dimensional limit for nanoparticle. In other words, the aggregation process can result in the situation when primordially supposed nanocomposite ceases to be the one. Therefore, at present, several methods exist, which allowed suppressing nanoparticles aggregation process. Proceeding from this, in the present paper, theoretical treatment of disperse nanofiller aggregation process in butadiene-styrene rubber matrix within the frameworks of irreversible aggregation models was carried out.

Chapter XXV - Based on modified Lagrangian equation for relative movement of two interacting material points, the notion of spatial-energy parameter (P-parameter), which is a complex characteristic of important atomic values responsible for interatomic interactions and directly connected with electron density in atom.

The value of relative difference of P–parameters of interacting atoms – components of α -coefficient of structural interactions was used as the main quantitative characteristic of structural interactions in condensed media

Chapter XXVI - In the last 25 years, an interest of physicists to the theory of polymer synthesis has sharply increased (for example, papers where the concept of a mean field was used). Simultaneously with the indicated papers, a number of publications concerned analytical study and computer simulation of reactions in different spaces, including fractal ones, has appeared. It has been clarified that the main factor, defining chemical reactions course, is a space connectivity degree, irrespective of its type. Also a large amount of theoretical and applied researches on irreversible aggregation models of different kinds offered for the description of such processes as a flocculation, coagulation and polymerization are carried out. These papers are intimately connected to the fractal analysis, intensively developing per last years, as the aggregation within the frameworks of the indicated models forms fractal aggregates. Nevertheless, the application of these modern physical concepts for the description of polymers synthesis still has unitary character.

However, the application of the fractal analysis methods to synthesis process for today becomes a vital problem. Such necessity is due not to the convenience of the fractal analysis as mathematical approach, which supposes the existence of approaches alternate to it. The necessity of the indicated problem solution is defined only by physical reasons. The basic object during synthesis of polymers in solutions is the macromolecular coil, which represents a fractal object. As it is known, the description of fractal objects within the frameworks of an Euclidean geometry is incorrect, which predetermines the necessity of fractal analysis application. Besides, practically all kinetic curves at synthesis of polymers represent curves with a decreasing reaction rate, which is typical designation for fractal reactions, i.e., reactions of fractal objects or reactions in fractal space. Therefore, the purpose of the present review is fractal analysis methods application for synthesis kinetics description and this process final characteristics determination for branched polymers on the example of polyhydroxyther.

Chapter XXVII - It is well known that in particulate-filled elastomeric nanocomposites (rubbers), nanofiller particles form linear spatial structures ("chains"). At the same time, in polymer composites, filled with disperse microparticles (microcomposites), particles (aggregates of particles) of filler form a fractal network, which defines polymer matrix structure (analog of fractal lattice in computer simulation). This results in different mechanisms of polymer matrix structure formation in micro- and nanocomposites. If in the first filler, particles (aggregates of particles) fractal network availability results in "disturbance" of polymer matrix structure that is expressed in the increase of its fractal dimension d_f, then in case of polymer nanocomposites at nanofiller, contents change the value d_f is not changed and equal to matrix polymer structure fractal dimension. As it has to been expected, composites indicated classes structure formation mechanism change defines their properties change, in particular, reinforcement degree.

At present, there are several methods of filler structure (distribution) determination in polymer matrix, both experimental and theoretical. All the indicated methods describe this distribution by fractal dimension D_n of filler particles network. However, correct determination of any object fractal (Hausdorff) dimension includes three obligatory conditions. The first from them is the above-indicated determination of fractal dimension numerical magnitude, which should not be equal to object topological dimension. As it is known, any real (physical) fractal possesses fractal properties within a certain scales range. And at last, the third condition is the correct choice of measurement scales range itself. As it has been shown in papers, the minimum range should exceed at any rate one self-similar iteration.

The present paper purpose is dimension D_n estimation, both experimentally and theoretically, and checking two above-indicated conditions fulfillment, i.e., obtaining of nanofiller particles (aggregates of particles) network ("chains") fractality strict proof in elastomeric nanocomposites on the example of particulate-filled butadiene-styrene rubber.

In: Handbook of Research on Nanomaterials, Nanochemistry ...   ISBN: 978-1-61942-525-5
Editors: A. K. Haghi and G. E. Zaikov   © 2013 Nova Science Publishers, Inc.

Chapter I

Lamination of Nanofiber Web into Nanostructured Fabric

A. K. Haghi[*]
University of Guilan, Iran

Introduction

Clothing is a person's second skin, since it covers great parts of the body and has a large surface area in contact with the environment. Therefore, clothing is proper interface between environment and human body and could act as an ideal tool to enhance personal protection. Over the years, growing concern regarding health and safety of persons in various sectors, such as industries, hospitals, research institutions, battlefields and other hazardous conditions, has led to intensive research and development in field of personal protective clothing. Nowadays, there are different types of protective clothing. The simplest and most preliminary of this equipment is made from rubber or plastic that is completely impervious to hazardous substances, air and water vapor. Another approach to protective clothing is laminating activated carbon into multilayer fabric in order to absorb toxic vapors from environment and prevent penetration to the skin [1]. The use of activated carbon is considered only a short-term solution because it loses its effectiveness upon exposure to sweat and moisture. The use of semi-permeable membranes as a constituent of the protective material is another approach. In this way, reactive chemical decontaminants encapsulates in microparticles [2] or fills in microporous hollow fibers [3]. The microparticle or fiber walls are permeable to toxic vapors but impermeable to decontaminants, so that the toxic agents diffuse selectively into them and neutralize. All of these equipments could trap such toxic pollutions but usually are impervious to air and water vapor and thus retain body heat. In other words, a negative relationship always exists between thermal comfort and protection performance for currently available protective clothing. For example, nonwoven fabrics with high air permeability exhibit low barrier performance, whereas microporous materials, laminated fabrics and tightly

[*] Haghi@Guilan.ac.ir.

constructed wovens offer higher level of protection but lower air permeability. Thus there still exists a very real demand for improved protective clothing that can offer acceptable levels of impermeability to highly toxic pollutions of low molecular weight, while minimizing wearer discomfort and heat stress [4].

Electrospinning provides an ultrathin membrane-like web of extremely fine fibers with very small pore size and high porosity, which makes them excellent candidates for use in filtration, membrane, and possibly protective clothing applications.

Preliminary investigations have indicated that the using of nanofiber web in protective clothing structure could present minimal impedance to air permeability and extreme efficiency in trapping aerosol toxic pollutions. Potential of electrospun webs for future protective clothing systems has been investigated [5-7]. Schreuder-Gibson et al. has shown an enhancement of aerosol protection via a thin layer of electrospun fibers. They found that the electrospun webs of nylon 66, polybenzimidazole, polyacrylonitrile, and polyurethane provided good aerosol particle protection, without a considerable change in moisture vapor transport or breathability of the system [5]. While nanofiber webs suggest exciting characteristics, it has been reported that they have limited mechanical properties [8, 9]. To compensate for this drawback in order to use of them in protective clothing applications, electrospun nanofiber webs could be laminated via an adhesive into a multilayer fabric system [10-11]. The protective clothing made of this multilayer fabric will provide both protection against toxic aerosol and thermal comfort for user.

The adhesives in the fabric lamination are as solvent/water-based adhesive or as hot-melt adhesive. At the first group, the adhesives are as solution in solvent or water and solidify by evaporating of the carrying liquid. Solvent-based adhesives could "wet" the surfaces to be joined better than water-based adhesives and also could solidify faster. But unfortunately, they are environmentally unfriendly, usually flammable and more expensive. Of course, it doesn't mean that the water-based adhesives are always preferred for laminating, since in practice, drying off water in terms of energy and time is expensive, too. Besides, water-based adhesives are not resistant to water or moisture because of their hydrophilic nature. At the second group, hot-melt adhesives are environmentally friendly, inexpensive, require less heat and energy, and so are now more preferred. Generally, there are two procedures to melt these adhesives: static hot-melt laminating that is accomplished by flat iron or Hoffman press and continuous hot-melt laminating that uses the hot calendars. In addition, these adhesives are available in several forms: as a web, as a continuous film, or in powder form. The adhesives in film or web form are more expensive than the corresponding adhesive powders. The web form are discontinuous and produce laminates that are flexible, porous and breathable, whereas, continuous film adhesives cause stiffening and produce laminates that are not porous and permeable to both air and water vapor. This behavior attributed to impervious nature of adhesive film and its shrinkage under the action of heat [12]. Thus, the knowledge of laminating skills and adhesive types is very essential to producing an appropriate multilayer fabric. Specifically, this subject becomes more highlighted as we will laminate the ultrathin nanofiber web into multilayer fabric, because the laminating process may be adversely influenced on the nanofiber web properties. Lee et al. [7], without disclosure of laminating details, reported that the hot-melt method is more suitable for nanofiber web laminating. In this method, laminating temperature is one of the most effective parameters. Incorrect selection of this parameter may lead to change or damage ultrathin nanofiber web. Therefore, it is necessary to find out a laminating temperature that has the least effect on nanofiber web during process. The purpose of this study is to consider the influence of laminating temperature on the nanofiber web/multilayer fabric properties to make protective fabric that is resistance against aerosol pollutions. Multilayer fabrics were made by laminating of nanofiber web into cotton fabric via hot-melt method at different temperatures. Effects of laminating temperature on the nanofiber web morphology, air transport properties and the adhesive force were discussed.

Experimental

Electrospinning and Laminating Process

The electrospinning conditions and layers properties for laminating are summarized in Table1. Polyacrylonitrile (PAN) of 70,000 g/mol molecular weight from Polyacryl Co. (Isfehan, Iran) has been used with N,N-dimethylformamide (DMF) from Merck to form a 12 %Wt polymer solution after stirring for 5h and exposing for 24h at ambient temperature. The yellow and ripened solution was inserted into a plastic syringe with a stainless steel nozzle, and then it was placed in a metering pump from WORLD PRECISION INSTRUMENTS (Florida, USA). Next, this set installed on a plate on which it could traverse to left-right direction along drum collector (Fig.1). The electrospinning process was carried out for 8h, and the nanofibers were collected on an aluminum-covered rotating drum, which was previously covered with a Poly-Propylene Spun-bond Nonwoven (PPSN) substrate. After removing of PPSN covered with nanofiber from drum and attaching another layer of PPSN on it, this set was incorporated between two cotton weft-warp fabrics as a structure of fabric-PPSN-nanofiber web- PPSN-fabric (Fig.2). Finally, hot-melt laminating performed using a simple flat iron for 1min, under a pressure of 9gf/cm^2 and at temperatures 85,110,120,140,150°C (above softening point of PPSN) to form the multilayer fabrics.

Table 1. Electrospinning conditions and layers properties for laminating

Electrospinning conditions		Layer properties	
		PPSN	
Polymer concentration	12% w/w	Thickness	0.19 mm
Flow rate	1 µl/h	Air permeability	824 cm^3/s/cm^2
		Melting point	140°C
Nozzle inner diameter	0.4 mm	Mass	25 g/m^2
Nozzle-Drum distance	7 cm	Nanofiber web	
		Mass	3.82 g/m^2
Voltage	11 KV	Fabric	
Drum speed	9 m/min	Thickness	0.24 mm
Spinning Time	8 hr	Warp-weft density	25×25 per cm

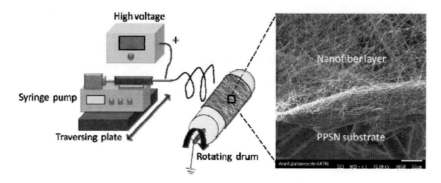

Figure 1. Electrospinning setup and an enlarged image of nanofiber layer on PPSN.

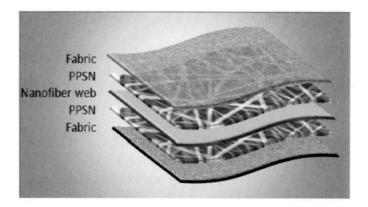

Figure 2. Multilayer fabric components.

Nanofiber Web Morphology

A piece of multilayer fabrics was freeze fractured in liquid nitrogen, and after sputter-coating with Au/Pd, a cross-section image of them was captured using a scanning electron microscope (Seron Technology, AIS-2100, Korea).

Also, to consider the nanofiber web surface after hot-melt laminating, other laminations were prepared by a non-stick sheet made of Teflon (0.25 mm thickness) as a replacement for one of the fabrics (fabric /PPSN/nanofiber web/PPSN/Teflon sheet). Laminating process was carried out at the same conditions, which mentioned to produce primary laminations. Finally, after removing of Teflon sheet, the nanofiber layer side was observed under an optical microscope (MICROPHOT-FXA, Nikon, Japan) connected to a digital camera.

Measurement of Air permeability

Air permeability of multilayer fabrics after laminating was tested by fabric air permeability tester (TEXTEST FX3300, Zürich, Switzerland). Five pieces of each sample under air pressure of 125 Pa were tested, at ambient condition (16°C, 70%RH) and then obtained average air permeability.

Results and Discussion

In electrospinning phase, PPSN was chosen as a substrate to provide strength to the nanofiber web and to prevent its destruction in removing from the collector. In Figure 1, an ultrathin layer of nanofiber web on PPSN layer is illustrated, which conveniently shows the relative fiber sizes of nanofibers web (approximately 380 nm) compared to PPSN fibers. Also, this Figure shows that the macropores of PPSN substrate is covered with numerous electrospun nanofibers, which will create innumerable microscopic pores in this system. But in laminating phase, this substrate acts as an adhesive and causes to bond the nanofiber web to the fabric.

In general, it is relatively simple to create a strong bond between these layers, which guarantees no delamination or failure in multilayer structures; the challenge is to preserve the original properties of the nanofiber web and fabrics to produce a laminate with the required appearance, handle, thermal comfort and protection.

In other words, the application of adhesive should have minimum affect on the fabric flexibility or on the nanofiber web structure. In order to achieve to this aim, it is necessary that: a) the least amount of a highly effective adhesive is applied, b) the adhesive correctly covers the widest possible surface area of layers for better linkage between them and c) the adhesive penetrates to a certain extent of the nanofiber web/fabric [12]. Therefore, we selected PPSN, which is a hot-melt adhesive in web form. As mentioned above, the perfect use of web form adhesive can be lead to produce multilayer fabrics, which are porous, flexible and permeable to both air and water vapor. On the other hand, since the melting point of PPSN is low, hot-melt laminating can perform at lower temperatures.

Hence, the probability of shrinkage that may happen on layers in effect of heat becomes smaller. Of course in this study, we utilized cotton fabrics and polyacrylonitrile nanofiber web for laminating, which intrinsically are resistant to shrinkage even at higher temperatures (above laminating temperature). By this description, laminating process was performed at five different temperatures to consider the effect of laminating temperature on the nanofiber web/multilayer fabric properties. Fig. 3(A-E) shows a SEM image of multilayer fabric cross-section after laminating at different temperatures. It is obvious that these images don't deliver any information about nanofiber web morphology in multilayer structure, so it becomes impossible to consider the effect of laminating temperature on nanofiber web.

Therefore, in a novel way, we decided to prepare a secondary multilayer by substitution of one of the fabrics (ref. Fig. 2) with Teflon sheet. By this replacement, the surface of nanofiber web will become accessible after laminating because Teflon is a non-stick material and easily separates from adhesive. Fig. 3(a-e) presents optical microscope images of nanofiber web and adhesive after laminating at different temperatures. It is apparent that the adhesive gradually flattened on nanofiber web (Fig. 3(a-c)) when laminating at a temperature increased to melting point of adhesive (140 °C). This behavior is attributed to increment in plasticity of adhesive because of temperature rise and the pressure applied from the iron weight. But, by selection of melting point as laminating temperature, the adhesive completely melted and began to penetrate into the nanofiber web structure instead of spread on it (Fig. 3(d)). This penetration, in some regions, was continued to some extent so that the adhesive was even passed across the web layer. The dark crisscross lines in Fig. 3(d) obviously show where this excessive penetration occurred.

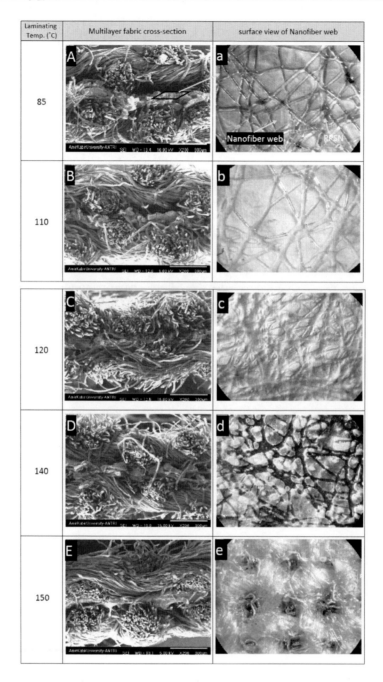

Figure 3. SEM images of multilayer fabric cross-section at 200 magnification (A-E) and optical microscope images of nanofiber web surface at 100 magnification (a-e).

The adhesive penetration could intensify by increasing of laminating temperature above melting point because the fluidity of melted adhesive increases by temperature rise. Fig. 3(e) clearly shows the amount of adhesive diffusion in the web, which was laminated at 150 °C. In this case, the whole diffusion of adhesive lead to creating a transparent film and to appearing the fabric structure under optical microscope. Also, to examine how laminating temperature affects the breathability of multilayer fabric, air permeability experiment was performed. The bar chart in Fig.4 indicates the effect of laminating temperature on air transport properties of multilayer fabrics. As might be expected, the air permeability decreased with increasing laminating temperature. This procedure means that the air permeability of multilayer fabric is related to adhesive's form after laminating because the polyacrylonitrile nanofiber web and cotton fabrics intrinsically are resistant to heat (ref. Fig 3). Of course, it is to be noted that the pressure applied during laminating can lead to compacting the web/fabric structure and to reducing the air permeability, too.

Nevertheless, this parameter didn't have an effective role on air permeability variations at this work, because the pressure applied for all samples had the same quantity. As discussed, by increasing of laminating temperature to melting point, PPSN was gradually flattened between layers so that it was transformed from web form to film-like. It is obvious in Fig.3 (a-c) that the pore size of adhesive layer becomes smaller in effect of this transformation. Therefore, we can conclude that the adhesive layer as a barrier resists to convective airflow during experiment and finally reduces the air permeability of multilayer fabric according to the pore size decrease. But this reason was not acceptable for the samples that were laminated at melting point (140 °C) since the adhesive missed self-layer form because of penetration into the web/fabric structures (Fig.3 d). In these samples, the adhesive penetration leads to block the pores of web/fabrics and to prevent of the air pass during experiment. It should be noted that the adhesive was penetrated into the web much more than the fabric, because PPSN structurally had more surface junction with the web (Fig. 3(A-E)). Therefore, here, the nanofiber web contained the adhesive itself could form an impervious barrier to air flow.

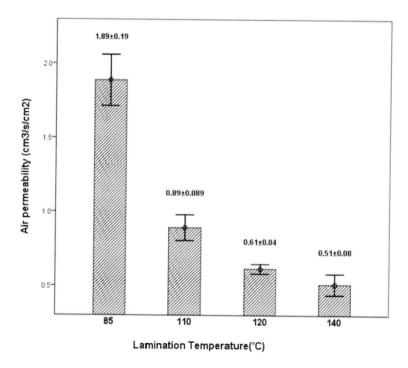

Figure 4. Air permeability of multilayer fabric after laminating at different temperatures.

Furthermore, we only observed that the adhesive force between layers was improved according to temperature rise.

For example, the samples laminated at 85°C exhibited very poor adhesion between the nanofiber web and the fabrics as much as they could be delaminated by light abrasion of thumb. Generally, it is essential that no delamination occurs during use of this multilayer fabric, because the nanofiber web might be destroyed due to abrasion of fabric layer. Before melting point, improving the adhesive force according to temperature rise is simultaneously attributed to the more penetration of adhesive into layers and the expansion of bonding area between them, as already discussed.

Also at melting point, the deep penetration of adhesive into the web/fabric leads to increase in this force.

Conclusion

In this study, the effect of laminating temperature on the nanofiber web/multilayer fabric properties was investigated to make next-generation protective clothing.

First, we demonstrated that it is impossible to consider the effect of laminating temperature on the nanofiber web morphology by a SEM image of multilayer fabric cross-section.

Thus, we prepared a surface image of nanofiber web after laminating at different temperature using an optical microscope. It was observed that nanofiber web was approximately unchanged when laminating temperature was below PPSN melting point.

In addition, to compare air transport properties of multilayer fabrics, air permeability tests were performed. It was found that by increasing laminating temperature, air permeability was decreased. Furthermore, it only was observed that the adhesive force between layers in multilayer fabrics was increased with temperature rise. These results indicate that laminating temperature is an effective parameter for laminating of nanofiber web into fabric structure.

Thus, varying this parameter could lead to developing fabrics with different levels of thermal comfort and protection depending on our need and use. For example, laminating temperature should be selected close to melting point of adhesive, if we would produce a protective fabric with good adhesive force and medium air permeability.

References

[1]   M. Ziabari, V. Mottaghitalab and A.K. Haghi, *Korean J. Chem. Eng.*, 25, 923 (2008).

[2]   A.K. Haghi and M. Akbari, *Phys. Stat. Sol. A*, 204, 1830 (2007).

[3]   M. Kanafchian, M. Valizadeh, A.K.Haghi, *Korean J. Chem. Eng.*, 28, 428 (2011).

[4]   M. Ziabari, V. Mottaghitalab, A. K. Haghi, *Korean J. Chem. Eng* ,25, 905 (2008).

[5]   M. Kanafchian, M. Valizadeh, A.K.Haghi, *Korean J. Chem. Eng.*, 28, 445(2011).

[6]   S. Lee and S. K. Obendorf, *J. Appl. Polym. Sci.*, 102, 3430 (2006).

[7]   Lee S, Kimura D, Lee KH, Park J C, Kim IS, , *Textile Res. J.,* 80, 99 (2010).

[8]   Pedicini, A., and Farris, R. J., *Polymer* 44, 6857 (2003).

[9]   Lee, K. H., Lee, B. S., Kim, C. H., Kim, H. Y., Kim, K. W., and Nah, C. W., *Macromol. Res.* 13, 441 (2005).

[10] Lee SM, Kimura D, Yokoyama A, Lee KH, Park JC, Kim IS, , *Textile Res. J.,* 79, 1085 (2009).

[11] Liu, L., Huang, Z. M., He, C. L., and Han, X. J., , *Mater. Sci. Eng. A* 435–436, 309(2006).

[12] Fung W, Materials and their properties. In: *Coated and laminated textiles, 1st ed*, Woodhead Publishing, 63-71, 2002.

In: Handbook of Research on Nanomaterials, Nanochemistry ...   ISBN: 978-1-61942-525-5
Editors: A. K. Haghi and G. E. Zaikov   © 2013 Nova Science Publishers, Inc.

Chapter II

Carbon Nanotube/Chitosan Nanocomposite

Z. Moridi, V. Mottaghitalab and A. K. Haghi[*]

University of Guilan, Iran

1. Introduction

Recently, the words "nanobiocomposites" or "biopolymer nanocomposites" are most frequently observed in environmentally friendly research studies. The synthetic polymers have been widely used in a various application of nanocomposites. However, they become a major source of waste after use due to their poor biodegradability. On the other hand, most of the synthetic polymers are without biocompatibility in *vivo* and *vitro* environments. Hence, scientists were interested in biopolymers as biodegradable materials [1]. Later, several groups of natural biopolymers such as polysaccharide, proteins, and nucleic acids were used in various applications [2]. Nevertheless, the use of these materials has been limited due to relatively poor mechanical properties. Therefore, researcher efforts have been made to improve the properties of biopolymers as a matrix by using of reinforcement [3].

Chitosan (CHT) is a polysaccharide biopolymer that has been widely used as a matrix in nanobiocomposites. Chitosan represents high biocompatibility and biodegradability properties, although these biopolymers have an essential requirement to additional material with high mechanical properties [4]. Following discovery of carbon nanotube, results of characterization represented unique electrical and mechanical properties. Thereby, many research studies have focused on improving the physical properties of biopolymer nanocomposites by use of the fundamental behavior of carbon nanotubes [5].

It is the aim of this review to summarize recent advances in the production of carbon nanotubes/chitosan nanocomposites by several methods. Specifically, we will discuss about our recent work in preparing CNTs/CHT nanofiber composites by using of electrospinning method.

---

[*] Haghi@Guilan.ac.ir.

2. Biopolymers

Biomaterial has been defined as biocompatibility materials with the living systems. The biocompatibility implies the chemical, physical (surface morphology), and biological suitability of an implant surface to the host tissues. S. Ramakrishna et al. reviewed various biomaterials and their applications over the last 30 years. They represented applications of biopolymers and their biocomposites in medical applications [6]. These materials can classify natural and synthetic biopolymers. Synthetic biopolymers have been provided more cheaply with high mechanical properties. The low biocompatibility of synthetic biopolymers compared with natural biopolymers such as polysaccharides, lipids, and proteins lead to having paid great attention to the natural biopolymers. On the other hand, the natural biopolymers usually have weak mechanical properties. Therefore, many efforts have been made for improving their properties by blending some filler [7].

Among the natural biopolymers, polysaccharides seem to be the most promising materials in various biomedical fields. These biopolymers have various resources including animal origin, plant origin, algal origin, and microbial origin. Among various polysaccharides, chitosan is the most usual due to its chemical structure [8].

2.1. Chitosan

Chitin (Fig 1.) is the second most abundant natural polymer in the world and extracted from various plants and animals [9]. However, derivations of chitin have been noticed because insolubility of chitin in aqueous media. Chitosan (Fig. 2.) is deacetylated derivation of chitin with the form of free amine. Unlike chitin, chitosan is soluble in diluted acids and organic acids. Polysaccharides contain 2-acetamido-2-deoxy-β-D-glucose and 2-amino-2-deoxy-β-D-glucose. Deacetylation of chitin converts acetamide groups to amino groups [10]. Deacetylation of degree (DD) is one of the important effective parameters in chitosan properties and has been defined as "the mole fraction of deacetylated units in the polymer chain" [11].

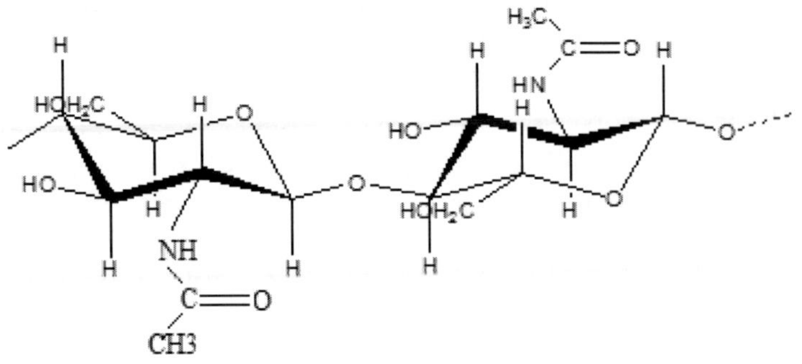

Figure 1. Structure of chitin.

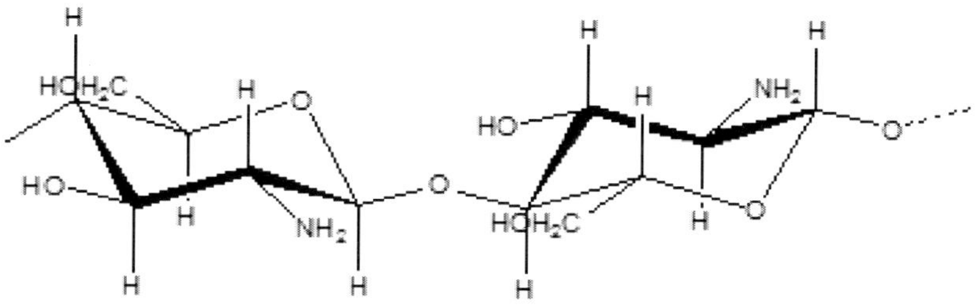

Figure 2. Structure of chitosan.

Chitosan could be suitably modified to impart desired properties due to the presence of the amino groups. Hence, a wide variety of applications for chitosan have been reported over the recent decades. Table 1. shows chitosan applications in variety fields and their principal characteristics. The high biocompatibility [12] and biodegradability [13] of chitosan yield most potential applications in biomedical [14].

Table 1. chitosan applications in variety fields and their principal characteristics

chitosan application		principal characteristics	ref
water engineering		metal ionic adsorption	[15]
biomedical application	biosensors and immobilization of enzymes and cells	biocompatibility, biodegradability to harmless products, nontoxicity, antibacterial properties, gel forming properties and hydrophilicity, remarkable affinity to proteins	[16]
	antimicrobial and wound dressing	wound healing properties	[17]
	tissue engineering	biocompatibility, biodegradable, and antimicrobial properties	[18]
	drug and gene delivery	biodegradable, nontoxicity, biocompatibility, high charge density, mucoadhesion	[19]
	orthopedic/periodontal application	antibacterial,	[20]
Photography		resistance to abrasion, optical characteristics, film forming ability	[21]
cosmetic application		fungicidal and fungi static properties	[22]
food preservative		biodegradability, biocompatibility, antimicrobial activity, non-toxicity	[23]
Agriculture		biodegradability, non-toxicity, antibacterial, cells activator, disease and insect resistant ability	[24]
textile industry		microorganism resistance, absorption of anionic dyes	[25]
paper finishing		high density of positive charge, non-toxicity, biodegradability, biocompatibility, antimicrobial and antifungal	[26]
solid-state batteries		ionic conductivity	[27]
chromatographic separations		the presence of free -NH2, primary -OH, secondary -OH	[28]
chitosan gel for LED and NLO applications		dye containing chitosan gels	[29]

2.2. Nanobiocomposites with Chitosan Matrix

Chitosan biopolymers have a great potential in biomedical applications due to their biocompatibility and biodegradability properties. However, the low physical properties of chitosan are most important challenge limiting their applications. The development of high-performance chitosan biopolymers has been received by incorporating fillers that display a significant mechanical reinforcement [30].

Polymer nanocomposites are polymers that have been reinforced by nano-sized particles with high surface area-to-volume ratio, including nanoparticles, nanoplatelet, nanofibers, and carbon nanotubes.

Nowadays, carbon nanotubes are considered to be highly potential fillers due to improving the materials properties of biopolymers [31]. Following these reports, researchers assessed the effect of CNTs fillers in chitosan matrix. Results of these research studies showed appropriate properties of CNTs/chitosan nanobiocomposites with high potential of biomedical science.

3. Carbon Nanotubes

Carbon nanotube, which is a tubular of Buckminster fullerene, was first discovered by Iijima in 1991 [32]. These are straight segments of tube with arrangements of carbon hexagonal units [33-34]. Scientists have greatly attended to CNTs during recent years due to the existence of superior electrical, mechanical and thermal properties [35].

Carbon nanotubes are classified as single-walled carbon nanotubes (SWNTs) formed by a single graphene sheet, and multiwalled carbon nanotubes (MWNTs) formed by several graphene sheets that have been wrapped around the tube core [36]. The typical range of diameters of carbon nanotubes are a few nanometers (~0.8-2 nm at SWNTs [37-38] and ~10-400 nm at MWNTs [39]), and their lengths are up to several micrometers [40].There are three significant methods for synthesizing CNTs including arc-discharge [41], laser ablation [42], chemical vapor deposition (CVD) [43]. The production of CNTs also can be realized by other synthesis techniques such as the substrate [44] the sol-gel [45] and gas phase metal catalyst [46] .

The C−C covalent bonding between the carbon atoms are similar to graphite sheets formed by sp^2 hybridization. As the result of this structure, CNTs exhibit a high specific surface area (about 10^3) [47] and thus a high tensile strength (more than 200 GPa) and elastic modulus (typically 1-5 TPa) [48].

Carbon nanotubes have also very high thermal and electrical conductivity. However, these properties are different in a variety of employed synthesis methods, defects, chirality, the degree of graphitization, and diameter [49]. For instance, the CNT can be metallic or semiconducting, depending on the chirality [50].

Preparation of CNTs solution is impossible due to their poor solubility. Also, a strong van der waals interaction of CNTs between several nanotubes leads to aggregation into bundle and ropes [51]. Therefore, the various chemical and physical modification strategies will be necessary for improving their chemical affinity [52].

There are two approaches to the surface modification of CNTs including the covalent (grafting) and non-covalent bonding (wrapping) of polymer molecule onto the surface of CNTs [53]. In addition, the reported cytotoxic effects of CNTs *in vitro* may be mitigated by chemical surface modification [54]. On the other hand, studies show that the end caps on nanotubes are more reactive than sidewalls. Hence, adsorption of polymers onto surface of CNTs can be utilized together with functionalization of defects and associated carbons [55].

The chemical modification of CNTs by covalent bonding is one of the important methods for improving their surface characteristics. Because of the extended π-network of the sp^2-hybridized nanotubes, CNTs have a tendency for covalent attachment, which introduces the sp^3-hibrydized C atoms [56].

These functional groups can be attached to termini of tubes by surface-bound carboxylic acids (grafting to) or direct sidewall modifications of CNTs that are based on the "in situ polymerization processing" (grafting from) [57]. Chemical functionalization of CNTs creates various activated groups (such as carboxyl [58], amine [59], fluorine [60], etc.) onto the CNTs surface by covalent bonds. However, there are two disadvantages for these methods. Firstly, the CNT structure may be decomposed due to functionalization reaction [61] and long ultrasonication process [62].

The disruption of π electron system is reduced as the result of these damages, leading to reduction of electrical and mechanical properties of CNTs. Secondly, the acidic and oxidation treatments that are often used for the functionalization of CNTs are environmentally unfriendly [63].

Thus, non-covalent functionalization of CNTs is greatly attended because of preserving their intrinsic properties while improving solubility and processability. In this method, non-covalent interaction between the π electrons of sp^2 hybridized structure at sidewalls of CNTs and other π electrons is formed by π-π stacking [64].

These non-covalent interactions can be raised between CNTs and amphiphilic molecules (surfactants) (Fig 3a). [65], polymers [66], and biopolymers such as DNA [67], polysaccharides [68], etc. In the first method, surfactants including non-ionic surfactants, anionic surfactants and cationic surfactants are applied for functionalization of CNTs.

The hydrophobic parts of surfactants are adsorbed onto the nanotubes surface, and hydrophilic parts interact with water [69]. Polymers and biopolymers can functionalize CNTs by using of two methods including endohedral (Fig 3b) and wrapping (Fig 3c). Endohedral method is a strategy for the functionalization of CNTs. In this method, nanoparticles such as proteins and DNA are entrapped in the inner hollow cylinders of CNTs [70]. In another technique, the van der waals interactions and π-π stacking between CNTs and polymer lead to the wrapping of polymer around the CNTs [71].

Various polymers and biopolymers such as polyaniline [72], DNA [73], and chitosan [74] interact physically through wrapping of nanotube surface and π-π stacking by solubilized polymeric chain. However, Jian et al. (2002) created a technique for the non-covalent functionalization of SWNTs most similar to π-π stacking by PPE without polymer wrapping [75].

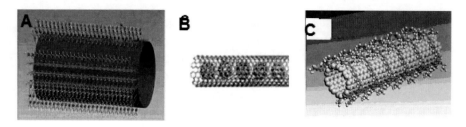

Figure 3. non-covalent functionalization of CNTs by (a) surfactants, (b) wrapping, (c) endohedral.

These functionalization methods can provide many applications of CNTs. In this context, one of the most important applications of CNTs is biomedical science such as biosensors [76], drug delivery [77], and tissue engineering [78].

3.1. Nanotube Composites

According to low physical properties of biopolymers, researchers would use some filler to the reinforcement of their electrical, mechanical, and thermal properties. Following discovery of CNTs, they have made many efforts to apply CNTs as filler in other polymers for improving properties of matrix polymer same to bulk materials [79]. The first time, Ajayan, in 1994, applied CNTs as filler in epoxy resin by the alignment method [80]. Later, many studies have focused on CNTs as excellent substitute for conventional nanofillers in the nanocomposites. Recently, many polymers and biopolymers have been reinforced by CNTs. As mentioned earlier, these nanocomposites have remarkable characteristics compared with bulk materials due to their unique properties [81].

There are several parameters affecting the mechanical properties of composites including proper dispersion, large aspect ratio of filler, interfacial stress transfer, well alignment of reinforcement, and solvent choice [82].

Uniformity and stability of nanotube dispersion in polymer matrixes are most important parameters for performance of composite. Good dispersion leads to efficient load transfer concentration centers in composites and uniform stress distribution [83]. Pemg-Cheng Ma et al. reviewed dispersion and functionalization techniques of carbon nanotubes for polymer-based nanocomposites and their effects on the properties of CNT/polymer nanocomposites. They demonstrated that the control of these two factors lead to uniform dispersion. Overall, the result showed that the proper dispersion enhanced a variety of mechanical properties of nanocomposites [71].

Fiber aspect ratio is defined as "the ratio of average fiber length to fiber diameter." This parameter is one of the main effective parameters on the longitudinal modulus [84]. Carbon nanotubes generally have high aspect ratio, but their ultimate performance in a polymer composite is different. The high aspect ratio of dispersed CNTs could lead to a significant load transfer [85]. However, aggregation of the nanotubes could lead to decrease of effective aspect ratio of the CNTs. Hence, properties of nanotube composites lower enhance than predictions. This is one of the processing challenges and poor CNTs dispersion [86].

The interfacial stress transfer has been performed by employing external stresses to the composites. The assessments showed that fillers take a significant larger share of the load due to CNTs-polymer matrix interaction. Also, the literature on mechanical properties of polymer nanotube composites represented enhancement of Young's modulus due to adding CNTs [87]. Wagner et al. investigated the effect of stress-induced fragmentation of multiwalled carbon nanotubes in a polymer matrix. The results showed that polymer deformation generates tensile stress and then transmits to CNTs [88].

The alignment CNT/polymer matrix in composite homogeneously is another effective parameter in properties of carbon nanotube composites. Researchers [89], for instance, assessed the effects of CNT alignment on electrical conductivity and mechanical properties of SWNT/epoxy nanocomposites. The electrical conductivity, Young's modulus and tensile strength of the SWNT/ epoxy composite rise with increasing SWNT alignment due to increase of interface bonding of CNTs in the polymer matrix.

Umar Khan et al., in 2007, examined the effect of solvent choice on the mechanical properties of CNTs–polymer composites. They were fabricated double-walled nanotubes and polyvinyl alcohol composites into the different solvents including water, DMSO and NMP. This work shows that solvent choice can have a dramatic effect on the mechanical properties of CNTs-polymer composites [90]. Also, a critical CNTs concentration has defined as optimum improvement of mechanical properties of nanotube composites where a fine network of filler formed [91].

There are other effective parameters in mechanical properties of nanotube composite such as size, crystallinity, crystalline orientation, purity, entanglement, and straightness. Generally, the ideal CNT properties depend on matrix and application [92].

The various functional groups on CNTs surface enable coupling with polymer matrix. A strong interface between coupled CNT/polymer creates efficient stress transfer. As a previous point, stress transfer is a critical parameter for control of mechanical properties of composite. However, covalent treatment of CNT reduces electrical [93], and thermal [94-95] properties of CNTs. These reductions affect on finally properties of nanotubes.

Matrix polymer can wrap around CNT surface by non-covalent functionalization. This process causes improvement in composite properties through various specific interactions. These interactions can improve properties of nanotube composites [96]. In this context, researchers [97] evaluated electrical and thermal conductivity in CNTs/epoxy composites. Figures 4. and 5. Show, respectively, electrical and thermal conductivity in various filler content including carbon black (CB), double-walled carbon nanotube (DWNT), and multi-functionalization. The experimental results represented that the electrical and thermal conductivity in nanocomposites improve by non-covalent functionalization of CNTs.

3.2. Mechanical and Electrical Properties of Carbon Nanotube/Natural Biopolymer Composites

Table 2. represents mechanical and electrical information of CNTs/natural polymer compared with neat natural polymer. These investigations show the higher mechanical and electrical properties of CNTs/natural polymers than neat natural polymers.

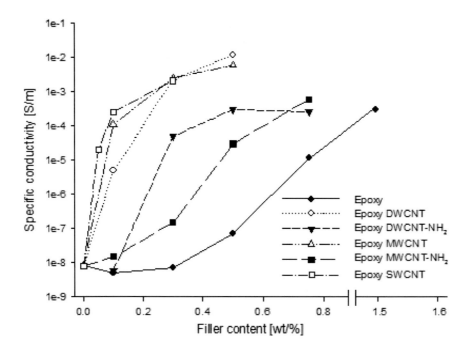

Figure 4. Electrical conductivity of the nanocomposites as function of filler content in weight percent [97].

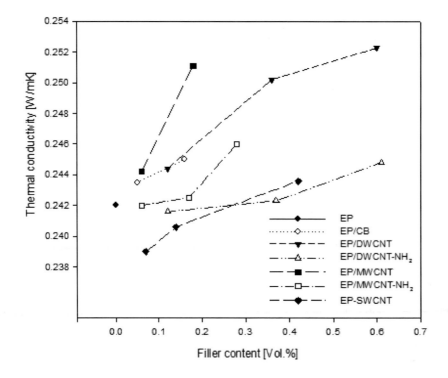

Figure 5. Thermal conductivity as function of the relative provided interfacial area per gram composite (m2/g) [97].

Table 2. mechanical and electrical information of neat biopolymers compared with their carbon nanotube nanocomposites

method	biopolymer	mechanical			comparison modulus (Pa)	stage modulus (Gpa)	conductivity	ref
		tensile modulus (Mpa)	tensile strength (Mpa)	strain to failure (%)				
polymerized hydrogel	neat collagen				1284±94		11.37ms±0.16	[98]
	collagen/CNTs				1127±73		11.85ms±0.67	
solution-evaporation	neat chitosan	1.08±0.04	37.7±4.5				0.021 nS/cm	[99-100]
	chitosan/CNTs	2.15±0.09	74.3±4.6				120 nS/cm	
wet spinning	neat chitosan	4250						[101]
	chitosan/CNTs	1025 0						
electrospinning	neat silk	140±2.21	6.18±0.3	5.78±0.65			0.028 S/cm	[102]
	silk/CNTs	4817.24±69.23	44.46±2.1	1.22±0.14			0.144 S/cm	
dry-jet wet spinning	neat cellulose	13100±1100	198±25	2.8±0.7		5.1	negligible	[103-104]
	cellulose/CNTs	14900±13 00	257±9	5.8±1.0		7.4	3000 S/cm	
electrospinning	neat cellulose	553±39	21.9±1.8	8.04±0.27				[105]
	cellulose/CNT	1144±37	40.7±2.7	10.46±0.33				

3.3. Carbon Nanotube Composite Application

Great attention has been paid in recent years to applying nanotube composites in various fields. Researchers [106] reviewed nanotubes composites based on gas sensors. These sensors play important roles for industry, environmental monitoring, biomedicine and so forth. The unique geometry, morphology, and material properties of CNTs led to apply them in gas sensors.

There are many topical studies for biological and biomedical applications of carbon nanotube composites due to its biocompatibility [107]. These components promoted biosensors [108], tissue engineering [95], and drug delivery [109] fields in biomedical technology.

On the other hand, light weight, mechanical strength, electrical conductivity, and flexibility are significant properties of carbon nanotubes for aerospace applications [110].

Researchers [111] represented an overview of carbon nanotube composite applications including electrochemical actuation, strain sensors, power harvesting, and bioelectronic sensors. They presented appropriate elastic and electrical properties for using nanoscale smart materials to synthesize intelligent electronic structures. In this context, we previously developed polyaniline/SWNTs composite fiber [112] and showed high strength, robustness, good conductivity and pronounced electroactivity of the composite. They presented new battery materials [113] and enhancement of performance artificial muscles [114] by using of these carbon nanotube composites.

Researchers [115] addressed sustainable environment and green technologies perspective for carbon nanotube applications. These contexts are including many engineering fields such as wastewater treatment, air pollution monitoring, biotechnologies, renewable energy technologies, and green nanocomposites.

Researchers [116] for the first time discovered photo-induced electron transfer from CNTs. Later, optical and photovoltaic properties of carbon nanotube composites have been studied by many groups. Results suggested the possible creation of photovoltaic devices due to hole-collecting electrode of CNTs [117].

Food packaging is another remarkable application of carbon nanotube composites. Usually, poor mechanical and barrier properties have limited applying biopolymers. Hence, appropriate filler is necessary for promotion of matrix properties. Unique properties of CNTs have been improved thermal stability, strength and modulus, and better water vapor transmission rate of applied composites in this industry [118].

4. Chitosan/Carbon Nanotube Composites

In recent decades, scientists became interested in creation of chitosan/CNTs composite due to providing unexampled properties of this composite. They attempted to create new properties by adding the CNTs to chitosan biopolymers. In recent years, several research articles were published in variety of applications. We summarized all of the applications of chitosan/CNTs nano-composites by these articles in a graph at Figure 6.

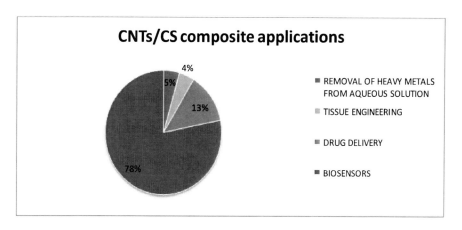

Figure 6. Graph of chitosan/CNTs nano-composites application.

4.1. Chitosan/Carbon Nanotube Nanofluids

Viscosity and thermal conductivity of nanofluids containing MWNTs stabilized by chitosan were investigated by some researchers [119]. The MWNTs fluid was stabilized by chitosan solution. Studies showed that thermal conductivity enhancements obtained were significantly higher than those predicted using the Maxwell's theory. Also, they observed that dispersing chitosan into deionized water increased the viscosity of nanofluid significantly and behaved as nan-Newtonian fluid.

4.2. Preparation Methods of Chitosan/CNTs Nano-Composites

There are several methods for creation of nanobiocomposites. Among them, researchers have studied some of these methods for preparation the chitosan/CNTs nano-composites. We represented these methods in the continuance of our review.

4.2.1. Solution-Casting-Evaporation

Researchers [120] assessed electrochemical sensing of carbon nanotube/chitosan system on dehydrogenase enzymes for preparing glucose biosensor first in 2004. They prepared the nanotube composite by use of solution-casting-evaporation method. In this method, the CNT/CHT films were prepared by casting of CNT/CHT solution on the surface of glassy carbon electrode and then dried. This CNT/CHT system showed a new biocomposite platform for development of dehydrogenase-based electrochemical biosensors due to providing a signal transduction of CNT. The great results of this composite in biomedical application led to many studies in this context.

The effect of CNT/CS matrix on direct electron transfer of glucose oxidase and glucose biosensor was examined by some researchers.[121]. They exhibited high sensitivity and better stability of CNT/CS composites compared with pure chitosan films. Furthermore, others [122] used the SWNT/CS films for preparation a new galactose biosensor with highly reliable detection of galactose. Researchers [123] immobilized lactate dehydrogenase within MWNT/CS nanocomposite for producing lactate biosensors. This proposed biosensor provided a fast response time and high sensitivity. Also, others [124] showed the immobilization of GOD molecules into chitosan-wrapped SWNT film is an efficient method for the development of a new class of very sensitive, stable, and reproducible electrochemical biosensors.

Several experiments were performed on DNA biosensor based on chitosan film doped with carbon nanotubes by others [125]. They found that CNT/CHT film can be used as a stable and sensitive platform for DNA detection. The results demonstrated improvement of sensor performance by adding CNT to chitosan film. Moreover, the analytical performance of glassy carbon electrodes modified with a dispersion of MWNT/CS for quantification of DNA was reported by others [126]. This new platform immobilized the DNA and opened the doors to new strategies for development of biosensors.

In other experiments, researchers [127] reported high sensitivity of glassy carbon electrode modified by MWNT-CHT for cathodic stripping voltammetric measurement of bromide (Br-).

Qian et al. [128] prepared amperometric hydrogen peroxide biosensor based on composite film of MWNT/CS. The results showed excellent electrocatalytical activity of the biosensor for H_2O_2 with good repeatability and stability.

Researchers [129] reported effect of CNT/CHT matrix on amperometric laccase biosensor. Results showed some major advantages of this biosensor involving detecting different substrates, possessing high affinity and sensitivity, durable long-term stability, and facile preparation procedure.

Others [130] paid particular attention to prepare SWNT/CHT film by solution-cast method and then characterized their drug delivery properties. They found that the SWNT/CS film has enhanced slowing down release of dexamethasone.

Growth of apatite on chitosan-multiwalled carbon nanotube composite membranes at low MWNT concentrations was reported by others [131]. Apatite was formed on the composites with low concentrations.

CNT/CHT nanobiocomposite for immunosensor was fabricated by some researchers [132]. Electron transport in this nanobiocomposite enhanced and improved the detection of ochratoxin-A, due to high electrochemical properties of SWNT. Also, CNT/CHT nanocomposite used for detection of human chorionic gonadotrophin antibody was performed by others [133] and displayed high sensitivity and good reproducibility.

4.2.1.1. Properties and Characterization

Researchers [134] represented that morphology and mechanical properties of chitosan have promoted by adding CNTs. Beside, others [135] proved that conducting direct electron is very useful for adsorption of hemoglobin in CNT/CHT composite film. These studies have demonstrated that this nanobiocomposite can used in many field such as biosensing and biofuel cell approaches.

Some researchers [136] evaluated water transport behavior of chitosan porous membranes containing MWNTs. They characterized two nanotube composites with low molecular weight CSP6K and high molecular weight CSP10K. Because of hollow nanochannel of MWNTs located among the pore network of chitosan membrane, the water transport results for CSP6K enhanced, when the MWNTs content is over a critical content.

But, for CSP10K series membranes, the water transport rate decreased with increase of MWNTs content due to the strong compatibilizing effect of MWNTs.

CNT/CHT nanocomposites were utilized by using of poly(styrene sulfonic acid)-modified CNTs by others [137]. Thermal, mechanical, and electrical properties of CNT/CHT composite film prepared by solution-casting have application potentials for separation membranes and sensor electrodes.

4.2.2. Cross-linking-Casting-Evaporation

In a new approach, MWNTs functionalized with –COOH groups at the end or at the sidewall defects of nanotubes by carbon nanotubes in nitric acid solvent. The functionalized carbon nanotubes immobilized into chitosan films by researchers [138]. This film applied in amperometric enzyme biosensors and resulted glucose detection and high sensitivity.

In a novel method, researchers [139] cross-linked chitosan with free –CHO groups by glutaraldehyde and then MWNTs were added to the mixture. The cross-linked MWNT-CHT composite immobilized acetylcholinesterase (AChE) for detecting of both acetylthiocholine and organophosphorous insecticides. On the other hand, researchers [140] created a new method for cross-linking CHT with carboxylated CNT. This new method was performed by adding glutaraldehyde to MWNT/CHT solution. They immobilized AChE on the composite for preparing an amperometric acetylthiocholine sensor. The suitable fabrication reproducibility, rapid response, high sensitivity, and stability could provide an amperometric detection of carbaryl and treazophos [141] pesticide. Results [142] showed the removal of heavy metals including copper, zinc, cadmium, and nickel ions from aqueous solution in MWNT/CHT nanocomposite film.

4.2.3. Surface Deposition Cross-linking

Researches [143] decorated carbon nanotube with chitosan by surface deposition and cross-linking process. In this new method, chitosan macromolecules as polymer cationic surfactants were adsorbed on the surface of the CNTs. In this step, CHT is capable of stably dispersing the CNT in acidic aqueous solution. The pH value of the system was increased by ammonia solution to become non-dissolvable of chitosan in aqueous media. Consequently, the soluble chitosan deposited on the surface of carbon nanotubes similar to chitosan coating. Finally, the surface-deposited chitosan was cross-linked to the CNTs by glutaraldehyde. They found potential applications in biosensing, gene and drug delivering for this composite.

4.2.4. Electrodeposition Method

Researchers [144] used nanocomposite film of CNT/CHT as glucose biosensor by a simple and controllable method. In this one-step electrodeposition method, a pair of gold electrodes was connected to a direct current power supply and then dipped into the CNT/CHT solution. Herein, the pH near the cathode surface increased thereby solubility of chitosan decreased. In pH about 6.3, chitosan become insoluble and the chitosan-entrapped CNT will be deposited onto the cathode surface.

Others [145] also characterized electrocatalytic oxidation and sensitive electroanalysis of NADH on a novel film of CS-DA-MWNTs and improved detection sensitivity. In this new method, glutaraldehyde cross-linked CHT DA with the covalent attachment of DA molecules to CHT chains formed by Schiff bases. Following, solution of MWNT dispersed in CHT-DA solution dropped on an Au electrode for preparing CHT-DA-MWNTs film and finally dried.

4.2.5. Covalently Grafting

Carboxylic acid (-COOH) groups were formed on the walls of CNTs by refluxing of CNTs in acidic solution. The carboxylated CNTs were added to aqueous solution of chitosan. Grafting reactions were accomplished by purging with N_2, and heated to 98 °C of CNTs/CS solution. Researchers [146] compared mechanical properties and water stability of CNTs-grafted-CS with the ungrafted CNTs. A significantly improved dispersion in chitosan matrix has resulted in an important improvement storage modulus and water stability of the chitosan nanocomposites.

Researchers [147] created another process for making a CHT-grafted MWNT composite. In this different method, after preparing oxidized MWNT (MWNT-COOH), they generated the acyl chloride functionalized MWNT (MWNT-COCl) in a solution of thionyl chloride. In the end, the MWNT-grafted-CS was synthesized by adding CHT to MWNT-COCl suspension in anhydrous dimethyl formamide. The covalent modification has improved interfacial bonding and resulted in high stability of CNT dispersion. Biosensors and other biological applications are evaluated as potential usage of this component. Also, others [148] prepared a similar composite by reacting CNT-COCl and chitosan with potassium persulfate, lactic acid, and acetic acid solution at 75 °C. They estimated that the CNT-grafted-CHT composite can be used in bone tissue engineering because the improvement of thermal properties.

4.2.5.1. Nucleophilic Substitution Reaction

Covalent modification of MWNT was accomplished with a low molecular weight chitosan (LMCS) [149]. In this method, the acyl chloride functionalized grafted to LMCS in DMF/Pyridine solution. This novel derivation of MWNTs can be solved in DMF, DMAc and DMSO but also in aqueous acetic acid solution.

4.2.6. Electrostatic Interaction

Furthermore [150], synthesized CHT nanoparticles-coated fMWNTs composite by electrostatic interactions between CHT particles and functionalized CNT. They prepared CHT nanoparticles and CHT microspheres by precipitation method and cross-linking method, respectively. The electrostatic interactions between CHT particles solution in distilled deionized water and the carboxylated CNTs were conformed by changing the pH solution. Results showed same surface charges in pH 2 (both were positively charged) and pH 8 (both were negatively charged). The electrostatic interactions can be caused at pH 5.5 due to different charges between CHT particles and fCNT with positive and negative surface charges, respectively. These CHT particles/CNT composite materials could be utilized for potential biomedical purposes.

Researchers[151] constructed SWNTs/ phosphotungstic acid modified SWNTs/CS composites using phosphotungstic acid as an anchor reagent to modify SWNTs. They succeeded in using PW_{12}-modified SWNT with a negative surface charge and, on the contrary, positively charged chitosan by electrostatic interaction. These strong interfacial interactions between SWNTs and chitosan matrix presented favorable cytocompatibility for the potential use as scaffolds for bon tissue engineering.

4.2.7. Microwave Irradiation

Researchers [152] created a new technique for synthesis of chitosan-modified carbon nanotube by using microwave irradiation. In this technique, MWNTs solution in nitric acid were placed under microwave irradiation and dried for purification of MWNTs. A mixture of purified MWNTs and chitosan solution was reacted in the microwave oven and then centrifuged.

The yielded black-colored solution was adjusted at pH 8 and centrifuged for precipitation of CNT/CHT composite. This facilitated technique is much more efficient than conventional methods.

4.2.8. Layer-by-Layer

Researchers [153] characterized MWNT/CHT composite rods with layer-by-layer structure were prepared *via in situ* precipitation method. Samples were prepared by coating CHT solution on internal surface of a cylindrical tube and then filling with MWNT/CHT solution in acetic acid.

They examined morphology, mechanical, and thermal properties of this composite rod. The excellent mechanical property of these new composite rods has made potential of bone fracture internal fixation application.

4.2.8.1. Layer-by-Layer Self Assembly

Researchers [154] produced a homogeneous multilayer film of MWNT/CHT by using layer-by-layer self assembly method. In this method, negatively charged substrates were dipped into poly (ethyleneimine) aqueous solution, MWNTs suspension, and CHT solution, respectively, and dried at the end. In this process, both of CHT and PEI solutions were contained NaCl for the LBL assembly. The films showed stable optical properties and were be appropriate for biosensor applications.

4.2.9. Freeze-Drying

Researchers [155] synthesized and characterized a highly conductive, porous, and biocompatible MWNT/CHT biocomposite film by freeze-drying technique. This process was performed by freezing MWNT/CHT dispersion into an aluminum mold and then drying. Such a composite permitted delivery of needed antibiotics with effect of increased antibiotic efficacy in a patent by others [156].

4.2.10. Wet-Spinning

Researchers [151] recently reported that chitosan is a good dispersing agent for SWNT. They also demonstrated several methods for preparing SWNT/CHT macroscopic structure in the form of films, hydrogels and fibers [148]. The CNT/CHT dispersion in acetic acid were spun into a ethanol:NaOH coagulation solution bath. They were demonstrated increasing mechanical properties of wet spun fibers by improving dispersion [156].

4.2.11. Electrospinning

In our recent work, the chitosan (CHT)/multi-walled carbon nanotubes (MWNTs) composite nanofiber were fabricated by using electrospinning. In our experimental researches, different solvents including acetic acid 1-90%, formic acid, and Tri-Fluoroacetic Acid TFA/DCM were tested for the electrospinning of chitosan/carbon nanotube. No jet was seen upon applying the high voltage, even above 25 kV, by using of acetic acid 1-30% and formic acid as the solvent for chitosan/carbon nanotube. When the acetic acid 30-90% was used as the solvent, beads were deposited on the collector. Therefore, under these conditions, nanofibers were not formed.

The TFA/DCM (70:30) solvent was the only solvent that resulted in electrospinability of chitosan/carbon nanotube. The scanning electron microscopic (Fig. images showed the homogenous fibers with an average diameter of 455 nm (306-672)) were prepared with chitosan/carbon nanotube dispersion in TFA/DCM 70:30. These nanofibers have a potential for biomedical applications.

Figure 8. electron micrographs of electrospun fibers at chitosan concentration 10 wt%, 24 kV, 5 cm, TFA/DCM: 70/30.

Conclusion

With less than ten years history, several tens of research studies have been created in chitosan biocomposites reinforcement using carbon nanotubes. In conclusion, much progress has been made in preparation and characterization of the CNTs/CHT nanocomposites. We reported several methods for preparing these nanobiocomposites. In addition, the CNTs/CHT applications have been classified including biomedicine (tissue engineering, biosensors, and drug delivery) and wastewater in this review.

Most importantly, the overriding results of electrospinning of CNTs/CHT nanocomposites in our recent paper have been discussed. There is expected a high potential application in tissue engineering and drug delivery by these nanobiocomposites. It is believed that with more attentions to the preparation methods of CNTs/CHT, nanocomposites and their characterization have a promising future in biomedicine science.

References

[1]   W. Praznik and C. V. Stevens, *Renewable Bioresources: Scope and Modification for Non-food Applications*, John Wiley & Sons, Ltd, Hoboken, p. 49 (2004)

[2]   D. A. D. Parry and E. N. Baker, *Rep. Prog. Phys.* 47, 1133 (1984)

[3]   S. Bhattacharyya, S. Guillot, H. Dabboue, J.-F. Tranchant and J.-P. Salvetat, *Biomacromolecules.* 9, 505 (2008)

[4]   M. Lavorgna, F. Piscitelli, P. Mangiacapra and G. G. Buonocore, *Carbohydr. Polym.* 82, 291 (2010)

[5]   X. Cao, Y. Chen, P. R. Chang and M. A. Huneault, *J. Appl. Polym. Sci.* 106, 1431 (2007)

[6]   S. Ramakrishna, J. Mayer, E. Wintermantel and K. W. Leong, *Compos. Sci. Technol.* 61, 1189 (2001)

[7]   D. Liang, B. S. Hsiao and B. Chu, *Adv. Drug Delivery Rev.* 59, 1392 (2007)

[8]   Z. Liu, Y. Jiao, Y. Wang, C. Zhou and Z. Zhang, *Adv. Drug Deliver Rev.* 60, 1650 (2008)

[9]   O. C. Agboh and Y. Qin, *Polym. Adv. Technol.* 8, 355 (1997)

[10]   I. Aranaz, M. Mengíbar, R. Harris, I. Paños, B. Miralles, N. Acosta, G. Galed and Á. Heras, *Current Chemical Biology.* 3, 203 (2009)

[11]   Y. Zhang, C. Xue, Y. Xue, R. Gao and X. Zhang, *Carbohydr. Res.* 340, 1914 (2005)

[12]   P. J. VandeVord, H. W. T. Matthew, S. P. DeSilva, L. Mayton, B. Wu and P. H. Wooley, *J. Biomed. Mater. Res.* 59, 585 (2002)

[13]   M. Ratajska, G. Strobin, M. Wiśniewska-Wrona, D. Ciechańska, H. Struszczyk, S. Boryniec, D. Biniaś and W. Biniaś, *Fibers Text East Eur.* 11, 75 (2003)

[14]   R. Jayakumar, M. Prabaharan, S. V. Nair and H. Tamura, *Biotechnol. Adv.* 28, 142 (2010)

[15]   J. T. Bamgbose, S. Adewuyi, O. Bamgbose and A. A. Adetoye, *Afr. J. Biotechnol.* 9, 2560 (2010)

[16]   B. Krajewska, *Enzyme Microb. Technol.* 35, 126 (2004)

[17]   H. Ueno, T. Mori and T. Fujinaga, *Adv. Drug Deliver Rev.* 52, 105 (2001)

[18]   I.-Y. Kim, S.-J. Seo, H.-S. Moon, M.-K. Yoo, I.-Y. Park, B.-C. Kim and C.-S. Cho, *Biotechnol. Adv.* 26, 1 (2008)

[19]   V. R. Sinha, A. K. Singla, S. Wadhawan, R. Kaushik, R. Kumria, K. Bansal and S. Dhawan, *Int. J. Pharm.* 274, 1 (2004)

[20]   A. D. Martino, M. Sittinger and M. V. Risbud, *Biomaterials.* 26, 5983-5990 (2005)

[21]   R. A. A. Muzzarelli, *Cell mol. life sci.* 53, 131 (1997)

[22] R. A. A. Muzzarelli, C. Muzzarelli, R. Tarsi, M. Miliani, F. Gabbanelli and M. Cartolari, *Biomacromolecules.* 2, 165 (2001)

[23] P. K. Dutta, S. Tripathi, G. K. Mehrotra and J. Dutta, *Food Chem.* 114, 1173 (2009)

[24] S. Boonlertnirun, C. Boonraung and R. Suvanasara, *Journal of Metals, Materials and Minerals.* 18, 47 (2008)

[25] K.-S. Huang, W.-J. Wu, J.-B. Chen and H.-S. Lian, *Carbohydr. Polym.* 73, 254 (2008)

[26] P. Lertsutthiwong, S. Chandrkrachang and W. F. Stevens, *Journal of Metals, Materials and Minerals.* 10, 43 (2000)

[27] R. H. Y. Subban and A. K. Arof, *Phys. Scr.* 53, 382 (1996)

[28] M. H. Ottøy, K. M. Vårum, B. E. Christensen, M. W. Anthonsen and O. Smidsrød, *Carbohydr. Polym.* 31, 253 (1996)

[29] P. K. Dutta, J. Dutta and V. S. Tripathi, *J. Sci. Ind. Res.* 63, 20 (2004)

[30] Q. Li, J. Zhou and L. Zhang, *J Polym Sci Pol Phys*, 47, 1069 (2009)

[31] E. T. Thostenson, C. Li and T.-W. Chou, *Compos. Sci. Technol.* 65, 491 (2005)

[32] S. Iijima, *Nature.* 354, 56 (1991)

[33] D. S. Benthune, C. H. Kiang, M. S. d. Vries, G. Gorman, R. Savoy, J. Vazquez and R. Beyers, *Nature.* 363, 605 (1993)

[34] S. Iijima and T. Ichihashi, *Nature*, 363, 603 (1993)

[35] M. Trojanowicz, *TrAC Trends Anal. Chem.* 25, 480 (2006)

[36] L. Duclaux, *Carbon.* 40, 1751 (2002)

[37] Y. Y. Wang, S. Gupta, J. M. Garguilo, Z. J. Liu, L. C. Qin and R. J. Nemanich, *Diamond and Related Materials.* 14, 714 (2005)

[38] J. Guo, S. Datta and M. Lundstrom, *IEEE Transaction On Electron Devices.* 51, 172 (2004)

[39] C.-S. Kuo, A. Bai, C.-M. Huang, Y.-Y. Li, C.-C. Hu and C.-C. Chen, *Carbon.* 43, 2760 (2005)

[40] R. L. Jacobsen, T. M. Tritt, J. R. Guth, A. C. Ehrlich and D. J. Gillespie, *Carbon.* 33, 1217 (1995)

[41] C. Journet, W. K. Maser, P. Bernier, A. Loiseau, M. L. l. Chapelle, S. Lefrant, P. Deniard, R. Leek and J. E. Fischer, *Nature.* 388, 756 (1997)

[42] A. Thess, R. Lee, P. Nikolaev, H. Dai, p. Petit, J. Robert, C. Xu, Y. H. Lee, S. G. Kim, A. G. Rinzler, D. T. Colbert, G. E. Scuseria, D. Tomanek, J. E. Fischer and R. E. Smalley, *Science.* 273, 483 (1996)

[43] A. M. Cassell, J. A. Raymakers, J. Kong and H. Dai, *J. Phys. Chem. B* 103, 6482 (1999)

[44] S. Fan, W. Liang, H. Dang, N. Franklin, T. Tombler, M. Chapline and H. Dai, *Physica E*, 8, 179 (2000)

[45] S. Xie, W. Li, Z. Pan, B. Chang and L. Sun, *Mater Sci. Eng. A*, 286, 11 (2000)

[46] Z. K. Tang, L. Zhang, N. Wang, X. X. Zhang, G. H. Wen, G. D. Li, J. N. Wang, C. T. Chan and P. Sheng, *Science.* 292, 2462 (2001)

[47] A. Peigney, C. Laurent, E. Flahaut, R. R. Bacsa and A. Rousset, *Carbon.* 39, 507 (2001)

[48] Z. W. Pan, S. S. Xie, L. Lu, B. H. Chang, L. F. Sun, W. Y. Zhou, G. Wang and D. L. Zhang, *Appl. Phys. Lett.* 74, 3152 (1999)

[49] L. Forro, J. P. Salvetat, J. M. Bonard, R. Basca, N. H. Thomson, S. Garaj, L. Thien-Nga, R. Gaal, A. Kulik, B. Ruzicka, L. Degiorgi, A. Bachtold, C. Schonenberger, S. Pekker and K. Hernadi, *Science and Application of Nanotubes.* 297 (2000)

[50] S. Frank, P. Poncharal, Z. L. Wang and W. A. d. Hccr, *Science.* 280, 1744-1746 (1998)

[51] D. A. Britzab and A. N. Khlobystov, *Chem. Soc. Rev.* 35, 637 (2006)

[52] R. Andrews and M. C. Weisenberger, *Curr. Opin. Solid State Mater Sci.* 8, 31 (2004)

[53] A. Hirsch, *Angew Chem. Int. Ed.* 41, 1853 (2002)

[54] C. P. Firme and R. B. Prabhakar, *Nanomed. Nanotechnol. Biol. Med.* 6, 245 (2010)

[55] S. Niyogi, M. A. Hamon, H. Hu, B. Zhao, P. Bhowmik, R. Sen, M. E. Itkis and R. C. Haddon, *Acc. Chem. Res.* 35, 1105 (2002)

[56] H. Kuzmany, A. Kukovecz, F. Simona, M. Holzweber, C. Kramberger and T. Pichler, *Synth. Met.* 141, 113 (2004)

[57] Z. Spitalsky, D. Tasis, K. Papagelis and C. Galiotis, *Prog Polym Sci*, 35, 357 (2010)

[58] R. Narain, A. Housni and L. Lane, *J. Polym. Sci. A Polym. Chem.* 44, 6558 (2006)

[59] M. Wang, K. P. Pramoda and S. H. Goh, *Carbon.* 44, 613 (2006)

[60] H. Touhara, A. Yonemoto, K. Yamamoto, S. Komiyama, S. Kawasaki, F. Okino, T. Yanagisawa and M. Endo, *Fluorine Chem.* 114, 181 (2002)

[61] W. Zhang, J. K. Sprafke, M. Ma, E. Y. Tsui, S. A. Sydlik, G. Rutledge and T. M. Swager, *J. Am. Chem. Soc.* 131, 8446 (2009)

[62] P. He, Y. Gao, J. Lian, L. Wang, D. Qian, J. Zhao, W. Wang, M. J. Schulz, X. P. Zhou and D. Shi, *Compos. Part A-Appl. S* 37, 1270 (2006)

[63] A. B. Sulong, C. H. Azhari, R. Zulkifli, M. R. Othman and J. Park, *Eur. J. Sci. Res.* 33, 295 (2009)

[64] C. Wang, Z. X. Guo, S. Fu, W. Wu and D. Zhu, *Prog. Polym. Sci.* 29, 1079 (2004)

[65] J. Rausch, R.-C. Zhuang and E. Mäder, *Compos. Part A-Appl. S.* 41, 1038 (2010)

[66] N. G. Sahoo, S. Rana, J. W. Cho, L. Li and S. H. Chan, *Prog. Polym. Sci.* 35, 837 (2010)

[67] D. Zheng, X. Li and J. Ye, *Bioelectrochemistry.* 74, 240 (2009)

[68] X. Zhang, L. Meng and Q. Lu, *ACS Nano.* 10, 3200 (2009)

[69] H. Wang, *Curr. Opin. Colloid In*, 14, 364 (2009)

[70] A. D. Schlüter, A. Hirsch and O. Vostrowsky. Functionalization of Carbon Nanotubes. *Functional Molecular Nanostructures.* 245: 193 (2005)

[71] P.-C. Ma, N. A. Siddiqui, G. Marom and J.-K. Kim, *Composites Part A.* 41, 1345 (2010)

[72] V. Mottaghitalab, G. M. Spinks and G. G. Wallace, *Synth. Met.* 152, 77 (2005)

[73] W. Cheung, F. Pontoriero, O. Taratula, A. M. Chen and H. He, *Adv. Drug Deliver Rev.* 62, 633 (2010)

[74] S. Piovesan, P. A. Cox, J. R. Smith, D. G. Fatouros and M. Roldo, *Phys. Chem. Chem. Phys.* 12, 15636 (2010)

[75] J. Chen, H. Liu, W. A. Weimer, M. D. Halls, D. H. Waldeck and G. C. Walker, *J. Am. Chem. Soc.* 124, 9034 (2002)

[76] V. Vamvakaki, M. Fouskaki and N. Chaniotakis, *Anal. Lett.* 40, 2271 (2007)

[77] Y. Kang, Y.-C. Liu, Q. Wang, J.-W. Shen, T. Wu and W.-J. Guan, *Biomaterials.* 30, 2807 (2009)

[78] B. S. Harrison and A. Atala, *Biomaterials.* 28, 344 (2007)

[79] M. Moniruzzaman and K. I. Winey, *Macromolecules.* 39, 5194 (2006)

[80] P. M. Ajayan, O. Stephan, C. Colliex and D. Trauth, *Science*. 265, 1212 (1994)

[81] P. Liu, *Eur. Polym. J.* 41, 2693 (2005)

[82] M. A. L. Manchado, L. Valentini, J. Biagiotti and J. M. Kenny, *Carbon*, 43, 1499 (2005)

[83] Q. Wang and V. K. Varadan, *Smart Mater Struct.* 14, 281 (2005)

[84] P. K. Mallick, *Fiber Reinforced Composites: Materials, Manufacturing, and Design*, Taylor & Francis Group, London, p. 43 (2008)

[85] M. S. P. Shaffer and J. K. W. Sandler, *Processing and properties of nanocomposites*, World Scientific Publishing Co. Pte. Ltd., Singapore, p. 31 (2006)

[86] C. W. Nan, Z. Shi and Y. Lin, *Chem. Phys. Lett.* 375, 666 (2003)

[87] J. N. Coleman, U. Khan, W. J. Blau and Y. K. Gun'ko, *Carbon*. 44, 1624 (2006)

[88] H. D. Wagner, O. Lourie, Y. Feldman and R. Tenne, *Appl. Phys. Lett.* 72, 188 (1998)

[89] Q. Wang, J. Dai, W. Li, Z. Wei and J. Jiang, *Compos. Sci. Technol.* 68, 1644 (2008)

[90] U. Khan, K. Ryan, W. J. Blau and J. N. Coleman, *Compos. Sci. Technol.* 67, 3158 (2007)

[91] A. Allaoui, S. Bai, H. M. Cheng and J. B. Bai, *Compos. Sci. Technol.* 62, 1993 (2002)

[92] A. M. K. Esawi and M. M. Farag, *Mater Design.* 28, 2394 (2007)

[93] F. H. Gojny, M. H. G. Wichmann, B. Fiedler, I. A. Kinloch, W. Bauhofer, A. H. Windle and K. Schulte, *Polymer.* 47, 2036 (2006)

[94] K. Kamaras, M. E. Itkis, H. Hu, B. Zhao and R. C. Haddon, *Science.* 301, 1501 (2003)

[95] S. Shenogin, A. Bodapati, L. Xue, R. Ozisik and P. Keblinski, *Appl. Phys. Lett.* 85, 2229 (2004)

[96] R. A. MacDonald, B. F. Laurenzi, G. Viswanathan, P. M. Ajayan and J. P. Stegemann, *J. Biomed. Mater Res. A.* 74A, 489 (2005)

[97] S. Bose, R. A. Khare and P. Moldenaers, *Polymer.* 51, 975 (2010)

[98] Z. Tosun and P. S. McFetridge, *J. Neural Eng.* 7, 1 (2010)

[99] S.-F. Wang, L. Shen, W.-D. Zhang and Y.-J. Tong, *Biomacromolecules.* 6, 3067 (2005)

[100] G. M. Spinks, S. R. Shin, G. G. Wallace, P. G. Whitten, S. I. Kim and S. J. Kim, *Sens Actuators B*, 115, 678 (2006)

[101] M. Gandhi, H. Yang, L. Shor and F. Ko, *Polymer.* 50, 1918 (2009)

[102] S. S. Rahatekar, A. Rasheed, R. Jain, M. Zammarano, K. K. Koziol, A. H. Windle, J. W. Gilman and S. Kumar, *Polymer.* 50, 4577 (2009)

[103] 103 H. Zhang, Z. Wang, Z. Zhang, J. Wu, J. Zhang and J. He, *Adv. Mater.* 19, 698 (2007)

[104] P. Lu and Y.-L. Hsieh, *ACS Appl. Mater Interfaces.* 2, 2413 (2010)

[105] Y. Wang and J. T. W. Yeow, *Journal of Sensors.* 2009, 1 (2009)

[106] W. Yang, P. Thordarson, J. J. Gooding, S. P. Ringer and F. Braet, *Nanotechnology.* 18, 1 (2007)

[107] J. Wang, *Electroanal.* 17, 7 (2005)

[108] M. Foldvari and M. Bagonluri, *Nanomed. Nanotechnol. Biol. Med.* 4, 183 (2008)

[109] S. Belluccia, C. Balasubramanianab, F. Micciullaac and G. Rinaldid, *J. Exp. Nanosci.* 2, 193 (2007)

[110] I. Kang, Y. Y. Heung, J. H. Kim, J. W. Lee, R. Gollapudi, S. Subramaniam, S. Narasimhadevara, D. Hurd, G. R. Kirikera, V. Shanov, M. J. Schulz, D. Shi, J. Boerio, S. Mall and M. Ruggles-Wren, *Composites Part B* 37, 382 (2006)

[111] V. Mottaghitalab, G. M. Spinks and G. G. Wallace, *Polymer.* 47, 4996 (2006)

[112] C. Y. Wang, V. Mottaghitalab, C. O. Too, G. M. Spinks and G. G. Wallace, *J. Power Sources.* 163, 1105 (2007)

[113] V. Mottaghitalab, B. Xi, G. M. Spinks and G. G. Wallace, *Synth. Met.* 156, 796 (2006)

[114] Y. T. Ong, A. L. Ahmad, S. H. S. Zein and S. H. Tan, *Braz. J. Chem. Eng.* 27, 227 (2010)

[115] N. S. Sariciftci, L. Smilowitz, A. J. Heeger and F. Wudi, *Science.* 258, 1474 (1992)

[116] P. J. F. Harris, *Int. Mater Rev.* 49, 31 (2004)

[117] H. M. C. d. Azeredo, *Food Res. Int.* 42, 1240 (2009)

[118] T. X. Phuoc, M. Massoudi and R.-H. Chen, *Int. J. Therm. Sci.* 50, 12 (2011)

[119] M. Zhang, A. Smith and W. Gorski, *Anal. Chem.* 76, 5045. (2004)

[120] Y. Liu, M. Wang, F. Zhao, Z. Xu and S. Dong, *Biosens Bioelectron.* 21, 984 (2005)

[121] J. Tkac, J. W. Whittaker and T. Ruzgas, *Biosens Bioelectron.* 22, 1820 (2007)

[122] Y.-C. Tsai, S.-Y. Chen and H.-W. Liaw, *Sens. Actuators. B*, 125, 474 (2007)

[123] Y. Zhou, H. Yang and H.-Y. Chen, *Talanta.* 76, 419 (2008)

[124] J. Li, Q. Liu, Y. Liu, S. Liu and S. Yao, *Anal. Biochem.* 346, 107 (2005)

[125] S. Bollo, N. F. Ferreyr and G. A. Rivasb, *Electroanal.* 19, 833 (2007)

[126] Y. Zeng, Z.-H. Zhu, R.-X. Wang and G.-H. Lu, *Electrochim. Acta.* 51, 649 (2005)

[127] L. Qian and X. Yang, *Talanta.* 68, 721 (2006)

[128] Y. Liu, X. Qu, H. Guo, H. Chen, B. Liu and S. Dong, *Biosens. Bioelectron.* 21, 2195 (2006)

[129] S. Naficy, J. M. Razal, G. M. Spinks and G. G. Wallace, *Sens Actuators A*, 155, 120 (2009)

[130] J. Yang, Z. Yao, C. Tang, B. W. Darvell, H. Zhang, L. Pan, J. Liu and Z. Chen, *Appl. Surf. Sci.* 255, 8551 (2009)

[131] A. Kaushik, P. R. Solanki, M. K. Pandey, K. Kaneto, S. Ahmad and B. D. Malhotra, *Thin Solid Films.* 519, 1160 (2010)

[132] H. Yang, R. Yuan, Y. Chai and Z. Ying, *Colloids Surf. B.* 82, 463 (2011)

[133] W. Zheng, Y. Q. Chen and Y. F. Zheng, *Appl. Surf. Sci.* 255, 571 (2008)

[134] C. Tang, Q. Zhang, K. Wang, Q. Fu and C. Zhang, *J. Membr. Sci.* 337, 240 (2009)

[135] Y.-L. Liu, W.-H. Chen and Y.-H. Chang, *Carbohydr. Polym.* 76, 232 (2009)

[136] M. E. Ghica, R. Pauliukaite, O. Fatibello-Filho and C. M. A. Brett, *Sens Actuators. B*, 142, 308 (2009)

[137] V. B. Kandimalla and H. Ju, *Chem. Eur. J.* 12, 1074 (2006)

[138] D. Du, X. Huang, J. Cai, A. Zhang, J. Ding and S. Chen, *Anal. Bioanal. Chem.* 387, 1059 (2007)

[139] D. Du, X. Huang, J. Cai and A. Zhang, *Sens. Actuators B*, 127, 531 (2007)

[140] M. A. Salam, M. S. I. Makki and M. Y. A. Abdelaal, *J. Alloys Compd.* 509, 2582 (2010)

[141] Y. Liu, J. Tang, X. Chen and J. H. Xin, *Carbon.* 43, 3178 (2005)

[142] X.-L. Luo, J.-J. Xu, J.-L. Wang and H.-Y. Chen, *Chem. Commun.* 16, 2169 (2005)

[143] B. Ge, Y. Tan, Q. Xie, M. Ma and S. Yao, *Sens. Actuators. B*, 137, 547 (2009)

[144] Y.-T. Shieh and Y.-F. Yang, *Eur. Polym. J.* 42, 3162 (2006)

[145] Z. Wu, W. Feng, Y. Feng, Q. Liu, X. Xu, T. Sekino, A. Fujii and M. Ozaki, *Carbon.* 45, 1212 (2007)

[146] L. Carson, C. Kelly-Brown, M. Stewart, A. Oki, G. Regisford, Z. Luo and V. I. Bakhmutov, *Mater Lett.* 63, 617 (2009)

[147] G. Ke, W. C. Guan, C. Y. Tang, Z. Hu, W. J. Guan, D. L. Zeng and F. Deng, *Chin. Chem. Lett.* 18, 361 (2007)

[148] S.-H. Baek, B. Kim and K.-D. Suh, *Colloids Surf. A*, 316, 292 (2008)

[149] Q. Zhao, J. Yin, X. Feng, Z. Shi, Z. Ge and Z. Jin, *J. Nanosci. Nanotechnol.* 10, 1 (2010)

[150] J.-G. Yu, K.-L. Huang, J.-C. Tang, Q. Yang and D.-S. Huang, *Int. J. Biol. Macromol.* 44, 316 (2009)

[151] Z.-k. Wang, Q.-l. Hu and L. Cai, *Chinese J. Polym. Sci.* 28, 801 (2010)

[152] X.-b. Li and X.-y. Jiang, *New Carbon Mater.* 25, 237 (2010)

[153] C. Lau and M. J. Cooney, *Langmuir.* 24, 7004 (2008)

[154] J. A. Jennings, W. O. Haggard and J. D. Bumgardner, "Chitosan/carbon nanotube composite scaffolds for drug delivery," United States Patent Application, 0266694.A1 (2010)

[155] J. M. Razal, K. J. Gilmore and G. G. Wallace, *Adv. Funct. Mater.* 18, 61 (2008)

[156] C. Lynam, S. E. Moulton and G. G. Wallace, *Adv. Mater.* 19, 1244 (2007)

In: Handbook of Research on Nanomaterials, Nanochemistry ...   ISBN: 978-1-61942-525-5
Editors: A. K. Haghi and G. E. Zaikov   © 2013 Nova Science Publishers, Inc.

Chapter III

Electrospun Biodegradable Nanofibers

A. K. Haghi[*]
University of Guilan, Iran

1. Introduction

Nanotechnology has become in recent years a topic of great interest to scientists and engineers and is now established as prioritized research area in many countries. The reduction of the size to the nano-meter range brings an array of new possibilities in terms of material properties, in particular with respect to achievable surface–to-volume ratios. Electrospinning of nanofibers is a novel process for producing superfine fibers by forcing a solution through a spinnerette with an electric field. An emerging technology of manufacturing of thin natural fibers is based on the principle of electrospinning process. In conventional fiber spinning, the mechanical force is applied to the end of a jet, whereas in the electrospinning process, the electric body force act on element of charged fluid. Electrospinning has emerged as a specialized processing technique for the formation of sub-micron fibers (typically between 100 nm and 1 μm in diameter), with high specific surface areas. Due to their high specific surface area, high porosity, and small pore size, the unique fibers have been suggested for wide range of applications. Electrospinning of nanofibers offers unique capabilities for producing novel natural nanofibers and fabrics with controllable pore structure.

Some 4-9% of cotton fiber is lost at textile mill in so-called opening and cleaning, which involves mechanically separating compressed clumps of fibers for removal of trapped debris. Another 1% is lost in drawing and roving-pulling lengths of fiber into longer and longer segments, which are then twisted together for strength. An average of 20% is lost during combing and yarn production. Typically, waste cotton is used in relatively low-value products such as cotton balls, yarn, and cotton batting. A new process for electrospinning waste cotton using a less harmful solvent has been developed.

Electrospinning is an economical and simple method used in the preparation of polymer fibers. The fibers prepared via this method typically have diameters much smaller than is

---
[*] Haghi@Guilan.ac.ir.

possible to attain using standard mechanical fiber-spinning technologies [1]. Electrospinning has gained much attention in the last few years as a cheap and straightforward method to produce nanofibers. Electrospinning differs from the traditional wet/dry fiber spinning in a number of ways, of which the most striking differences are the origin of the pulling force and the final fiber diameters. The mechanical pulling forces in the traditional industrial fiber spinning processes lead to fibers in the micrometer range and are contrasted in electrospinning by electrical pulling forces that enable the production of nanofibers. Depending on the solution properties, the throughput of single-jet electrospinning systems ranges around 10 ml/min. This low fluid throughput may limit the industrial use of electrospinning.

A stable cone-jet mode followed by the onset of the characteristic bending instability, which eventually leads to great reduction in the jet diameter, necessitate the low flow rate [2]. When the diameters of cellulose fiber materials are shrunk from micrometers (e.g., 10–100 mm) to submicrons or nanometers, there appear several amazing characteristics such as very large surface area-to-volume ratio (this ratio for a nanofiber can be as large as 103 times of that of a microfiber), flexibility in surface functionalities and superior mechanical performance (e.g., stiffness and tensile strength) compared with any other known form of the material.

These outstanding properties make the polymer nanofibers optimal candidates for many important applications [3]. These include filter media, composite materials, biomedical applications (tissue engineering scaffolds, bandages, drug release systems), protective clothing for the military, optoelectronic devices and semi-conductive materials, biosensor/chemosensor [4]. Another biomedical application of electrospun fibers that is currently receiving much attention is drug delivery devices. Researchers have monitored the release profile of several different drugs from a variety of biodegradable electrospun membranes. Another application for electrospun fibers is porous membranes for filtration devices.

Due to the inter-connected network type structure that electrospun fibers form, they exhibit good tensile properties, low air permeability, and good aerosol protection capabilities. Moreover, by controlling the fiber diameter, electrospun fibers can be produced over a wide range of porosities. Research has also focused on the influence of charging effects of electrospun non-woven mats on their filtration efficiency. The filtration properties slightly depended on the surface charge of the membrane; however the fiber diameter was found to have the strongest influence on the aerosol penetration. Electrospun fibers are currently being utilized for several other applications as well. Some of these include areas in nanocomposites.

Figure 1 compares the dimensions of nanofibers, micro fibers and ordinary fibers. When the diameters of polymer fiber materials are shrunk from micrometers (for example, 10-100 μm) to sub-microns or nanometers (for example, 10×10^{-3} -100×10^{-3} μm), there appear several amazing characteristics such as very large surface area-to-volume ratio, flexibility in surface functionalities, and superior mechanical performance compared with any other known form of material.

These outstanding properties make the polymer nanofibers optimal candidates for many important applications [4].

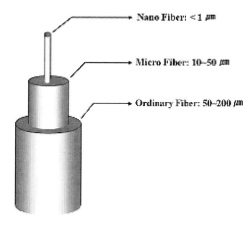

Figure 1.Classifications of fibers by the fiber diameter.

1.1. Electrospinning Set up

A schematic diagram to interpret electrospinning of nanofibers is shown in Figure 2.

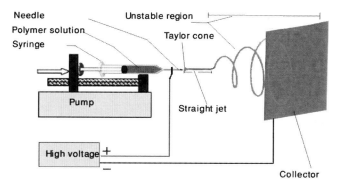

Figure 2. Schematic of electrospinning set up.

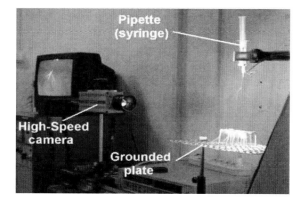

Figure 3. Electrospinning process.

There are basically three components to fulfill the process: a high voltage supplier, a capillary tube with a pipette or needle of small diameter, and a metal collecting screen. In the electrospinning process (Figure 3), a high voltage is used to create an electrically charged jet of polymer solution or melt out of the pipette. Before reaching the collecting screen, the solution jet evaporates or solidifies and is collected as an interconnected web of small fibers [5]. One electrode is placed into the spinning solution/melt or needle, and the other is attached to the collector. In most cases, the collector is simply grounded. The electric field is subjected to the end of the capillary tube that contains the solution fluid held by its surface tension. This induces a charge on the surface of the liquid. Mutual charge repulsion and the contraction of the surface charges to the counter electrode cause a force directly opposite to the surface tension [6]. As the intensity of the electric field is increased, the hemispherical surface of the fluid at the tip of the capillary tube elongates to form a conical shape known as the Taylor cone [7]. Further increasing the electric field, a critical value is attained with which the repulsive electrostatic force overcomes the surface tension and the charged jet of the fluid is ejected from the tip of the Taylor cone [8]. The jet exhibits bending instabilities due to repulsive forces between the charges carried with the jet. The jet extends through spiraling loops, as the loops increase in diameter, the jet grows longer and thinner until it solidifies or collects on the target [9]. A schematic diagram to interpret electrospinning of polymer nanofibers is shown in Figure 4 and Figure 5. There are basically three components to fulfill the process: a high voltage supplier, a capillary tube with a pipette or needle of small diameter, and a metal collecting screen. In the electrospinning process, a high voltage is used to create an electrically charged jet of polymer solution or melt out of the pipette. Before reaching the collecting screen or drum, the solution jet evaporates or solidifies and is collected as an interconnected web of small fibers. When an electric field is applied between a needle capillary end and a collector, surface charge is induced on a polymer fluid deforming a spherical pendant droplet to a conical shape. As the electric field surpasses a threshold value where electrostatic repulsion force of surface charges overcome surface tension, the charged fluid jet is ejected from the tip of the Taylor cone, and the charge density on the jet interacts with the external field to produce instability.

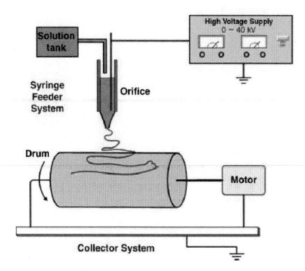

Figure 4. Schematic diagram of the electrospinning set up by drum collector.

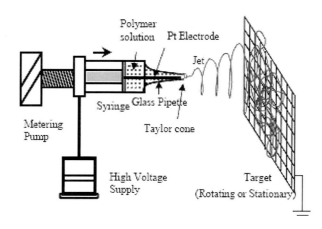

Figure 5. Schematic diagram of the electrospinning set up by screen collector.

It has been found that morphology such as fiber diameter and its uniformity of the electrospun polymer fibers are dependent on many processing parameters. These parameters can be divided into three groups as shown in Table 1. Under certain conditions, not only uniform fibers but also bead-like formed fibers can be produced by electrospinning. Although the parameters of the electrospinning process have been well analyzed in each of polymers, this information has been inadequate enough to support the electrospinning of ultrafine nanometer scale polymer fibers. A more systematic parametric study is hence required to investigate.

Table 1. Processing parameters in electrospinning

Solution properties	Viscosity
	Polymer concentration
	Molecular weight of polymer
	Electrical conductivity
	Elasticity
	Surface tension
Processing conditions	Applied voltage
	Distance from needle to collector
	Volume feed rate
	Needle diameter
Ambient conditions	Temperature
	Humidity
	Atmospheric pressure

2. Effect of Systematic Parameters on Electrospun Nanofibers

It has been found that morphology such as fiber diameter and its uniformity of the electrospun nanofibers are dependent on many processing parameters. These parameters can be divided into three main groups: a) solution properties, b) processing conditions, and c) ambient conditions. Each of the parameters has been found to affect the morphology of the electrospun fibers.

2.1. Solution Properties

Parameters such as viscosity of solution, solution concentration, molecular weight of solution, electrical conductivity, elasticity and surface tension have an important effect on morphology of nanofibers.

2.1.1. Viscosity

The viscosity range of a different nanofiber solution, which is spinnable, is different. One of the most significant parameters influencing the fiber diameter is the solution viscosity. A higher viscosity results in a large fiber diameter. Figure 6 shows the representative images of beads formation in electrospun nanofibers. Beads and beaded fibers are less likely to be formed for the more viscous solutions. The diameter of the beads become bigger, and the average distance between beads on the fibers longer as the viscosity increases.

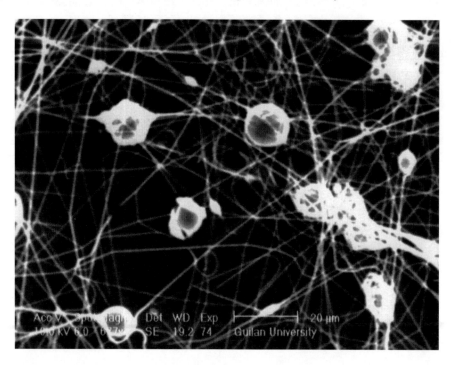

Figure 6. Electron micrograph of beads formation in electrospun nanofibers.

2.1.2. Solution Concentration

In electrospinning process, for fiber formation to occur, a minimum solution concentration is required. As the solution concentration increases, a mixture of beads and fibers is obtained (Figure 7). The shape of the beads changes from spherical to spindle-like when the solution concentration varies from low to high levels. It should be noted that the fiber diameter increases with increasing solution concentration because the higher viscosity resistance. Nevertheless, at a higher concentration, viscoelastic force, which usually resists rapid changes in fiber shape, may result in uniform fiber formation. However, it is impossible to electrospin if the solution concentration or the corresponding viscosity become too high due to the difficulty.

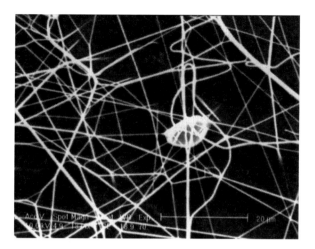

Figure 7. Electron micrograph of beads and fibers formation in electrospun nanofibers.

2.1.3. Molecular weight

Molecular weight also has a significant effect on the rheological and electrical properties such as viscosity, surface tension, conductivity and dielectric strength. It has been reported that too low molecular weight solution tends to form beads rather than fibers, and high molecular weight nanofiber solution gives fibers with larger average diameter (Figure 8).

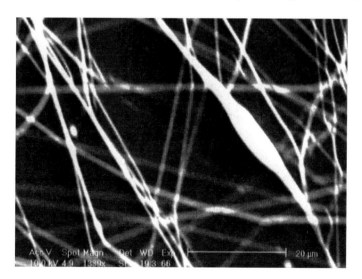

Figure 8. Electron micrograph of variable diameter formation in electrospun nanofibers.

2.1.4. Surface Tension

The surface tension of a liquid is often defined as the force acting at right angles to any line of unit length on the liquid surface. However, this definition is somewhat misleading, since there is no elastic skin or tangential force as such at the surface of a pure liquid. It is more satisfactory to define surface tension and surface free energy as the work required to increase the area of a surface isothermally and reversibly by unit amount. As a consequence of surface tension, there is a balancing pressure difference across any curved surface, the pressure being greater on the concave side. By reducing surface tension of a nanofiber solution, fibers could be obtained without beads (Figures 9-10). This might be correct in some sense, but it should be applied with caution. The surface tension seems more likely to be a function of solvent compositions, but it is negligibly dependent on the solution concentration. Different solvents may contribute different surface tensions. However, not necessarily a lower surface tension of a solvent will always be more suitable for electrospinning. Generally, surface tension determines the upper and lower boundaries of electrospinning window if all other variables are held constant. The formation of droplets, bead and fibers can be driven by the surface tension of solution, and lower surface tension of the spinning solution helps electrospinning to occur at lower electric field.

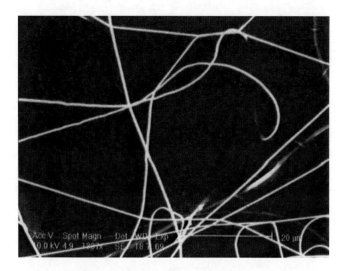

Figure 9. Electron micrograph of electrospun nanofiber without beads formation.

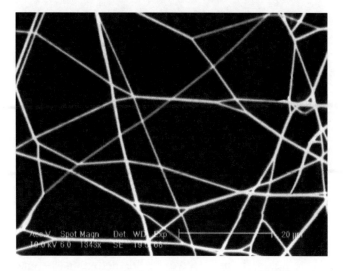

Figure 10. Electron micrograph of electrospun nanofiber without beads formation.

2.1.5. Solution conductivity

There is a significant drop in the diameter of the electrospun nanofibers when the electrical conductivity of the solution increases. Beads may also be observed due to low conductivity of the solution, which results in insufficient elongation of a jet by electrical force to produce uniform fiber. In general, electrospun nanofibers with the smallest fiber diameter can be obtained with the highest electrical conductivity. This interprets that the drop in the size of the fibers is due to the increased electrical conductivity.

2.2. Processing Condition

2.2.1. Applied Voltage

In the case of electrospinning, the electric current due to the ionic conduction of charge in the nanofiber solution is usually assumed small enough to be negligible. The only mechanism of charge transport is the flow of solution from the tip to the target. Thus, an increase in the electrospinning current generally reflects an increase in the mass flow rate from the capillary tip to the grounded target when all other variables (conductivity, dielectric constant, and flow rate of solution to the capillary tip) are held constant.

With the increase of the electrical potential, the resulting nanofibers became rougher. It is sometimes reported that a diameter of electrospun fibers is not significantly affected by an applied voltage. This voltage effects is particularly diminished when the solution concentration is low. Applied voltage may affect some factors such as mass of solution fed out from a tip of needle, elongation level of a jet by an electrical force, morphology of a jet (a single or multiple jets), etc. A balance among these factors may determine a final diameter of electrospun fibers. It should be also noted that beaded fibers may be found to be electrospun with too high level of applied voltage. Although voltage effects show different tendencies, the voltage generally does not have a significant role in controlling the fiber morphology.

Nevertheless, increasing the applied voltage (i.e., increasing the electric field strength) will increase the electrostatic repulsive force on the fluid jet that favors the thinner fiber formation. On the other hand, the solution will be removed from the capillary tip more quickly as jet is ejected from Taylor cone. This results in the increase of the fiber diameter.

2.2.2. Feed Rate

The morphological structure can be slightly changed by changing the solution flow rate, as shown in Figure 11. At the flow rate of 0.3 ml/h, a few big beads were observed on the fibers. When the flow rate exceeded a critical value, the delivery rate of the solution jet to the capillary tip exceeds the rate at which the solution was removed from the tip by the electric forces. This shift in the mass balance resulted in sustained but unstable jet and fibers with big beads formation.

The solution's electrical conductivity was found as a dominant parameter to control the morphology of electrospun nanofibers [9]. In the case of low-molecular-weight liquid, when a high electrical force is applied, formation of droplets can occur. A theory proposed by Rayleigh explained this phenomenon. As the evaporation of a droplet takes place, the droplet decreases in size. Therefore the charge density of its surface is increased. This increase in charge density due to Coulomb repulsion overcomes the surface tension of the droplet and causes the droplet to split into smaller droplets. However, in the case of a solution with high molecular weight liquid, the emerging jet does not break up into droplets but is stabilized and forms a string of beads connected by a fiber. As the concentration is increased, a string of connected beads is seen, and with further increase, there is reduced bead formation until only smooth fibers are formed. And sometimes spindle-like beads can form due to the extension causing by the electrostatic stress. The changing of fiber morphology can probably be attributed to a competition between surface tension and viscosity. As concentration is increased, the viscosity of the solution increases as well. The surface tension attempts to reduce surface area per unit mass, thereby causing the formation of beads/spheres. Viscoelastic forces resisted the formation of beads and allowed for the formation of smooth fibers. Therefore formation of beads at lower solution concentration (low viscosity) occurs where surface tension had a greater influence than the viscoelastic force. However, bead formation can be reduced and finally eliminated at higher solution concentration, where viscoelastic forces had a greater influence in comparison with surface tension. But when the concentration is too high, high viscosity and rapid evaporation of solvent makes the extension of jet more difficult, thicker, and un-uniform fibers will be formed.

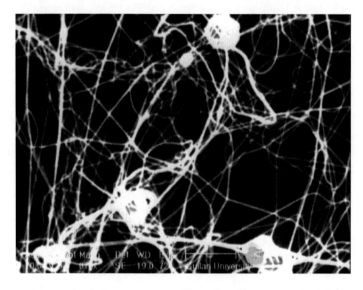

Figure 11. Electron micrograph of electrospun nanofiber when flow rate exceeded the critical value.

Suitable level of processing parameters must be optimized to electrospin solutions into nanofibers with desired morphology, and the parameters' levels are dependent on properties of solution and solvents used in each of electrospinning process. Understanding of the concept how each of processing parameter affects the morphology of the electrospun nanofibers is essential. All the parameters can be divided into two main groups; i.e., one with parameters that affect the mass of solution fed out from a tip of needle, and the other with parameters that affect an electrical force during electrospinning. Solution concentration, applied voltage and volume feed rate are usually considered to affect the mass. Increased solution concentration and feed rate tend to bring more mass into the jet. High applied voltage reflects to force to pull a solution out from the needle; hence higher applied voltage causes more solution coming out. On the other hand, it should be noted that solution electrical conductivity and applied voltage affect a charge density, thus an electrical force, which acts to elongate a jet during electrospinning.

3. Experimental

Polyacrylonitrile (PAN) fiber (Dolan) was from Hoechst .NMP (N-methyl-2- pyrolidon) was from Riedel-de Haën. The polyaniline used was synthesized in our laboratory. Polyaniline (PANi) was synthesized by the oxidative polymerization of aniline in acidic media. Three ml of distilled aniline was dissolved in 150 ml of 1N HCl and kept at 0 °C, 7. 325g of (NH4)2S2O8 was dissolved in 35 ml of 1N HCl and added dropwise under constant stirring to the aniline/HCl solution over a period of 20 mins. The resulting dark green solution was maintained under constant stirring for four hrs, filtered and washed with methanol and then with water; then it was dried before being added to 150 mL of 1N (NH4)OH solution. After an additional four hrs, the solution was filtered, and a deep blue emeraldine base form of polyaniline was obtained (PANiEB). Then it was dried and crushed into fine powder and then passed through a 100 mesh.

The polymer solutions were prepared by first dissolving exact amount of PANi in NMP. The PANi was slowly added to the solvent with constant stirring at room temperature. This solution was then allowed to stir for one hour in a sealed container. PAN/NMP solution was prepared separately and added dropwise to the well-stirred PANi solution, and the blend solution was allowed to stir with a mechanical stirrer for an additional one hour.

By mixing different solution ratios (0/100, 50/50, 60/40, 75/25) of 5% PANi solution and 20% PAN solution, various polymer blend solutions were prepared with the concentration of polyaniline ranging from 5 wt% to 42 wt%. The fiber diameter and polymer morphology of the electrospun polyaniline/polyacrylonitrile NMP solution were determined using an optical microscope Nikon Microphot-FXA. A small section of the non-woven mat was placed on the glass slide and placed on the microscope sample holder. A scanning electron microscope (SEM) Philips XL-30 was used to take the SEM photographs to do more precise characterization. A small section of the web was placed on SEM sample holder and coated with gold (BAL-TEC SCD 005 sputter coater).

4. Result and Discussion

In our first experiment, we tried to find out whether electrospinning of PANi pure solution can result in a web formation or not. Without the addition of PAN to PANi dissolved in NMP, no web formation occurred, because the concentration and viscosity of the solution was not high enough to form a stable drop at the end of the needle and just some dispersed drops were formed on the collector. Adding more polyaniline cannot increase the solution viscosity and just resulted in gelation of the solution.

Based on these results, the blend solution was electrospun; the initial results showed that in room temperature, fine fibers are formed. In these blends, the concentration of PANi ranged from 5 wt% to 42 wt%. The potential difference between the needle tip and the electrode was 20 Kv. Optical microscope photomicrographs (Figure 12) showed that the fibers are formed, but they are entangled with each other; also the bead forming is observed. In order to get more uniform webs, we tried to obtain the webs at the gap between two metal stripes that were placed on the collector plate, and it was seen that (Figure 13) webs got more uniform but yet the entanglement and beads were observed. For examining the web formation of PAN, it was electrospun from NMP at different concentrations. (10% and 15%) webs were formed at voltages between 17 and 20. PAN web showed an excellent web forming behavior, and the resulted webs were uniform (Figure 14).

After that, optical microscope micrographs confirmed that the fibers are formed; higher concentrations of PANi were used, and the resulted webs were examined by SEM for studying their diameter and morphology more precisely. The SEM photomicrograph revealed that the diameter of fibers in non-woven mat ranged from minimum 160 nm to maximum 560 nm, with an average fiber diameter of 358 nm. It was noticed that by increasing amount of PAN, the fiber formation enhanced and more uniform fibers were obtained; also the fiber diameter variation is smaller; it can be related to intrinsic fiber forming behavior of PAN as it is used widely as the base material for producing fibers and yarns. By increasing PANi, ratio amount of beading increased and fibers twisted to each other before reaching the collector; in some parts a uniform web was formed. It seemed that the fibers were wet, and it is the cause of sticking of fibers together. In 42% of PANi, the web forming was not seen and instead we had an entangled bulk with some polymer drops. Actually with increasing the PANi ratio, the fiber diameter decreased as it can be noticed clearly from the results indicated in Table 2.

Table 2 Fiber diameters in different PANi ratios

PANi percent (blending ratio)	0% (0/100)	20% (50/50)	27% (60/40)
Fiber diameter	445 nm	372 nm	292 nm

By increasing the temperature of electrospinning environment to 75č in order to let the solvent evaporate more rapidly, the problem of twisted fiber was overcome, and more uniform webs and finer nanofibers were formed, but yet the beading problem was seen. More research is in progress to enhance the web characterization and decrease the fiber diameter to real nanometer size.

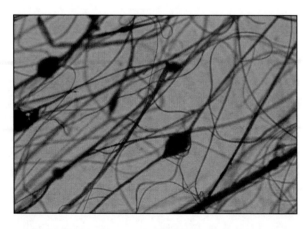

Figure 12. Optical microscope micrograph of 16% PANi blend solution, beads can be observed clearly.

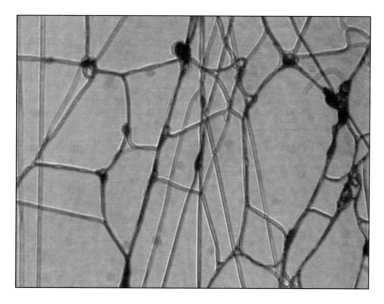

Figure 13. Optical microscope micrograph of 16% PANi blend solution caught in air.

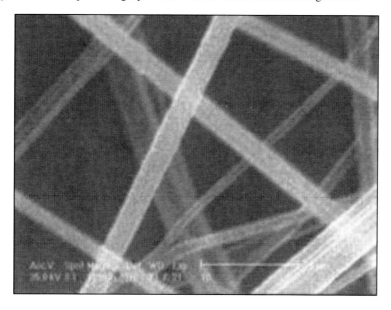

Figure 14. SEM photomicrograph of pure PAN.

Conclusion

Nanofibers of pure PAN dissolved in NMP was prepared, but pure PANi /NMP solution did not show the web forming. By adding PAN, fiber forming was observed. Different PANI/PAN blends were electrospun; the average diameters of nanofibers were 385 nm. It was seen that by increasing the PANi amount, the resulted nanofibers diameter decreased; also, with increasing the amount of PANi, the web becomes more irregular and nonuniform.

The electrospinning technique provides an inexpensive and easy way to produce nanofibers on low basis weight, small fiber diameter and pore size. It is hoped that this chapter will pave the way toward a better understanding of the application of electrospinning of nanofibers. There are three categories of variables that influence the electrospun fiber diameter, including (1) polymer solution variables, (2) process variables, and (3) environmental variables. Examples of solution variables are viscosity or polymer concentration, solvent volatility, conductivity, and surface tension. Process variables consist of electric field strength, fluid flow rate, and distance between electrodes. Low molecular weight fluids form beads or droplets in the presences of an electric field, while high molecular weight fluids generate fibers. However, an intermediate process is the occurrence of the "beads on a string" (Figures 15, 16) morphology. In many instances, bead formation is also observed in addition to fiber growth. This morphology is a result of capillary break-up of the spinning jet caused by the surface tension. Solution conductivity is another polymer solution property that greatly influences electrospun fiber diameter. The addition of salts to polymer solutions has been shown to increase the resulting net charge density of the electrospinning jet. The surface tension of the polymer solution also influences the resulting fiber morphology because large surface tensions promote the formation of polymer droplets. The surface tension of the fluid must be overcome by the electrical voltage in order for emission of an electrified jet from the syringe. Process variables also control the morphology of fibers during the electrospinning process. In general, fiber diameter is rather insensitive to process conditions when compared to varying the polymer solution properties; however extensive work has been published on the influence of voltage, flow rate, and working distance on electrospun fiber morphology. The distance between the electrodes or the working distance influences the electrospinning process. Generally as the working distance decreases, the time for the flight of path for the fluid jet decreases.

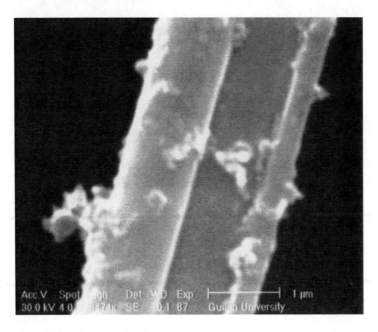

Figure 15. Formation of "beads on a string."

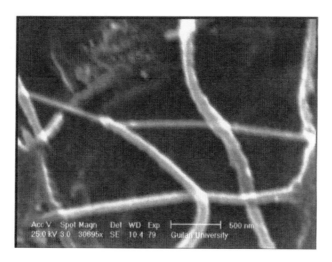

Figure 16. Formation of "beads on a string."

Moreover, it should be noted that temperature is a convoluted variable when attempting to discern its influence on electrospun fiber formation. Increasing the solution temperature causes (1) a change in chain conformation in solution, (2) a decrease in solution viscosity, and (3) an increase in rate of solvent evaporation. Thus, quantifying the effect of temperature on electrospinning proves difficult since all of the above can influence fiber morphology. Humidity has been shown to control the surface morphology of electrospun fibers.

References

[1]   Y. Wan, Q. Guo, N. Pan. 2004. Thermo-electro-hydrodynamic model for electrospinning process. *Int. J. Nonlinear Sci. Num. Simul.* 5.pp5–8.

[2]   J. He, Y. Wan, J. Yu. 2004. Allometric scaling and instability in electrospinning. *Int. J. Nonlinear Sci. Num. Simul.* 5(3).pp.243–52.

[3]   J. He, YQ. Wan, JY. Yu. 2004.Application of vibration technology to polymer electrospinning. *Int. J. Nonlinear Sci. Num. Simul.* 5(3).pp.253–61.

[4]   J. He, YQ Wan. 2004. Allometric scaling for voltage and current in electrospinning. *Polymer.* 45(19).pp-6731–4.

[5]   X-H .Qin, YQ ,Wan, JH .He. 2004. Effect of LiCl on electrospinning of PAN polymer solution: theoretical analysis and experimental verification. *Polymer.* 45(18).pp.6409–13.

[6]   S. Therona, E. Zussmana, AL. Yarin. 2004. Experimental investigation of the governing parameters in the electrospinning of polymer solutions. *Polymer.* 45.pp.2017–30.

[7]   M .Demir, I .Yilgor, E. Yilgor, B Erman. 2002. Electrospinning of polyurethane fibers. *Polymer.* 43.pp.3303–9.

[8]   A. Ganan-Calvo. 1999. The surface charge in electrospraying: its nature and its universal scaling laws. *J. Aerosol. Sci.* 30(7):863–72.

[9]   J. Feng. 2003. Stretching of a straight electrically charged viscoelastic jet. *J. Non-Newtonian Fluid Mech.* 116.pp.55–70.

In: Handbook of Research on Nanomaterials, Nanochemistry ...   ISBN: 978-1-61942-525-5
Editors: A. K. Haghi and G. E. Zaikov   © 2013 Nova Science Publishers, Inc.

Chapter IV

Drug Delivery Systems Based on Chitosan Nanoparticles

M. R. Saboktakin

Baku State University, Azerbaijan

Introduction

Natural polymers have potential pharmaceutical applications because of their low toxicity, biocompatibility, and excellent biodegradability. In recent years, biodegradable polymeric systems have gained importance for design of surgical devices, artificial organs, drug delivery systems with different routes of administration, carriers of immobilized enzymes and cells, biosensors, ocular inserts, and materials for orthopedic applications (BrOndsted & Kope˘cek, 1990). These polymers are classified as either synthetic (polyesters, polyamides, polyanhydrides) or natural (polyamino acids, polysaccharides) (Giammona, Pitarresi, Cavallora, & Spadaro, 1999; Krogars et al., 2000). Polysaccharide-based polymers represent a major class of biomaterials, which includes agarose, alginate, carageenan, dextran, and chitosan. Chitosan [(1,4)2- amino-2-d-glucose] is a cationic biopolymer produced by alkaline N-deacetylation of chitin, which is the main component of the shells of crab, shrimp, and krill (Chiu, Hsiue, Lee, & Huang, 1999; Jabbari & Nozari, 2000). Chitosan is a functional linear polymer derived from chitin, the most abundant natural polysaccharide on the earth after cellulose, and it is not digested in the upper GI tract by human digestive enzymes (Fanta & Doane, 1986; Furda, 1983). Chitosan is a copolymer consisting of 2-amino-2-deoxyd-glucose and 2-acetamido-2-deoxy-d-glucose units linked with β-(1-4) bonds. It should be susceptible to glycosidic hydrolysis by microbial enzymes in the colon because it possesses glycosidic linkages similar to those of other enzymatically depolymerized polysaccharides. Among diverse approaches that are possible for modifying polysaccharides, grafting of synthetic polymer is a convenient method for adding new properties to a polysaccharide with minimum loss of its initial properties (Saboktakin, Maharramov, & Ramazanov, 2007; Peppas, 1987). Graft copolymerization of vinyl monomers onto polysaccharides using free radical initiation, has attracted the interest of many scientists. Up to now, considerable works have been devoted to the grafting of vinyl monomers onto the substrates, especially starch and cellulose (Jabbari & Nozari, 2000; Xu & Li, 2005). Existence of polar functionally groups as carboxylic acid are needed not only for bio-adhesive properties but also for pH-sensitive properties of polymer (Ratner, 1989; Thierry, Winnik, Mehri, & Tabrizian, 2003), because the increase of MAA content in the hydrogels provides more hydrogen bonds at low pH and more electrostatic repulsion at high pH. It is as a part of our research program on chitosan modification to prepare materials with pH-sensitive properties for use as drug delivery (Mahfouz, Hamm, & Taupitz, 1997; Schmitz et al., 2000; Bloembergen & Pershan, 1967).

In this study, we hypothesized that the absorption of paclitaxel could be enhanced by administration with chitosan-polymethacrylic acid nanoparticles because of their greater permeability properties (Puttpipatkhachorn, Nunthanid, & Yamamato, 2001).

Materials and Methods

Chitosan with 1:1 molar ratios of methacrylic acid were polymerized at 60–70 °C in a thermostatic water bath, bis-acrylamide as a cross-linking agent (CA), using persulfate as an initiator ([I] = 0.02M) and water as the solvent (50 mL). The polymeric system was stirred by mechanical stirrer to sticky nanoparticles, and it was separated from medium without solvent addition. All the experiments were carried out in Pyrex glass ampoules. After the specific time (48 h), the precipitated network polymer was collected and dried in vacuum. Chitosan-methacrylic acid copolymer suspensions of 0.2% (w/v) were prepared in 1% acetic acid. Sodium tripolyphosphate (TPP, 1.0%) was added dropwise to 6mL of chitosan with stirring, followed by sonication with a dismembrator for 10 s at a power setting of 3W.The resulting chitosan particle suspension was centrifuged at 10,000×g for 10 min. The pelleted particles were re-suspended in deionized water with 10 s sonication and lyophilized. The mean size and zeta potential of the chitosan-methacrylic acid nanoparticles were determined by photon correlation spectroscopy using ZetaPlus particle analyzer. The major amount of drug was adsorbed by 2mg of chitosan nanoparticles in a certain time period. Chitosan-methacrylic acid nanoparticle suspensions (4 mg/mL) were mixed with paclitaxel solutions (0.5 and 1mg/mL), vortexed, and incubated at 37 °C for 1, 6, 12 and 18 h. After adsorption, the suspensions were centrifuged at 10,000×g for 10 min, and free drug was measured in the supernatant by a colorimetric method using periodic acid/Schiff (PAS) staining. Schiff reagent was prepared by diluting pararosaline solution (40 g/L in 2MHCl, Sigma) with water to give a final concentration of 1.0% sodium bisulfate (80 mg) was added to 5mL of Schiff reagent and the resultant solution was incubated at 37 °C until it became colorless or pale yellow. Periodic acid solution was freshly prepared by adding 10_L of 50% periodic acid to 7mL of 7% acetic acid. Supernatants were mixed with 100_L of dilute periodic acid and incubated for two h at 37 °C. Then, 100_L of Schiff reagent was added at room temperature, and after 30 min, the absorbance was measured at 560 nm.

The following procedure was used to assess the stability of paclitaxel during the bead preparation process. The prepared nanoparticles were extracted twice with a solvent mixture of 1:1 acetonitrile and ethanol (v/v), the extract was evaporated, and the residue was injected onto HPLC column. Stability-indicating chromatographic method was adopted for this purpose. The method consisted of aSymmetryC18 column (254mm×4.6mm;5_m)run using a mobile phase of composition methanol:water (70:30, v/v) at a flow rate of 0.5 mL/min, a waters pump (600E), and eluants. A definite weight range of 10–15mg of nanoparticles were cut and placed in a 1.5mL capacity microcentrifuge tube containing 1mL of release medium of the following composition at 37 ∘C: phosphate buffered saline (140mM,pH7.4) with 0.1% sodium azide and 0.1% Tween 80. At predetermined time points, 100_L of release medium was sampled with replacement to which 3mL of scintillation cocktail was added and vortexed before liquid scintillation counting. The cumulative amount of paclitaxel released as a function of time was calculated. To study the molecular properties of paclitaxel and chitosan-PMAA, the solid-state characterization was done by the application of thermal, X-ray diffraction, and microscopy techniques. During these studies, the solid characteristics of paclitaxel and chitosan- PMAA were compared with those of nanoparticles to reveal any changes occurring as a result of nanoparticle preparation. Differential scanning calorimetry (DSC) studies were performed with a Mettler Toledo 821 thermal analyzer (Greifensee, Switzerland) calibrated with indium as standard. For thermogram acquisition, sample sizes of 1–5mg were scanned with a heating rate of 5 ∘C/min over a temperature range of 25–300 ∘C. In order to check the reversibility of transition, samples were heated to a point just above the corresponding transition temperature, cooled to room temperature, and reheated up to 300 ∘C. Paclitaxel samples and chitosan-PMAA beads were viewed using a Philips XL-30 E SEM scanning electron microscope (SEM) at 30 kV (max.) for morphological examination. Powder samples of paclitaxel and beads were mounted onto aluminum stubs using double-sided adhesive tape and then sputter coated with a thin layer of gold at 10 Torr vacuum before examination. The specimens were scanned with an electron beam of 1.2 kV acceleration potential, and images were collected in collected in secondary electron mode. Molecular arrangement of paclitaxel and chitosan-PMAA in powder as well as in nanoparticles were compared by powder X-ray diffraction patterns acquired at room temperature on a Philips PW 1729 diffractometer (Eindhoven, Netherlands) using CuK radiation.

X-ray diffraction is a proven tool to study crystal lattice arrangements and yields very useful information on degree of sample crystallinity.

Results

In the present study, nanoparticles were prepared by the classical method, which involves spreading a uniform layer of polymer dispersion followed by a drying step for removal of solvent system. Since bead preparation methodology involved a heating step, it may have had a detrimental effect on the chemical stability of drug. Hence, stability assessment of paclitaxel impregnated in bead was done using stability-indicating method. For this purpose, paclitaxel was extracted from bead and analyzed by HPLC. A single peak at 21.2 min representing paclitaxel (with no additional peaks) was detected in the chromatogram, suggesting that the molecule was stable during preparation of beads Paclitaxel was extracted from different regions of chitosan- PMAA nanoparticles using acetonitrile:ETOH (1:1, v/v) solvent system. After normalization of amount of paclitaxel on weight basis of nanoparticles, the results indicated that the variation in distribution of paclitaxel in different regions of nanoparticles were <16%. The composition of the polymer defines its nature as a neutral or ionic network and, furthermore, its hydrophilic/hydrophobic characteristics. Ionic hydrogels, which could be cationic, containing basic functional groups or anionic, containing acidic functional groups, have been reported to be very sensitive to changes in the environmental pH. The swelling properties of the ionic hydrogels are unique due to the ionization of their pendent functional groups. The equilibrium swelling behavior of ionic hydrogels containing acidic and/or basic functional groups is illustrated in Fig. 1. Hydrogels containing basic functional groups are found increased swelling activity in acidic conditions and reduced in basic conditions but pH-sensitive anionic hydrogels shows low swelling activity in acidic medium and very high activity in basic medium.

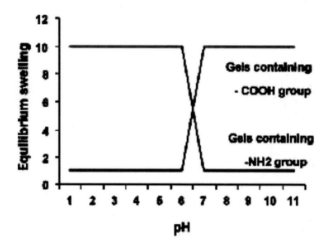

Figure 1. Equilibrium degree of swelling in response to pH.

As shown in Fig. 2, an increase in the content of MAA in the freed monomer mixtures resulted in less swelling in simulated gastric fluid but greater swelling in and simulated intestinal fluids. This is because the increase of MAA content in the hydrogels provides more hydrogen bonds at low pH and more electrostatic repulsion at high pH. Fig. 4 shows scanning electron microscope (SEM) of graft chitosan copolymer with polymethacrylic acid and nano-polymer bonded drug. Nano- and micro-polymer bonded drugs (50 mg) were poured into 3mL of aqueous buffer solution (SGF: pH 1 or SIF: pH 7.4). The mixture was introduced into a cellophane membrane dialysis bag. The bag was closed and transferred to a flask containing 20mL of the same solution maintained at 37 °C. The external solution was continuously stirred, and 3mL samples were removed at selected intervals. The removed volume was replaced with SGF or SIF. Triplicate samples were used. The sample of hydrolyzate was analyzed by UV spectrophotometer, and the quantity of paclitaxel was determined using a standard calibration curve obtained under the same conditions (Fig. 3). It appears that the degree of swelling depends on their particle size. As shown in Fig. 2, a decrease in the molecular size of carriers increased the swelling rate. The thermal behavior of a polymer is important in relation to its properties for controlling the release rate in order to have a suitable drug dosage form. The glass transition temperature (Tg) was determined from the DSC thermograms. The higher Tg values probably related to the introduction of cross-links, which would decrease the flexibility of the chains and the ability of the chains to undergo segmental motion, which would increase the Tg values. On the other hand, the introduction of a strongly polar carboxylic acid group can increase the Tg value because of the formation of internal hydrogen bonds between the polymer chains (Fig. 4). X-ray diffraction is also used to study the degree of crystallinity of pharmaceutical drugs and excipients. A lower 2Θ value indicates larger d-spacings, while an increase in the number of high-angle reflections indicates higher molecular state order.

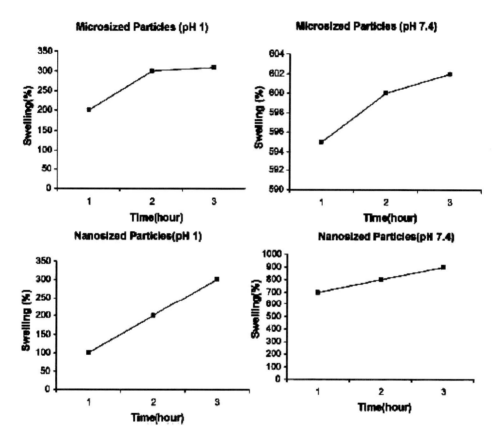

Figure 2. Time-dependent swelling behavior of micro- and nano-carriers for paclitaxel drug model as a function of time at 37°C.

In addition, broadness of reflections, high noise, and low peak intensities are characteristics of a poorly crystalline material. A broad hump in the diffraction pattern of chitosan hydrogel extending over a large range of 2Θ suggests that chitosan is present in amorphous state in the film.

X-ray diffraction patterns of paclitaxel and chitosan- PMAA hydrogel film were obtained and compared, which revealed marked differences in the molecular state of paclitaxel (Fig. 5). X-ray diffractogram of paclitaxel and chitosan-PMAA hydrogel film shows that several high-angle diffraction peaks were observed at the following $2\square$ values: 24.1°, 27.4°, 29.2°, 36.4°, 40.3°, and 44.6°. The 29.6° 2Θ peak had the highest intensity as observed for hydrogel film.

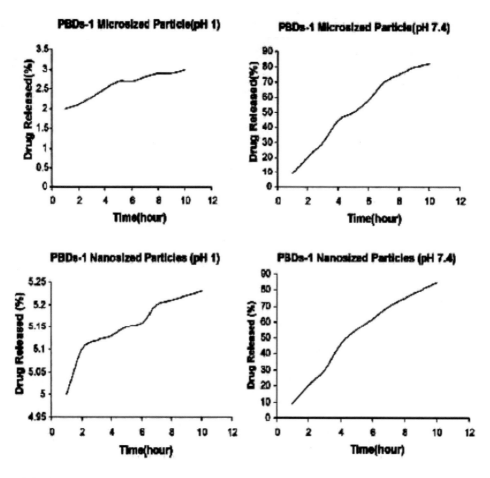

Figure 3. Release of paclitaxel drug from micro- and nano-polymeric carriers as a function of time at 37°C.

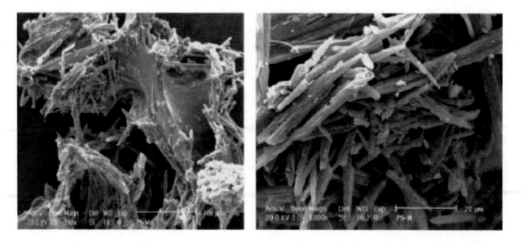

Figure 4. SEM of paclitaxel- chitosan nanoparticles.

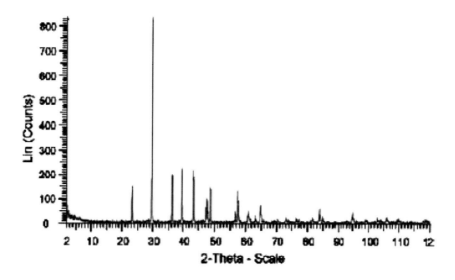

Figure 5. XRD pattern of chitosan nanoparticles.

Discussion

In order to develop a local delivery system for paclitaxel, a biodegradable N-chitosan-PMAA bead was attempted, which to our knowledge was the first effort of its kind. Potential formulation problems were anticipated since chitosan is only soluble in aqueous acidic solutions, whereas paclitaxel, being a hydrophobic drug, is insoluble under similar conditions. In early stages of formula optimization studies, it was observed that paclitaxel was incorporated into bead. The primary mechanisms for release of drugs from matrix systems in vitro are swelling, diffusion, and disintegration. In vitro degradation of chitosan-PMAA nanoparticles prepared by solution casing method occurred less rapidly as the degree 73% deacetylated showed slower biodegradation. Since the grade of chitosan used in the present study was of high molecular weight with a degree of deacetylation ≥85%, significant retardation of release of paclitaxel from nanoparticles is attributed to the polymer characteristics. In addition, diffusion of paclitaxel may have been hindered by increased tortuosity of polymer accompanied by a swelling mechanism.

Fig. 4 shows the SEM of paclitaxel and chitosan-PMAA hydrogel film that synthesized by chemical reaction. This hydrogel is very sensitive to the temperature, due to the interaction of electron and sample. Scanning electron micrography images were obtained from a diluted solution of the paclitaxel particle. The white spots are paclitaxel nanoparticles. The SEM image shows the presence of paclitaxel spherical particles in hydrogel matrix, which are homogeneously distributed throughout the hydrogel, which is also confirmed from 1H NMR studies. As observed from SEM photomicrographs, the crystals of paclitaxel have a different appearance than recrystallized paclitaxel. These nanoparticles do not have clearly defined crystal morphological features in the SEM photomicrographs. Hence, it appears that the irregularly shaped particles are surface deposited with poloxamer, which gives them an appearance resembling that of coated particles. X-ray diffraction is also used to study the degree of crystallinity of pharmaceutical drugs and excipients. A lower 2Θ value indicates larger d-spacings, while an increase in the number of high-angle reflections indicates higher molecular state order. In addition, broadness of reflections, high noise, and low peak intensities are characteristic of a poorly crystalline material. A broad hump in the diffraction pattern of chitosan extending over a large range of $2\square$ suggests this.

Conclusion

The swelling and hydrolytic behavior of the hydrogel beads was dependent on the content of MAA groups and caused a decrease in gel swelling in SGF or an increase in gel swelling in SIF. Modified chitosan with different contents of MAA and CA by graft copolymerization reactions were carried out under microwave-radiation. The swelling of the hydrogels was dependent on the content of MAA groups and caused a decrease in gel swelling in SGF or an increase in gel swelling in SIF. Incorporation of MAA made the hydrogels pH-dependent, and the transition between the swollen and the collapsed states occurred at high and low pH. The swelling ratios of the hydrogels beads increased at pH 7.4 but decreased at pH 1 with increasing incorporation of MAA.

References

Bloembergen, N., & Pershan, P. S. (1967). Model catalysis of ammonia synthesis and iron–water interfaces – a sum frequency generation vibrational spectroscopic study of solid–gas interfaces and anion photoelectron spectroscopic study of selected anion clusters. *Physical Review*, 128(2), 606.

BrOndsted, H., & Kope˘cek, J. (1990). Hydrogels for site-specific oral delivery. In *Proceeding of the International Synposium on Controlled Release of Bioactive Materials*, vol. 17 (pp. 128–129).

Chiu, H. C., Hsiue, G. H., Lee, Y. P., & Huang, L. W. (1999). Synthesis and characterization of pH-sensitive dextran hydrogels as a potential colon-specific drug delivery system. *Journal of Biomaterials Science. Polymer Edition*, 10, 591–608.

Fanta, G. F., & Doane, W. N. (1986). Grafted starches. In O. B. Wurzburg (Ed.), *Modified Starches: Properties and Uses.* (pp. 149–178). Boca Raton, FL: CRC.

Furda, I. (1983). Aminopolysaccharides – their potential as dietary fiber. In I. Furda (Ed.), *Unconventional Sources of Dietary Fiber, Physiological and In Vitro Functional Properties* (pp. 105–122). Washington, DC: American Chemical Society.

Giammona, G., Pitarresi, G., Cavallora, G., & Spadaro, G. (1999). New biodegradable hydrogels based on an acryloylated polyaspartamide cross-linked bygamma irradiation. *Journal of Biomedical Science. Polymer Edition*, 10, 969–987.

Jabbari, E., & Nozari, S. (2000). Swelling behavior of acrylic acid hydrogels prepared by γ-radiation cross-linking of polyacrylic acid in aqueous solution. *European Polymer Journal*, 36, 2685–2692.

Krogars, K., Heinamaki, J., Vesalahti, J., Marvola, M., Antikainen, O., & Yliruusi, J. (2000). Extrusion-spheronization of pH-sensitive polymeric matrix pellets for possible colonic drug delivery. *International Journal of Pharmacy,* 199, 187–194.

Mahfouz, A., Hamm, B., & Taupitz, M. (1997). *Contrast agents for MR imaging of the liver: Clinical overview. European Radiology,* 7, 507.

Peppas, N. A. (1987). Hydrogels in Medicine and Pharmacy. Boca Raton, FL: CRC Press. Puttpipatkhachorn, S., Nunthanid, J., & Yamamato, K. (2001). Drug physical state and drug–polymer interaction on drug release from chitosan matrix films. *Journal of Control Release,* 75, 143–153.

Ratner, B. D. (1989). Comprehensive polymer science – the synthesis, characterisation, reactions & applications of polymers. In S. K. Aggarwal (Ed.), *Comprehensive Polymer Science – The Synthesis, Characterization, Reactions and Applications of Polymers.* (pp. 201–241). Oxford: Pergamon Press.

Saboktakin, M. R., Maharramov, A., & Ramazanov, M. A. (2007). Synthesis and characterization of aromatic polyether dendrimer/mesalamine (5-ASA) nanocomposite as drug carrier system. *Journal of American Science,* 3(4), 45.

Schmitz, S. A., Winterhalter, S., Schiffler, S., Gust, R., Wagner, S., Kresse, M., et al. (2000). Superparamagnetic iron oxide nanoparticles functionalized polymers. *Investigative Radiology,* 35, 460.

Thierry, B., Winnik, F. M., Mehri, Y., & Tabrizian, M. (2003). A new Y3Al5O12 phase produced by liquid-feed flame spray. *Journal of American Chemical Society,* 125, 7494.

Xu, H., & Li, T. (2005). The analysis of boundary functions of CMS reaction factors. *Journal of Nature and Science,* 3(2), 25–28.

In: Handbook of Research on Nanomaterials, Nanochemistry ...   ISBN: 978-1-61942-525-5
Editors: A. K. Haghi and G. E. Zaikov   © 2013 Nova Science Publishers, Inc.

Chapter V

Application of Nano-SiO$_2$ in Cementitious Composites

A. K. Haghi[*]
University of Guilan, Iran

1. Introduction

The use of nanoparticles in developing materials has been introduced in the last few years. It has been observed that the extremely fine size of nanoparticles strongly affects their physical and chemical properties and enables them to fabricate new materials with novelty function. Thus integrating nanoparticles with existing cement-based building materials may enable us to fabricate new products with some outstanding properties [1,2]. Among the nanoparticles, nano silica has been used to improve the properties of cement-based materials. Some efforts on excellent mechanical properties and microstructure of cement composites with nano-SiO$_2$ have been reported [3,4,5]. Gengig Li[6] showed that incorporating nanoparticles to high-volume fly ash concrete can significantly increase the initial pozzolanic activity of fly ash. He also concluded that nano-SiO$_2$ can enhance the short-term and long-term strength of high-volume, high-strength concrete. K. Lin [7] reported that nano-SiO$_2$ particles could potentially improve the negative influences caused by enhanced the compressive and flexural strength of cement concrete and mortar [8,9]. As the rate of pozzolanic reaction is proportional to the amount of surface available for reaction and owing to the high specific surface of nanoparticles, they possess high pozzolanic activity. Nano-SiO$_2$ effectively consumes calcium hydroxide crystals (CH), which array in the interfacial transition zone between hardened cement paste and aggregates and produces hydrated calcium silicate (CSH), which enhances the strength of cement paste [10]. In addition, due to nano-scale size of particles, nano-SiO$_2$ can fill the ultrafine pores in cement matrix. This physical effect leads to reduction in porosity of transition zone in the fresh concrete. This mechanism strengthens the bond between the matrix and the aggregates and improves the

[*] Haghi@Guilan.ac.ir.

cement paste properties. Furthermore, it has been found that when the ultrafine particles of nano-SiO_2 uniformly disperse in the paste, thanks to high activity, they generate a large number of nucleation sites for the precipitation of the hydration products, which accelerates cement hydration [11].

It has been demonstrated that rice husk ash (RHA) can be added to concrete mixtures to substitute for the more expensive Portland cement to lower the construction cost. Compared to the other agricultural by-products, RHA is high in ash and similar to silica fume; it contains considerable amounts of SiO_2 [12]. Hence, RHA is not just a cheap alternative but a well-burned and a well-ground RH with most of its silica in an amorphous form and with enough specific surface is very active and considerably improves the strength and durability of cement and concrete [13,14]. Researchers have applied RHA and nano-SiO_2 to improve cement-based materials and have achieved great successes, However, considering nano-SiO_2 particles in RHA, mortar is an innovative approach. The main purpose of this research is to study the mechanical properties of RHA mortar containing nano-SiO_2 in order to further improvements of Portland cement-based materials.

2. Materials and Methods

In this study, ordinary Portland cement type I, standard graded sand, rice husk ash, nano-SiO_2 and tap water were used. RHA used in this experiment contained 92.1% SiO_2 with average particles size of 15.83 µm. The chemical compositions of RHA and cement were analyzed using an X-ray microprobe analyzer (Table 1). In order to achieve desired fluidity and better dispersion of nanoparticles, the polycarboxylate ether-based superplastisizer was incorporated into all mixes. The content of superplastisizer was adjusted for each mixture to keep constant the fluidity of mortars. Natural river sand was used with the fraction of sand passing through 1.18 mm sieve and retaining on 0.2mm. (conforming to ASTM C778). The specific gravity of sand was 2.51 gr/cm3. Basic material properties of nano-SiO_2 are given in Table 2.

Table 1. Chemical composition of cement and rice husk ash

Items	Chemical compositions (%)	
	OPC	RHA
SiO_2	21	92.1
AL_2O_3	4.6	0.41
Fe_2O_3	3.2	0.21
CaO	64.5	0.41
MgO	2	0.45
SO_3	2.9	-
L.O.I	1.5	-

Table 2.The properties of nano-SiO₂

Item	Diameter (nm)	Specific surface (m²/gr)	Density (g/cm³)	Purity (%)
Target	50	50	1.03	99.9

2.1. Mix Proportions

Ten different combinations as listed in Table 3 were cast. The compositions A-1 to A-6 were used to investigate the effect of nano-SiO$_2$ in substitution of cement in mortar and to find the optimum content of nano-SiO$_2$ in cement mortar. The compositions A-7 to A-10 were made to study the effect of nano-SiO$_2$ particles on properties of cement mortar containing RHA. The amount of RHA replacement in mortar was 20% by weight of cement, which is an acceptable range and is most often used [15]. The water/binder ratio for all mixtures was 0.5, where the binder weight is the total weight of cement, RHA and nano-SiO$_2$. The cement-sand ratio was 1:2.75 for all the mixtures. 50×50×50 mm cubes were considered for compressive and water absorption tests and 50×50×200 mm beams for flexural and shrinkage tests. The fresh mortar placed into the molds and tamped using a hard rubber mallet. After 24 hours, specimens were removed from the molds and cured in water at 23±2 ^{oC} until they were tested. The samples were tested using hydraulic testing machine under load control at 1350 N/S for compressive test (as per ASTM C 109) and 44N/S for flexural test (as per ASTM C348). After the mechanical tests, the crushed specimens were selected for scanning electronic microscope (SEM) test. The absorption test was carried out on two 50 mm cubes. Saturated surface dry specimens were kept in an oven at 110^{oC} for 72 h. After measuring the initial weight, specimens were immersed in water for 72h. Then the final weight was measured, and the final absorption was reported to assess the mortar permeability.

Table 3. Mix proportion of the specimens

Batch No	Sa/Bb	Wc/B	% Content (by weight)		
			O.C	RHA	N.S
A-1	2.75	0.5	100	-	0
A-2	2.75	0.5	99	-	1
A-3	2.75	0.5	97	-	3
A-4	2.75	0.5	95	-	5
A-5	2.75	0.5	93	-	7
A-6	2.75	0.5	91	-	9
A-7	2.75	0.5	80	20	0
A-8	2.75	0.5	79	20	1
A-9	2.75	0.5	77	20	3
A-10	2.75	0.5	75	20	5

a: Sand, b: Binder (Cement +RHA +Nano-SiO$_2$).c: Water.

3. Results and Discussion

3.1. Compressive Strength

Fig.1 shows the variation in the compressive strength of ordinary and RHA cement mortars at various contents of nano-SiO$_2$. It can be seen that the compressive strength of cement mortars with nano-SiO$_2$ are all higher than that of plain cement mortar. Results indicate that the optimal content of nano-SiO$_2$ for reinforcing concrete/mortar purpose should be about 7%. It is clear that increasing the amount of the nano-SiO$_2$ from 7% to 9% doesn't have considerable effect on compressive strength. Moreover, larger amounts of nano-SiO$_2$

actually reduce the strength of composites instead of improving it. It was found that when the content of nanoparticles is large, they are more difficult to disperse uniformly. Therefore, they create weak zone in form of voids; consequently the homogeneous hydrated microstructure cannot be formed, and low strength will be expected. From the results, it is clear that nano-SiO_2 is more effective in reinforcement of cement mortar than that of RHA. Also, it can be observed that the nano-SiO_2 improved the compressive strength of RHA mortars and incorporating nano-SiO_2 particles by RHA in mortar can lead to further improvements in compressive strength and likely other properties of the cement mortar.

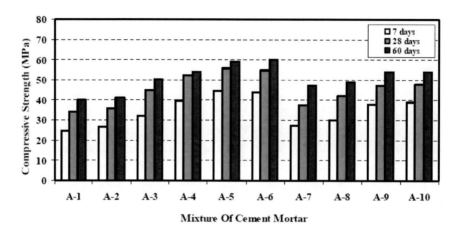

Figure 1. Compressive strength of the ordinary and RHA cement mortars at different content of nano-SiO_2.

3.2. Flexural Strength

The flexural strength of ten mortar mixtures at different ages is shown in Fig.2. It can be seen that the flexural strengths of the specimens with substitution of cement by nano-SiO_2 are all higher than that of plain mortar with the same water-to-binder ratio. The greatest increase among all ages is observed for the batch A4 with the 5% nano-SiO_2. At 5% nano-SiO_2, the flexural strength at 28 days was 7.5 MP, whereas it decreased to 6.2 MP with 9% nano-SiO_2. It indicates that high amounts of nano-SiO_2 (especially in excess of 7%) have negative effect on flexural strength. It is clear that the nano-SiO_2 particles are more effective in developing flexural strength than that of RHA, and incorporating nano-SiO_2 in cement mortars containing RHA can further increase the flexural strength. Two fundamental mechanisms can be deduced for strength enhancement by nano-SiO_2. The first strengthening mechanism is the filler effect. The micro-filling effect of nano-SiO_2 is one of the important factors for the development of dense concrete/mortar with very high strength, because it has been found that the small amounts of air content significantly decrease the strength of the mortar [16]. It has been reported that the size ratio between filler and the aggregates is one of the main parameters that strongly affects the strengthening caused by filling effect. Thanks to the high size ratio between nano-SiO_2 and aggregates, the filling effect of nano-SiO_2 particles is more obvious. Furthermore, the microstructure of the transition zone between aggregates and cement paste strongly influences the strength and durability of concrete [17]. Absence of nano-SiO₂ particles reduces the wall effect in the transition zone between the paste and the aggregates and strengthens this weaker zone due to the higher bond between those two phases. This mechanism also leads to an improvement in microstructure and properties of the mortars/concretes [18]. The second strengthening mechanism is the pozzolanic activity. Two major products of cement hydration are calcium silicate hydrate (CSH) and calcium hydroxide (CH), respectively. Calcium silicate hydrate, which is produced by hydration of C_3S and C_2S, plays a vital role in mechanical characteristics of cement paste, whereas calcium hydrate, which is also formed by hydration of cement, has not any cementing property. It contains about 20-25% of the volume of the hydration products. Calcium hydrates due to their morphology are relatively weak and brittle. Cracks can easily propagate through regions populated by them, especially at the aggregate /cement paste interface [16]. Nano-SiO₂ particles react with Calcium Hydrates formed during hydration of cement rapidly and produce calcium silicate hydrate with cementitious properties, which is beneficial for enhancement of strength in concrete/mortar.

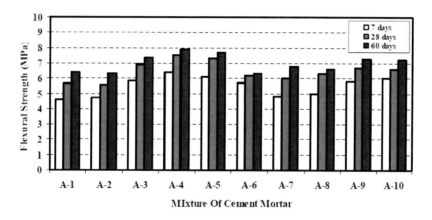

Figure 2. Flexural strength of the ordinary and RHA cement mortars at different content of nano-SiO₂.

3.3. Water Absorption

The absorption characteristics indirectly represent the porosity through an understanding of the permeable pore volume and its connectivity [19]. In order to investigate the effect of nano-SiO₂ particles on cement mortar permeability, water absorption test was carried out on mixes A-1 (plain cement mortar), A-5 (cement mortar incorporated 7% nano-SiO₂), A-7 (cement mortar with 20% RHA replacement), A9 (cement mortar with 20%RHA replacement incorporated 3% nano- SiO₂).The final absorption of mixes above are presented in Table 4. It can be seen that mixture A-5(cement mortar incorporated 7% nano-SiO₂) showed the lowest absorption between all the mixtures, which shows that nano-SiO₂ is more effective in reduction of permeability than that of RHA. Integrating nano-SiO₂ into RHA mortar reduced the water absorption from 5.42% to 4.45%. Results showed that the presence of nano-SiO₂ particles in cement mortar could decrease the water absorption and likely permeability of cement mortar. This impermeability increase can be attributed to two concomitant phenomena: 1. Nano-SiO₂ particles generate a large number of nucleation sites for hydration products and induce a more homogenous distribution of CSH and hence less pore structure. 2.

Nano-SiO$_2$ particles block the passages connecting capillary pores and water channels in cement paste [20].

Table 4. Water absorption values of different mixes

Batch No	Absorption (%)
A-1	6.12
A-5	4.23
A-7	5.421
A-9	4.458

3.4. Shrinkage

Prismatic specimens with 50×50×200 mm dimensions were prepared. The specimens were cured in the laboratory environment. The average temperature in the laboratory was 27±3 °C, and the relative humidity was 70%. The first measurement was taken using a length comparator with a precision of 2μm after 24 h of mixing, while the rest of measurements were taken at different ages of 3, 7, 14, 21, 28, 35, 42 days. The shrinkage behavior of mortars containing nano-SiO$_2$ is presented in Fig.3. From the results, it can be seen that drying shrinkage of mortars with nano-SiO$_2$ is apparently higher than that of control mortar and increases with increasing nano-SiO$_2$ content. Fig.4 shows the influence of nano-SiO$_2$ on shrinkage behavior of the RHA mortar. Results showed that RHA mortar experienced higher shrinkage than that of ordinary cement mortar. An increase was observed in RHA mortar containing nano-SiO$_2$ in comparison with the RHA mortar. The increase in the drying shrinkage of mortar containing nano-SiO$_2$ might be due mainly to refinement of pore size and increase of mesopores, which is directly related with the shrinkage due to self-desiccation. Moreover, it has been found that Nano-SiO$_2$, due to its high specific surface, serves additional nucleation sites for hydration products whereby chemical reactions are accelerated. Therefore, the degree of hydration increases as the amount of nano-SiO$_2$ increases and the autogenous shrinkage related to chemical shrinkage also increases [21].

3.5. Microstructure

Cement paste characteristics, for instance, strength and permeability, significantly depended on its nanostructure features in particular nanoporosity [22]. In recent years, the electron microscopy has demonstrated to be a very valuable method for determination of microstructure. Numerous studies on the influence of nano-SiO$_2$ on microstructure of plain cement mortar have been carried out. The results showed that nano-SiO$_2$ particles formed very dense and compact texture of hydrate products and decreased the size of big crystals such as Ca(OH)$_2$. In this paper in order to study the microstructure of RHA mortar, with and without nano-SiO$_2$, a XL30-type scanning electron microscope produced by Philips Company was used. The microstructure of RHA mortar with 3% replacement of nano-SiO$_2$ and without nano-SiO$_2$ at curing age of seven days are presented in Figures 5 and 6, respectively. Results showed that nano-SiO$_2$ particles improved the microstructure of RHA mortar on dense and compact form and generated more homogenous distribution of hydrated products.

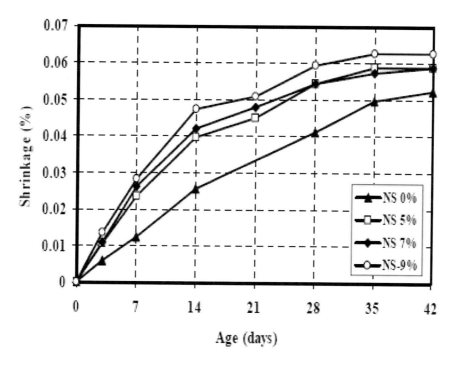

Figure 3. Shrinkage of mortars containing nano-SiO$_2$ versus time.

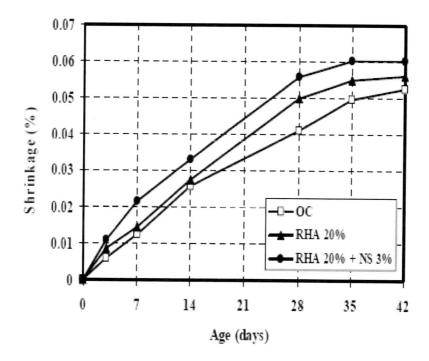

Figure 4. Shrinkage of RHA mortar with and without nano-SiO$_2$ versus time.

Figure 5. SEM micrograph of RHA mortar.

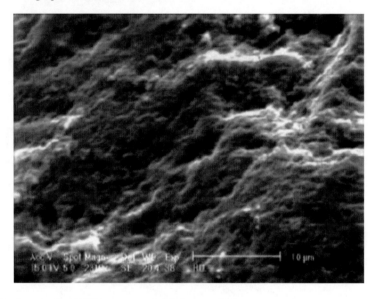

Figure 6. SEM micrograph of RHA mortar with 3% nano-SiO$_2$.

Conclusion

Noticeable increase was observed in compressive and flexural strength of ordinary cement mortars upon adding nano-SiO$_2$. Compressive and flexural strength of RHA cement mortars improved with the incorporation of nano-SiO$_2$. Integrating nano-SiO$_2$ with cement mortar containing RHA improved the microstructure of products on dense and compact form. Nano-SiO$_2$ had significant impact on drying shrinkage of mortars. The mortar samples with nano-SiO$_2$ experienced higher drying shrinkage. This effect was more prominent for larger amounts of nano-SiO$_2$. According to the results, there was a significant improvement in water absorption of mortars with intigerating nano-SiO$_2$. Nano-SiO$_2$ particles decreased the water absorption of cement composite by pore filling and pozzolanic effects. Also, it was observed that nano-SiO$_2$ particles were more effective in reduction of permeability than that of RHA.

References

[1]   Maile, A.; Huang, C.P. The chemistry and physics of nano-cement, Research experience for under graduates in bridge engineering, University of Delaware, USA, 2006

[2]   Li, Hui; Xiao, Hui-gang; Ou, Jin-ping A study on mechanical and pressure-sensitive properties of cement mortar with nanophase materials, *Cement and Concrete Research.* 2004, 34, 435-438.

[3]   Qing, Ye; Zenan, Zhang; Deyu, Kong; Rongshen, Chen Influence of nano-$_{SiO2}$ addition on properties of hardened cement paste as compared with silica fume, *Construction and Building Materials.* 2007, 21, 539-545.

[4]   Li, Hui; Zhang, Mao-hua; Ou, Jin-ping Abrasion resistance of concrete containing nano-particles for pavement, *Wear.* 2006, 260, 1262-1266.

[5]   Lin, K.L.; Chang, W.C.; Lin, D.F.; Luo, H.L.; Tsai, M.C. Effects of nano-$_{SiO2}$ and different ash particle sizes on sludge ash–cement mortar, *Journal of Environmental Management*, doi:10.1016./j.jenvman.2007. 03.036, 2007.

[6]   Li, Gengying; Properties of high-volume fly ash concrete incorporating nano-$_{SiO2}$, *Cement and Concrete Research.* 2004, 34, 1043-1049.

[7]   Lin, D.F.; Lin, K.L.; Chang, W.C.; Luo, H.L.; Cai, M.Q. Improvements of nano-$_{SiO2}$ on sludge/fly ash mortar, *Waste Management* doi: 10.1016/j.wasman.2007.03.023, 2007.

[8]   Jo, Byung-Wan; Kim, Chang-Hyun; Tae, Ghi-ho; Park, Jong-Bin Characteristics of cement mortar with nano-$_{SiO2}$ particles, *Construction and Building Materials.* 2007,21, 1351-1355.

[9]   Shih, Jeng-Ywan; Chang, Ta-Peng; H., Tien-Chin Effect of nanosilica on characterization of Portland cement composite, *Materials Science and Engineering* A, 2006, 424, 266-274.

[10] Ji, Tao Preliminary study on the water permeability and microstructure of concrete incorporating nano-$_{SiO2}$, *Cement and Concrete Research*, 2005, 35, 1943-1947.

[11] Li, Hui; Xiao, Hui-gang; Yuan, Jie; Ou, Jin-ping Microstructure of cement mortar with nano-particles, *Composite: part B*, 2004, 35, 185-189.

[12] Nair, D.G.; Alex, Fraaij; Klaassen, A.K.; Kentgens, A.P.M. A structural investigation relating to the pozzolanic activity of rice husk ashes, *Cement and Concrete Research* doi:10.1016/jcemconres.2007.10.004, 2007.

[13] Yu, Q.; Sawayama, K.; Sugita, S.; Shoya, M.; Isojima, Y. The reaction between rice husk ash and Ca(OH)$_2$ solution and the nature of its product, *Cement and Concrete Research* 1999, 29, 37-43.

[14] Feng, Q.; Yamamichi, H.; Shoya, M.; Sugita, S. Study on the pozzolanic properties of rice husk ash by hydrochloric acid pretreatment, *Cement and Concrete Research.* 2004, 34, 521-526.

[15] Sadrmomtazi, A.; Alidoust, O.; Hatami, F.; Haghi, A.K. A study on improvement of "Glasscrete" properties , Part 2: Fundamental Investigation, International Conference on Recent Advances in Composite Materials, New Dehli, India, 2007, Feb 20-23.

[16] Rao, G.A. Investigations on the performance of silica fume-incorporated cement pastes and mortars, *Cement and Concrete Research.* 2003, 33, 1765-1770.

[17] Yue, Li; Shuguang, Hu The microstructure of the interfacial transition zone between steel and cement paste, *Cement and Concrete Research.* 2001, 31, 385-388.

[18] Isaia, C.; Gastaldini, A.L.G.; Moraes, R. Physical and pozzolanic action of mineral addition on the mechanical strength oh high-performance concrete, *Cement and Concrete Research* 2003, 25, 69-76.

[19] Babu, K.G.; Babu, D.S. Performance of fly ash concrete containing lightweight EPS aggregates, *Cement &Concrete Composites.* 2004, 26, 605-611.

[20] Benachour, Y; Davy, C.A.; Skoczylas, F.; Houari, H. Effect of high calcite filler addition upon microstructural, mechanical, Shrinkage and transport properties of a mortar, *Cement and Concrete Research.* 2008, 38, 727-736.

[21] Melo Neto, A.A.; Cincotto, A.M.; Repette, W. Drying and autogenous shrinkage of pastes and mortars with activated slag cement, *Cement and Concrete Research.* 2008, 38, 565-574.

[22] Tanaka, K.; Kurmumisawa, K. Development of technique for observing pores in hardened cement paste, *Cement and Concrete Research.* 2002, 32, 1435-1441.

In: Handbook of Research on Nanomaterials, Nanochemistry ...   ISBN: 978-1-61942-525-5
Editors: A. K. Haghi and G. E. Zaikov   © 2013 Nova Science Publishers, Inc.

Chapter VI

Free Vibration Analysis of Double-Walled Carbon Nanotubes

R. Ansari, R. Rajabiehfard and B. Arash
Department of Mechanical Engineering,
University of Guilan, Iran

1. Introduction

During the past two decades, researchers from different disciplines have advanced the emerging field of nanotechnology at a phenomenal pace. In particular, the discovery of carbon nanotubes (CNTs) in 1991, by Iijima [1], has speeded the development of nanotechnology because of superior mechanical, electronic and other physical and chemical properties of CNTs over other materials known to man.

A single-walled carbon nanotube (SWCNT) is best described as a rolled-up tubular shell of graphene sheet, while a multiwalled nanotube (MWCNT) is a rolled-up stack of graphene sheets into concentric shells of carbons with adjacent shells separation of 0.34 nm. Mechanical properties of carbon nanotubes have shown to exceed those of other existing materials known to man, with high elastic modulus of greater than 1 TPa, comparable to that of diamonds, and strengths many times higher than the strongest steel at a fraction of the weight. Carbon nanotubes can be either metallic or semiconducting depending upon their structure characterized by chirality and also have very high thermal conductivity.

Generally, the theoretical modeling and analysis of nanostructured materials can be classified into two main categories. One category is atomic modeling, which includes techniques such as the classical molecular dynamics (MD), tight-binding molecular dynamics, and the ab initio method [4–7].

In this respect, a comprehensive review on these methods can be found in [8]. The other category is continuum modeling, which is increasingly being viewed as an alternative way of modeling materials on the atomistic scale. Although the MD simulations have generated abundant results for understanding the behavior of nanostructured materials, they are limited by the size and time scales of such atomic systems. Moreover, performing controlled experiments at the nanoscale is very difficult and prohibitively expensive. Consequently, continuum mechanics as a computationally efficient technique has still been the dominant tool for modeling large-scale systems at the nanometer scale. In the classical (local) continuum models, CNTs are taken as linear elastic thin shells [9, 10]. This continuum approximation is appropriate when the radius of CNTs is considerably larger than the interlayer spacing as remarked by Peng et al. [11].

As the size of nanomaterials is scaled down into very small scales, lattice spacing between individual atoms becomes increasingly important so that the discrete structure of the material can be no longer homogenized into a continuum. It is concluded that the applicability of classical continuum models at very small scales is questionable. Therefore, continuum models need to be further extended to consider the small-scale effects in nanomaterials studies. Many attempts have been made to develop more sophisticated types of continuum models in order to better accommodate the results from the MD simulations.

One type is a hybrid atomistic/continuum model, which allows one to directly incorporate interatomic potential into the continuum analysis.

This can be accomplished by equating the molecular potential energy of a nanostructured material with the mechanical strain energy of the representative volume element of a continuum model. The membrane version of this type of the atomistic-based theories was initially proposed by [12-15], and its bending theory was later developed by Wu [16]. The other type of continuum models is based on the theory of nonlocal continuum mechanics developed by Eringen [17, 18]. In this theory, which contains information about the long-range forces between atoms, the small-scale effect is simply introduced into the constitutive equations as a material parameter. The first application of the nonlocal continuum theory to nanotechnology was proposed by Peddieson et al. [19], followed by many other researchers.

A survey of the literature reveals that the vibrational aspects of CNTs have been the focus of considerable research [20-53]. Many of these studies have been conducted based on the classical continuum mechanics including the Bernoulli–Euler/ Timoshenko beam models [20-32] and shell models [33-36]. Most of the previous nonlocal vibration-related articles on CNTs have been established on the basis of the nonlocal Bernoulli–Euler/ Timoshenko beam models [37-50]. Unlike the Euler beam model, the Timoshenko beam model considers the effects of transverse shear deformation and rotary inertia.

The discrepancy becomes more pronounced for CNTs of low aspect ratios, which might be encountered in applications such as nanoprobes. This makes the Timoshenko beam model generally more preferable than the Euler counterpart. Besides, in a recent work by R. Li et al. [52] on vibrations of MWCNTs with simply support boundary conditions by a nonlocal shell model, it is concluded that the beam models may be inadequate for the study of the dynamics of nanotubes and may not yield proper results due to not taking the circumferential mode into account.

Hence, the transition from the nonlocal beam models to the nonlocal shell model in order to capture shell-like vibration modes becomes a necessity especially when the length-to-radius ratio of CNTs decreases. The present study is undertaken to accomplish this goal.

2. Governing Equations for CNTs

2.1. Shell Equations Based on Nonlocal Elasticity

The concept of non-locality is inherent in solid-state physics where the nonlocal attractions of atoms are prevalent. According to Eringen [54, 55], unlike the conventional local elasticity, in the nonlocal continuum theory, it is assumed that the stress at a point is a function of strains at all points in the continuum. To bring the non-locality into formulation, we employ the nonlocal constitutive equation given by Eringen [55]

$$(1 - (e_0 a)^2 \nabla^2)\boldsymbol{\sigma} = \mathbf{t} \tag{1}$$

where $e_0 a$ is the nonlocal parameter or characteristic length; t is the macroscopic stress tensor at a point. In the limit when the characteristic length goes to zero, nonlocal elasticity reduces to classical (local) elasticity. The stress tensor is related to strain by generalized Hooke's law as

$$\mathbf{t} = \mathbf{S} : \boldsymbol{\epsilon} \tag{2}$$

here S is the fourth order elasticity tensor and ":" denotes the double dot product. Hooke's law for the stress and strain relation is thus expressed by

$$
\begin{Bmatrix} \sigma_{xx} \\ \sigma_{\theta\theta} \\ \sigma_{x\theta} \\ \sigma_{\theta z} \\ \sigma_{xz} \end{Bmatrix} - (e_0 a)^2 \nabla^2 \begin{Bmatrix} \sigma_{xx} \\ \sigma_{\theta\theta} \\ \sigma_{x\theta} \\ \sigma_{\theta z} \\ \sigma_{xz} \end{Bmatrix} = \begin{bmatrix} \frac{E}{1-v^2} & \frac{vE}{1-v^2} & 0 & 0 & 0 \\ \frac{vE}{1-v^2} & \frac{E}{1-v^2} & 0 & 0 & 0 \\ 0 & 0 & 2G & 0 & 0 \\ 0 & 0 & 0 & G & 0 \\ 0 & 0 & 0 & 0 & G \end{bmatrix} \begin{Bmatrix} \varepsilon_{xx} \\ \varepsilon_{\theta\theta} \\ \gamma_{x\theta} \\ \gamma_{\theta z} \\ \gamma_{xz} \end{Bmatrix} \tag{3}
$$

Based on the FOSD theory, the three-dimensional displacement components u_x, u_y and u_z in the x, θ and z directions, respectively, as shown in Fig. 1, are assumed to be

$$
\begin{aligned}
u_x(x, \theta, z, t) &= u(x, \theta, t) + z\psi_x(x, \theta, \text{t}) \\
u_y(x, \theta, z, t) &= v(x, \theta, t) + z\psi_\theta(x, \theta, \text{t}) \\
u_z(x, \theta, z, t) &= w(x, \theta, z, t)
\end{aligned}
\tag{4}
$$

where u, v, w are the reference surface displacements and ψ_x, ψ_θ are the rotations of transverse normal about the x-axis and y-axis, respectively.

The middle surface strains $\epsilon_{xx}, \epsilon_{\theta\theta}$ and $\gamma_{x\theta}$, the shear strains γ_{xz} and $\gamma_{\theta z}$ and the middle surface curvatures k_x, k_θ and $k_{x\theta}$ are given by

$$\epsilon_{xx} = \frac{\partial u}{\partial x}, \epsilon_{\theta\theta} = \frac{1}{R}\frac{\partial v}{\partial \theta} + \frac{w}{R}, \gamma_{xy} = \frac{\partial v}{\partial x} + \frac{1}{R}\frac{\partial u}{\partial \theta}$$

$$k_x = \frac{\partial \psi_x}{\partial x}, k_\theta = \frac{1}{R}\frac{\partial \psi_\theta}{\partial \theta}, k_{x\nu} = \frac{\partial \psi_\theta}{\partial x} + \frac{1}{R}\frac{\partial \psi_x}{\partial \theta}$$

$$\gamma_{xz} = \frac{\partial w}{\partial x} + \psi_x, \gamma_{\theta z} = \frac{1}{R}\frac{\partial w}{\partial \theta} - \frac{v}{R} + \psi_\theta$$

$$(5)$$

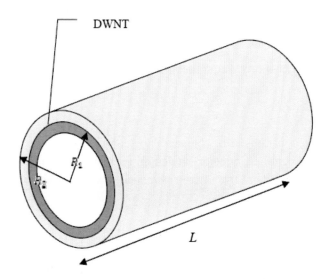

Figure 1. A double-walled CNT embedded in an elastic medium.

In the nonlocal elastic shell theory, the stress and moment resultants are defined based on the stress components in Eq. (3) and thus can be expressed as follows by referencing the kinematic relations in coupled Donnell theory

$$N_{xx} = \int_{-h/2}^{h/2} \sigma_{xx} dz \; i.e. \, N_{xx} - (e_0 a)^2 \nabla^2 N_{xx} = \frac{Eh}{1-v^2}\frac{\partial u}{\partial x} + \frac{vEh}{1-v^2}\left(\frac{1}{R}\frac{\partial v}{\partial \theta} + \frac{w}{R}\right)$$

$$(6\text{-}1)$$

$$N_{\theta\theta} = \int_{-h/2}^{h/2} \sigma_{\theta\theta} dz \; i.e. \, N_{\theta\theta} - (e_0 a)^2 \nabla^2 N_{\theta\theta} = \frac{vEh}{1-v^2}\frac{\partial u}{\partial x} + \frac{Eh}{1-v^2}\left(\frac{1}{R}\frac{\partial v}{\partial \theta} + \frac{w}{R}\right)$$

$$(6\text{-}2)$$

$$N_{x\theta} = \int_{-h/2}^{h/2} \sigma_{x\theta} dz \; i.e. \, N_{x\theta} - (e_0 a)^2 \nabla^2 N_{x\theta} = Gh\left(\frac{\partial v}{\partial x} + \frac{1}{R}\frac{\partial u}{\partial \theta}\right)$$

$$(6\text{-}3)$$

$$M_{xx} = \int_{-h/2}^{h/2} z\sigma_{xx} dz \; i.e. \, M_{xx} - (e_0 a)^2 \nabla^2 M_{xx} = D\left(\frac{\partial \psi_x}{\partial x} + \frac{v}{R}\frac{\partial \psi_\theta}{\partial \theta}\right)$$

$$(6\text{-}4)$$

$$M_{\theta\theta} = \int_{-h/2}^{h/2} z\sigma_{\theta\theta}\,dz \ \ i.e.\ M_{\theta\theta} - (e_0 a)^2 \nabla^2 M_{\theta\theta} = D\left(\frac{1}{R}\frac{\partial \psi_\theta}{\partial \theta} + v\frac{\partial \psi_x}{\partial x}\right)$$

(6-5)

$$M_{x\theta} = \int_{-h/2}^{h/2} z\sigma_{x\theta}\,dz \ \ i.e.\ M_{x\theta} - (e_0 a)^2 \nabla^2 M_{x\theta} = \frac{1}{2}D(1-v)\left(\frac{\partial \psi_\theta}{\partial x} + \frac{1}{R}\frac{\partial \psi_x}{\partial \theta}\right)$$

(6-6)

$$Q_{xx} = \int_{-h/2}^{h/2} \sigma_{xz}\,dz \ \ i.e.\ Q_{xx} - (e_0 a)^2 \nabla^2 Q_{xx} = Gh\left(\frac{\partial w}{\partial x} + \psi_x\right)$$

(6-7)

$$Q_{\theta\theta} = \int_{-h/2}^{h/2} \sigma_{\theta z}\,dz \ \ i.e.\ Q_{\theta\theta} - (e_0 a)^2 \nabla^2 Q_{\theta\theta} = Gh\left(\frac{1}{R}\frac{\partial w}{\partial \theta} - \frac{v}{R} + \psi_\theta\right)$$

(6-8)

In the multiple-shell model used herein, each tube of DWCNTs is described as an individual cylindrical shell of radius R, length L, thickness h, as in Fig. 1. If x and θ denote the longitudinal and circumferential coordinates, respectively, the governing equations on the basis of the Donnell shell theory are given as

$$\frac{\partial N_{xx}}{\partial x} + \frac{1}{R}\frac{\partial N_{x\theta}}{\partial \theta} = I_1\ddot{u} + I_2\ddot{\psi}_x$$
$$\frac{\partial N_{x\theta}}{\partial x} + \frac{1}{R}\frac{\partial N_{\theta\theta}}{\partial \theta} + \frac{Q_{\theta\theta}}{R} = I_1\ddot{v} + I_2\ddot{\psi}_\theta$$
$$\frac{\partial Q_{xx}}{\partial x} + \frac{1}{R}\frac{\partial Q_{\theta\theta}}{\partial \theta} - \frac{N_{\theta\theta}}{R} + p = I_1\ddot{w}$$
$$\frac{\partial M_{xx}}{\partial x} + \frac{1}{R}\frac{\partial M_{x\theta}}{\partial \theta} - Q_{xx} = I_2\ddot{u} + I_3\ddot{\psi}_x$$
$$\frac{\partial M_{x\theta}}{\partial x} + \frac{1}{R}\frac{\partial M_{\theta\theta}}{\partial \theta} - Q_{\theta\theta} = I_2\ddot{v} + I_3\ddot{\psi}_\theta$$

(7)

p is the pressure exerted on the tube i through the vdW interaction forces and/or the interaction between the tube and the surrounding elastic medium. To confine the effect of the surrounding medium on the outermost layers in the calculation of p_i, the Kronecker delta δ is introduced as

$$p = p_{vdW} + p_{Winkler}$$

(8)

The vdW model employed captures the effects of the interlayer vdW interactions of all layers in a MWNT and accounts for the curvature dependence of the vdW interactions, which is proposed by [56]

$$(p_{vdw})_i = w_i \sum_{j=1}^{N} c_{ij} - \sum_{j=1}^{N} c_{ij} \, w_j \quad (i = 1,2,...,N)$$

(9)

in which the vdW coefficients c_{ij} representing the pressure increment contributing to layer i from layer j are given by

$$c_{ij} = \left[\frac{1001\pi\varepsilon\sigma^{12}}{3a^4} E_{ij}^{13} - \frac{1120\pi\varepsilon\sigma^6}{9a^4} E_{ij}^{7} \right] R_j$$

(10)

where $a = 1.42 \, A°$ is the C-C bond length, ε the depth of the potential, σ a parameter that is determined by the equilibrium distance and R_j is the radius of jth layer and E_{ij}^m denotes the elliptic integral defined as

$$E_{ij}^m = (R_j + R_i)^{-m} \int_0^{\pi/2} \frac{d\theta}{[1 - K_{ij} \cos^2 \theta]^{m/2}}$$

(11)

here m is an integer and the coefficient K_{ij} is given by

$$K_{ij} = \frac{4R_j R_i}{(R_j + R_i)^2}$$

(12)

The surrounding elastic medium interacts with the outermost layer of the DWCNT under consideration. To include this interaction, the Winkler foundation model is employed as

$$(p_{Winkler})_i = -K_w w_i \quad (i = 2)$$

(13)

where K_w is the Winkler foundation modulus, which depends on the material properties of the elastic medium.

2.2. Field Equations

For the ith tube of a DWCNT, by the use of Eqs. (5), the Eqs. (6) can be stated in terms of the five field variables

$$\frac{Eh}{1-v^2}u_{,xx}^{(i)} - \frac{1}{2}\left(\frac{1}{R_i}\right)^2 \frac{Eh}{2(1+v)}u_{,\theta\theta}^{(i)} + \frac{1}{R_i}\left(\frac{vEh}{1-v^2} + \frac{1}{2}\frac{Eh}{2(1+v)}\right)v_{,x\theta}^{(i)} + \frac{1}{R_i}\frac{vEh}{1-v^2}w_{,x}^{(i)}$$

$$= I_1\ddot{u}^{(i)} + I_2\ddot{\psi}_x^{(i)} - (e_0 a)^2\left[I_1\left(\ddot{u}_{,xx}^{(i)} + \frac{1}{R_i^2}\ddot{u}_{,\theta\theta}^{(i)}\right) + I_2\left(\ddot{\psi}_{x,xx}^{(i)} + \frac{1}{R_i^2}\ddot{\psi}_{x,\theta\theta}^{(i)}\right)\right]$$

$$\frac{1}{R_i}\left(\frac{vEh}{1-v^2} + \frac{1}{2}\frac{Eh}{2(1+v)}\right)u_{,x\theta}^{(i)} + \frac{1}{2}\frac{Eh}{2(1+v)}v_{,xx}^{(i)} - \left(\frac{1}{R_i}\right)^2\frac{Eh}{1-v^2}v_{,\theta\theta}^{(i)} - \frac{Gh}{R_i^2}v^{(i)}$$

$$- \left(\frac{1}{R_i}\right)^2\left(\frac{Eh}{1-v^2} + Gh\right)w_{,\theta}^{(i)} + \frac{Gh}{R_i}\psi_\theta^{(i)}$$

$$= I_1\ddot{v}^{(i)} + I_2\ddot{\psi}_\theta^{(i)} - (e_0 a)^2\left[I_1\left(\ddot{v}_{,xx}^{(i)} + \frac{1}{R_i^2}\ddot{v}_{,\theta\theta}^{(i)}\right) + I_2\left(\ddot{\psi}_{\theta,xx}^{(i)} + \frac{1}{R_i^2}\ddot{\psi}_{\theta,\theta\theta}^{(i)}\right)\right]$$

$$- \frac{1}{R_i}\frac{vEh}{1-v^2}u_{,x}^{(i)} + \left(\frac{1}{R_i}\right)^2\left(\frac{Eh}{1-v^2} + Gh\right)v_{,\theta}^{(i)} + Ghw_{,xx}^{(i)} - \left(\frac{1}{R_i}\right)^2 Ghw_{,\theta\theta}^{(i)} - \left(\frac{1}{R_i}\right)^2\frac{Eh}{1-v^2}w^{(i)}$$

$$+ Gh\psi_{x,x}^{(i)} + \frac{Gh}{R_i}\psi_{\theta,\theta}^{(i)}$$

$$= I_1\ddot{w}^{(i)} + w^{(i)}\sum_{j=1}^{N}c_{ij} - \sum_{j=1}^{N}c_{ij}w^{(j)} - \delta_{iN}K_w w^{(i)}$$

$$- (e_0 a)^2\left[I_1\left(\ddot{w}_{,xx}^{(i)} + \frac{1}{R_i^2}\ddot{w}_{,\theta\theta}^{(i)}\right) + \left(w_{,xx}^{(i)} + \frac{1}{R_i^2}w_{,\theta\theta}^{(i)}\right)\sum_{j=1}^{N}c_{ij}\right.$$

$$\left. - \sum_{j=1}^{N}c_{ij}\left(w_{,xx}^{(j)} + \frac{1}{R_i^2}w_{,\theta\theta}^{(j)}\right) - \delta_{iN}K_w\left(w_{,xx}^{(i)} + \frac{1}{R_i^2}w_{,\theta\theta}^{(i)}\right)\right]$$

$$-Ghw_{,x}^{(i)} + D\psi_{x,xx}^{(i)} + \frac{1}{2}\left(\frac{1}{R_i}\right)^2\frac{(1-v)D}{2}\psi_{x,\theta\theta}^{(i)} - Gh\psi_x^{(i)} + \frac{1}{R_i}\left(vD + \frac{1}{2}\frac{(1-v)D}{2}\right)\psi_{\theta,x\theta}^{(i)}$$

$$= I_2\ddot{u}^{(i)} + I_3\ddot{\psi}_x^{(i)} - (e_0 a)^2\left[I_2\left(\ddot{u}_{,xx}^{(i)} + \frac{1}{R_i^2}\ddot{u}_{,\theta\theta}^{(i)}\right) + I_3\left(\ddot{\psi}_{x,xx}^{(i)} + \frac{1}{R_i^2}\ddot{\psi}_{x,\theta\theta}^{(i)}\right)\right]$$

$$\frac{Gh}{R_i}v^{(i)} + \frac{Gh}{R_i}w_{,\theta}^{(i)} + \frac{1}{R_i}\left(vD + \frac{1}{2}\frac{(1-v)D}{2}\right)\psi_{x,x\theta}^{(i)} + \frac{1}{2}\frac{(1-v)D}{2}\psi_{\theta,xx}^{(i)} + \left(\frac{1}{R_i}\right)^2 D\psi_{\theta,\theta\theta}^{(i)}$$

$$- Gh\psi_\theta^{(i)}$$

$$= I_2\ddot{v}^{(i)} + I_3\ddot{\psi}_\theta^{(i)} - (e_0 a)^2\left[I_2\left(\ddot{v}_{,xx}^{(i)} + \frac{1}{R_i^2}\ddot{v}_{,\theta\theta}^{(i)}\right) + I_3\left(\ddot{\psi}_{\theta,xx}^{(i)} + \frac{1}{R_i^2}\ddot{\psi}_{\theta,\theta\theta}^{(i)}\right)\right]$$

$$(i = 1,2) \tag{14}$$

2.3. Solution Procedure

   For DWCNT, the displacement field components are assumed to be functions of circumferential wave number, n, and the axial wave number, m. It is a simple task to indicate that for simply supported boundary conditions, the field equations admit solutions of the form

$$u^{(i)}(x,\theta,t) = \mathcal{A}_i \cos\left(\frac{n\pi x}{L}\right)\cos(m\theta)\,e^{j\omega t}$$

$$v^{(i)}(x,\theta,t) = \mathcal{B}_i \sin\left(\frac{n\pi x}{L}\right)\sin(m\theta)\,e^{j\omega t}$$

$$w^{(i)}(x,\theta,t) = \mathcal{C}_i \sin\left(\frac{n\pi x}{L}\right)\cos(m\theta)\,e^{j\omega t}$$

$$\psi_x^{(i)}(x,\theta,t) = \mathcal{D}_i \cos\left(\frac{n\pi x}{L}\right)\cos(m\theta)\,e^{j\omega t}$$

$$\psi_\theta^{(i)}(x,\theta,t) = \mathcal{E}_i \sin\left(\frac{n\pi x}{L}\right)\sin(m\theta)\,e^{j\omega t}$$

$$(15)$$

After introducing Eq. (15) into Eqs (14), these equations can be written in the matrix form

$$([\mathbf{K}] - \omega^2[\mathbf{M}])\{\mathbf{X}\} = 0 \tag{16}$$

where

$$[\mathbf{K}] = \begin{bmatrix} \left[k_{stiff}\right]^{(1)} + \left[k_{vdw}\right]^{(1)}_{(1,2)} & -\left[k_{vdw}\right]^{(2)}_{(1,2)} \\ -\left[k_{vdw}\right]^{(1)}_{(2,1)} & \left[k_{stiff}\right]^{(2)} + \left[k_{vdw}\right]^{(2)}_{(2,1)} \end{bmatrix}$$

where

$$\left[k_{stiff}\right]^{(i)} = \begin{bmatrix} L_{11} & L_{12} & L_{13} & 0 & 0 \\ L_{12} & L_{22} & L_{23} & 0 & L_{25} \\ L_{13} & L_{23} & L_{33} & L_{34} & L_{35} \\ 0 & 0 & L_{34} & L_{44} & L_{45} \\ 0 & L_{25} & L_{35} & L_{45} & L_{55} \end{bmatrix}, \left[k_{vdw}\right]_{(i,j)} = \begin{bmatrix} 0 & 0 & 0 & 0 & 0 \\ 0 & 0 & 0 & 0 & 0 \\ 0 & 0 & c_{ij} & 0 & 0 \\ 0 & 0 & 0 & 0 & 0 \\ 0 & 0 & 0 & 0 & 0 \end{bmatrix}$$

where the operators $L_{pq} = (p,q = 1,2,\ldots,5)$ are given in the Appendix, and

$$[\mathbf{M}] = \begin{bmatrix} [M]^{(1)} & \mathbf{0} \\ \mathbf{0} & [M]^{(2)} \end{bmatrix}$$

where

$$[M]^{(i)} = \begin{bmatrix} M_{11} & 0 & 0 & M_{14} & 0 \\ 0 & M_{22} & 0 & 0 & M_{25} \\ 0 & 0 & M_{33} & 0 & 0 \\ M_{14} & 0 & 0 & M_{44} & 0 \\ 0 & M_{25} & 0 & 0 & M_{55} \end{bmatrix},$$

where the operators $M_{pq} = (p,q = 1,2,\ldots,5)$ are given in the Appendix, and

$$\{\mathbf{X}\} = [\{\mathcal{A}_1 \ \mathcal{B}_1 \ \mathcal{C}_1 \ \mathcal{D}_1 \ \mathcal{E}_1\} \ \{\mathcal{A}_2 \ \mathcal{B}_2 \ \mathcal{C}_2 \ \mathcal{D}_2 \ \mathcal{E}_2\}]^T,$$

3. Results and Discussion

The geometries and the mechanical properties of each layer of CNTs are

$$E = 1\,Tpa, D = 0.85\,ev\ h = 0.34\,nm\,, v = 0.27\,, \rho = 1340\frac{Kg}{m^3}.$$

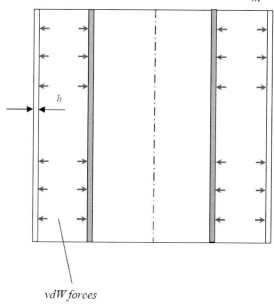

Figure 2. Cross-sectional view of a double-walled CNT under the vdW interactions.

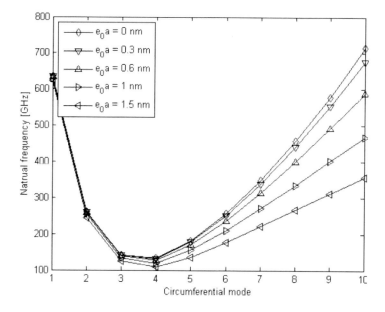

Figure 3. Variation of natural frequency with circumferential wave number for a double-walled CNT with simply supported end conditions ($R_1 = 8.5\,nm, \frac{L}{R_1} = 5, K_w = 0\,\frac{Pa}{m}$).

Presented graphically in Fig. 3 is the natural frequency of a double-walled CNT versus circumferential mode number for several values of the small length scale ranging from $e_0 a = 0$ (corresponding to the classical/ local continuum model) to $e_0 a = 1.5$. One can observe from this Figure that the lowest natural frequency decreases as the small length scale increases. It physically means that the small-scale effects in the nonlocal model make nanotubes more flexible. It is further observed that the magnitude of decrease in natural frequencies corresponding to higher circumferential modes is considerably higher than those corresponding to lower ones.

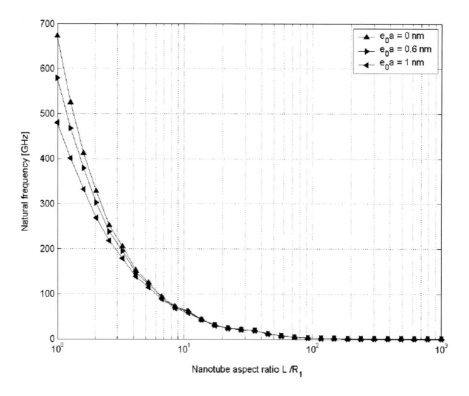

Figure 4. Variation of natural frequencies of a double-walled CNT over a wide range of its aspect ratio (

$R_1 = 8.5 \ nm, K_w = 0 \frac{Pa}{m}$).

The variation of natural frequencies of a double-walled CNT over a wide range of its aspect ratio for various nonlocal parameters is plotted in Fig. 4. The values of nonlocal parameter are assumed to be varied from $e_0 a = 0 \ nm^2$ to $e_0 a = 1 \ nm^2$. The profound effects of the small length scale on the natural frequencies of the CNT are seen from Fig. 4, especially for shorter CNTs and higher values of nonlocal parameter. As the ratio of length-to-innermost radius increases, natural frequencies tend to decrease, and the effects of small length scale diminish so that the frequency envelopes tend to converge. This observation means that the classical continuum model would give a reasonable prediction in the study of nanotubes of high aspect ratios for which the whole structure can be homogenized into a continuum.

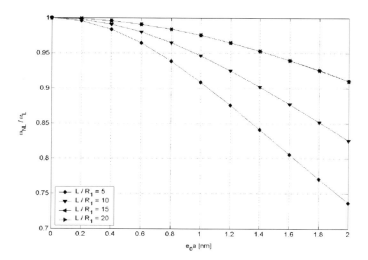

Figure 5. Effect of the small length scale on the natural frequencies ratio for double-walled CNTs with various length-to-innermost radius ($R_1 = 8.5 \; nm, K_w = 0 \frac{Pa}{m}$).

To further investigate the influence of the small length scale on the natural frequencies of nanotubes, the ratio of nonlocal frequency to local frequency will be discussed later on. The frequency ratios corresponding to various length-to-innermost radius ratios for a double-walled CNT are graphed in Fig. 5. It is observed that the effects of the small length scale are more prominent for shorter length CNTs as the name implies. In the prediction of natural frequencies of a CNT of $L/R_1 = 5$ via a classical continuum model, for instance, a relative error of 26.5% for the nonlocal parameter $e_0 a = 2 \; nm^2$ is introduced. This relative error reduces to about 9% when the aspect ratio of the CNT is increased by $L/R_1 = 20$.

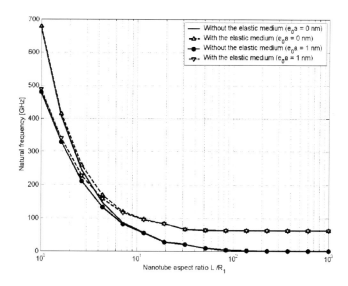

Figure 6. Effect of the surrounding medium on the natural frequencies of a double walled CNT ($R_1 = 8.5 \; nm$).

The local and nonlocal natural frequencies over a broad range of L/R_1 for a triple-walled CNT with and without being embedded in the surrounding medium are indicated in Figure 6. As seen from this Figure, the effect of the surrounding elastic medium on natural frequencies of CNTs is negligible for $L/R_1 \leq 2$. However, the frequency difference due to this effect becomes more pronounced when the ratio of length-to-innermost radius increases until it reaches a maximum value corresponding to infinitely long CNTs. It is also seen that the effect of small length scale on the natural frequency becomes smaller as the aspect ratio of nanotube increases, regardless of the medium effect.

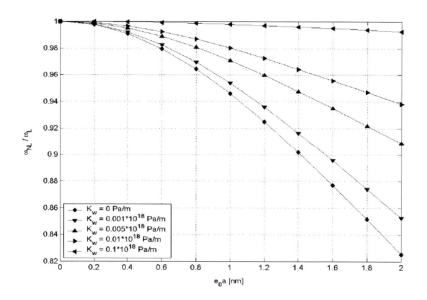

Figure 7. Effect of the small length scale on the frequency ratios corresponding to various foundation moduli for a double-walled CNT with all edges simply-supported ($R_1 = 8.5\ nm\ and\ \frac{L}{R_1} = 10$).

Fig. 7 is presented to further examine the role of the surrounding medium in the small size dependence of the natural frequencies of a double-walled simply supported CNT. It is observed that the relative error deceases when the modulus of the elastic medium becomes larger. Physically speaking, in the study of vibration characteristics of CNTs without being embedded in the surrounding medium, more care must be taken over the dependence of the natural frequencies on the small length scale. For the given nonlocal parameter $e_0 a = 2\ nm^2$, the relative error in predicting the frequencies varies from about 17.5% to less than 1%, which correspond to $K_w = 0$ and $K_w = 10^{17} Pa/m$, respectively.

Conclusion

On the basis of the theory of nonlocal continuum mechanics, the free vibration characteristics of double-walled carbon nanotubes for a simply supported boundary condition were studied. The equations of motion of a double-walled carbon nanotube were derived based on the coupled Donnell theory and the Eringen nonlocal elasticity. The following findings are summarized:

- The significance of the small size effects on the natural frequencies of double-walled carbon nanotubes is shown to be dependent on the geometric sizes of CNT and the elastic surrounding medium.
- The small-scale effects in the nonlocal continuum model make small-size CNTs more flexible. In other words, the classical continuum model tends to overestimate the natural frequencies of small size nanotubes, and one must recourse to the nonlocal version to reduce the relative error. As the small-scale parameter increases, the frequencies obtained for the nonlocal shell become smaller than those for its local counterpart.
- The natural frequencies corresponding to higher vibration modes are more sensitive to the small length scale.
- The existence of the elastic medium significantly enhances the values of natural frequencies particularly for lower ones.
- The small size effect becomes more pronounced when the modulus of the elastic medium becomes smaller.

Appendix

$$L_{11} = -\frac{Eh}{1-v^2}\left(\frac{n\pi}{L}\right)^2 - \frac{1}{2}\frac{Eh}{2(1+v)}\left(\frac{m}{R_i}\right)^2$$

$$L_{12} = \left(\frac{vEh}{1-v^2} + \frac{1}{2}\frac{Eh}{2(1+v)}\right)\left(\frac{n\pi}{L}\right)\left(\frac{m}{R_i}\right)$$

$$L_{13} = \frac{vEh}{1-v^2}\left(\frac{1}{R_i}\right)\left(\frac{n\pi}{L}\right)$$

$$L_{22} = -\frac{1}{2}\frac{Eh}{2(1+v)}\left(\frac{n\pi}{L}\right)^2 - \frac{Eh}{1-v^2}\left(\frac{m}{R_i}\right)^2 - \frac{K_s}{2}\frac{Eh}{2(1+v)}\left(\frac{1}{R_i}\right)^2$$

$$L_{23} = \left(-\frac{K_s}{2}\frac{Eh}{2(1+v)} - \frac{Eh}{1-v^2}\right)\left(\frac{m}{R_i}\right)^2$$

$$L_{25} = \frac{K_s}{2}\frac{Eh}{2(1+v)}\frac{1}{R_i}$$

$$L_{33} = -\frac{K_s}{2}\frac{Eh}{1+v}\left[\left(\frac{n\pi}{L}\right)^2 + \left(\frac{m}{R_i}\right)^2\right] - \frac{Eh}{1-v^2}\left(\frac{1}{R_i}\right)^2$$

$$L_{34} = -\frac{1}{2}\frac{K_s}{2}\frac{Eh}{1+v}\frac{n\pi}{L}$$

$$L_{35} = \frac{1}{2}\frac{K_s}{2}\frac{Eh}{1+v}\frac{m}{R_i}$$

$$L_{45} = \left(vD + \frac{1}{2}\frac{(1-v)D}{2} \right) \left(\frac{n\pi}{L} \right) \left(\frac{m}{R_i} \right)$$

$$L_{44} = -D \left(\frac{n\pi}{L} \right)^2 - \frac{1}{2}\frac{(1-v)D}{2} \left(\frac{m}{R_i} \right)^2 - \frac{1}{2}\frac{K_s}{2}\frac{Eh}{1+v}$$

$$L_{55} = -\frac{1}{2}\frac{(1-v)D}{2} \left(\frac{n\pi}{L} \right)^2 - D \left(\frac{m}{R_i} \right)^2 - \frac{1}{2}\frac{K_s}{2}\frac{Eh}{1+v}$$

$$M_{11} = M_{22} = M_{33} = \left[1 + (e_0 a)^2 \left(\left(\frac{n\pi}{L} \right)^2 + \left(\frac{m}{R_i} \right)^2 \right) \right] I_1$$

$$M_{44} = M_{55} = \left[1 + (e_0 a)^2 \left(\left(\frac{n\pi}{L} \right)^2 + \left(\frac{m}{R_i} \right)^2 \right) \right] I_3$$

$$M_{14} = M_{25} = \left[1 + (e_0 a)^2 \left(\left(\frac{n\pi}{L} \right)^2 + \left(\frac{m}{R_i} \right)^2 \right) \right] I_2$$

References

[1]   Kroto H W, Heath J R, O'Brien S C, Curl R F 1985 Smalley RE, C60: buckminsterfullerene *Nature*. 318 162-163.

[2]   Iijima S 1991 Helical microtubes of graphitic carbon *Nature* 8 354-356.

[3]   Kong X Y, Ding Y, Yang R, Wang Z L 2004 Single-Crystal Nanorings Formed by Epitaxial Self-Coiling of Polar Nanobelts. *Science*. 303 1348-1351.

[4]   Iijima S, Brabec C, Maiti A and Bernholc J 1996 *Chem. Phys.* 104 2089

[5]   Yakobson B I, Campbell M P, Brabec C J and Bernholc J 1997 *Comput. Mater. Sci.* 8 241

[6]   Hernandez E, Goze C, Bernier P and Rubio A 1998 *Phys. Rev. Lett.* 80 4502

[7]   Sanchez-Portal D et al. 1999 *Phys. Rev.* B 59 12678

[8]   Qian D, Wagner J G, Liu W K, Yu M F and Ruoff R S 2002 *Appl. Mech. Rev.* 55 495

[9]   Yakobson B I, Brabec C J, Bernholc J 1996 Nanomechanics of carbon tubes: instability beyond linear response *Phys. Rev. Lett.* 76 2511.

[10]   Ru C Q 2001 Axially compressed buckling of a double-walled carbon nanotube embedded in an elastic medium *J. Mech. Phys. Solids* 49 1265-1279.

[11]   Peng J, Wu J, Hwang K C, Song J, Huang Y 2008 Can a single-wall carbon nanotube be modeled as a thin shell? *J. Mech. Phys. Solids* 56 2213–2224.

[12]   Belytschko, T., Xiao, S.P., Schatz, G.C., Ruoff, R.S., 2002. Atomistic simulations of nanotube fracture. *Phys. Rev.* B 65, 235430.

[13]   Zhang, P., Huang, Y., Gao, H., Hwang, K.C., 2002a. Fracture nucleation in single-wall carbon nanotubes under tension: a ontinuum analysis incorporating interatomic potentials. *J. Appl. Mech.* 69, 454–458.

[14]   Zhang, P., Huang, Y.G., Geubelle, P.H., Hwang, K.C., 2002b. On the continuum modeling of carbon nanotubes. *Acta Mech. Sin.* 18, 528–536.

[15]   Zhang, P., Huang, Y., Geubelle, P.H., Klein, P.A., Hwang, K.C., 2002c. The elastic modulus of single-wall carbon nanotubes: a continuum analysis incorporating interatomic potentials. *Int. J. Solids Struct.* 39, 3893–3906.

[16]   Wu, J., Hwang, K.C., Huang, Y., 2008. An atomistic-based finite-deformation shell theory for single-wall carbon nanotubes. *J. Mech. Phys. Solids.* 56, 279–292.

[17]   Eringen A C 1983 On differential equations of nonlocal elasticity and solutions of screw dislocation and surface waves *J. Appl. Phys.* 54 4703–4710.

[18]   Eringen A C 2002 *Nonlocal Continuum Field Theories.* (Springer NewYork).

[19]   Peddieson J, Buchanan G R, McNitt R P 2003 Application of nonlocal continuum models to nanotechnology. *Int. J. Eng. Sci.* 41 305–312.

[20]   Yoon J, Ru C Q, Mioduchowski A 2003 Vibration of an embedded multiwall carbon nanotube. *Compos. Sci. Tech.* 63 1533–1542.

[21]   Yoon J, Ru C Q, Mioduchowski A 2005 Vibration and instability of carbon nanotubes conveying fluid. *Compos. Sci. Tech.* 65 1326–1336.

[22]   Zhang Y, Liu G, Han X 2005 Transverse vibrations of double-walled carbon nanotubes under compressive axial load. *Phys. Lett. A* 340 258–266.

[23]   Fu Y M, Hong J W, Wang X Q 2006 Analysis of nonlinear vibration for embedded carbon nanotubes. *J. Sound Vib.* 296 746–756.

[24]   Wang C M, Tan V B C, Zhang Y Y 2006 Timoshenko beam model for vibration analysis of multiwalled carbon nanotubes. *J. Sound Vib.* 294 1060–1072.

[25]   Wang Q, Varadan V K, 2006 Wave characteristics of carbon nanotubes. *Int. J. Solids Struct.* 43 254–265.

[26]   Wang L, Ni Q, Li M, Qian Q 2008 The thermal effect on vibration and instability of carbon nanotubes conveying fluid. *Physica E.* 40 3179-3182.

[27]   Aydogdu M 2008 Vibration of multiwalled carbon nanotubes by generalized shear deformation theory. *Int. J. Mech. Sci.* 50 837–844.

[28]   Xu K Y, Aifantis E C, Ya Xu n Y H 2008 Vibrations of Double-Walled Carbon Nanotubes With Different Boundary Conditions Between Inner and Outer Tubes. *J. Appl. Mech.* 75 021013.

[29]   Kuang Y D, He X Q, Chen C Y, Li G Q 2009 Analysis of nonlinear vibrations of double-walled carbon nanotubes conveying fluid. *Comput. Mat. Sci.* 45 875–880.

[30]   Ansari R., Hemmatnezhad M., Ramezannezhad H. 2009 Application of HPM to the Nonlinear Vibrations of Multiwalled Carbon Nanotubes. *Numerical Methods for Partial Differential Equations Journal* DOI 10.1002/num.20499.

[31]   Elishakoff I, Pentaras D 2009 Fundamental natural frequencies of double-walled carbon nanotubes. *J. Sound Vib.* 322 652–664.

[32]   Chang W J, Lee H L 2009 Free vibration of a single-walled carbon nanotube containing a fluid flow using the Timoshenko beam model. *Phys. Lett. A* 373 982–985.

[33]   Natsuki T, Endo M 2006 Vibration analysis of embedded carbon nanotubes using wave propagation approach. *J. Appl. Phys.* 99 034311.

[34]   Liew K M, Wang Q 2007 Analysis of wave propagation in carbon nanotubes via elastic shell theories. *International Journal of Engineering Science* 45 227–241.

[35]   Sun C, Liu K 2007 Vibration of multiwalled carbon nanotubes with initial axial loading. *Solid State Commun.* 143 202–207.

[36]   Yan Y, Wang W Q, Zhang L X 2009 Noncoaxial vibration of fluid-filled multiwalled carbon nanotubes. *Appl. Math. Modell.* doi:10.1016/j.apm.2009.03.031

[37]   Wang Q, Zhou G Y, Lin K C 2006 Scale effect on wave propagation of double-walled carbon nanotubes. *Int. J. Solids Struct.* 43 6071–6084.

[38] Wang Q, Varadan V K 2006 Vibration of carbon nanotubes studied using nonlocal continuum mechanics. *Smart Mater. Struct.* 15 659–666.

[39] Lu P, Lee H P, Lu C, Zhang P Q 2007 Application of nonlocal beam models for carbon nanotubes. *Int. J. Solids Struct.* 44 5289–5300.

[40] Reddy J N 2007 Nonlocal theories for bending, buckling and vibration of beams. *International Journal of Engineering Science* 45 288–307.

[41] Wang Q, Wang C M 2007 The constitutive relation and small-scale parameter of nonlocal continuum mechanics for modeling carbon nanotubes *Nanotechnology* 18 075702.

[42] Wang C M, Zhang Y Y, He X Q 2007 Vibration of nonlocal Timoshenko beams. *Nanotechnology.* 18 105401 (9pp)

[43] Khosravian N, Rafii-Tabar H 2008 Computational modelling of a non-viscous fluid flow in a multiwalled carbon nanotube modelled as a Timoshenko beam. *Nanotechnology.* 19 275703.

[44] Heireche H, Tounsi A, Benzair A 2008 Scale effect on wave propagation of double-walled carbon nanotubes with initial axial loading. *Nanotechnology.* 19 185703 (11pp)

[45] Heireche H, Tounsi A, Benzair A, Maachou M, Adda Bedia E A 2008 Sound wave propagation in single-walled carbon nanotubes using nonlocal elasticity. *Physica. E* 40 2791–2799.

[46] Aydogdu M 2009 Axial vibration of the nanorods with the nonlocal continuum rod model. *Physica. E* 41 861–864.

[47] Wang L 2009 Dynamical behaviors of double-walled carbon nanotubes conveying fluid accounting for the role of small length scale. *Computational Materials Science.* 45 584–588.

[48] Murmu T, Pradhan S C, 2009 Small-Scale Effect on the Vibration of Non-uniform Nanocantilever based on Nonlocal Elasticity Theory *Physica E.* doi:10.1016/j.physe.2009.04.015

[49] Murmu T, Pradhan S C 2009 Thermo-mechanical vibration of a single-walled carbon nanotube embedded in an elastic medium based on nonlocal elasticity theory. *Computational Materials Science.*

[50] Lee H L, Chang W J 2009 Vibration analysis of a viscous-fluid-conveying single-walled carbon nanotube embedded in an elastic medium. *Physica. E* 41 529–532.

[51] Wang Q, Varadan V K 2007 Application of nonlocal elastic shell theory in wave propagation analysis of carbon nanotubes. *Smart Mater. Struct.* 16 178–190.

[52] Li R, Kardomateas G A 2007 Vibration Characteristics of Multiwalled Carbon Nanotubes Embedded in Elastic Media by a Nonlocal Elastic Shell Model. *J. Appl. Mech.* 74 1087-1094.

[53] Hu Y G, Liew K M, Wang Q, He X Q, Yakobson B I 2008 Nonlocal shell model for elastic wave propagation in single- and double-walled carbon nanotubes. *J. Mech. Phys. Solids* 56 3475–3485.

[54] Eringen A C 1983 On differential equations of nonlocal elasticity and solutions of screw dislocation and surface waves. *J. Appl. Phys.* 54 4703–4710.

[55] Eringen A C 2002 *Nonlocal Continuum Field Theories.* (Springer NewYork).

[56] He X Q, Kitipornchaia S, Liew K M 2005 Buckling analysis of multiwalled carbon nanotubes: a continuum model accounting for van der Waals interaction *J. Mech. Phys. solids* 53 303-326.

In: Handbook of Research on Nanomaterials, Nanochemistry ...   ISBN: 978-1-61942-525-5
Editors: A. K. Haghi and G. E. Zaikov   © 2013 Nova Science Publishers, Inc.

Chapter VII

Local Delivery Systems based on Biodegradable Chitosan Beads

M. R. Saboktakin
Baku State University, Azerbaijan

Introduction

Natural polymers have potential pharmaceutical applications because of their low toxicity, biocompatibility, and excellent biodegradability. In recent years, biodegradable polymeric systems have gained importance for design of surgical devices, artificial organs, drug delivery systems with different routs of administration, carriers of immobilized enzymes and cells, biosensors, ocular inserts, and materials for orthopedic applications (BrOndsted & Kopeˇcek, 1990). These polymers are classified as either synthetic (polyesters, polyamides, polyanhydrides) or natural (polyamino acids, polysaccharides) (Giammona et al., 1999; Krogars et al., 2000). Polysaccharide-based polymers represent a major class of biomaterials, which includes agarose, alginate, carageenan, dextran, and chitosan. Chitosan [_(1,4)2-amino-2-d-glucose] is a cationic biopolymer produced by alkaline N-deacetylation of chitin, which is the main component of the shells of crab, shrimp, and krill (Chiu et al., 1999; Jabbari & Nozari, 2000). Chitosan is a functional linear polymer derived from chitin, the most abundant natural polysaccharide on the earth after cellulose, and it is not digested in the upper GI tract by human digestive enzymes (Fanta & Doane, 1986; Furda, 1983). Chitosan is a copolymer consisting of 2-amino-2-deoxy-d-glucose and 2-acetamido-2-deoxy-d-glucose units links with β-(1-4) bonds. It should be susceptible to glycosidic hydrolysis by microbial enzymes in the colon because it possesses glycosidic linkages similar to those of other enzymatically depolymerized polysaccharides. Among diverse approaches that are possible for modifying polysaccharides, grafting of synthetic polymer is a convenient method for adding new properties to a polysaccharide with minimum loss of its initial properties (Peppas, 1987; Saboktakin et al., 2007). Graft copolymerization of vinyl monomers onto polysaccharides using free radical initiators, has attracted the interest of many scientists. Up to now, considerable works have been devoted to the grafting of vinyl monomers onto the substrates, especially Starch and cellulose (Honghua & Tiejing, 2005; Jabbari & Nozari, 2000). Existence of polar functionally groups as carboxylic acid are needed not only for bio-adhesive properties but also for pH-sensitive properties of polymer (Ratner, 1989; Thierry et al., 2003). The increase of MAA content in the hydrogels provides more hydrogen bonds at low pH and more electrostatic repulsion at high pH. A part of our research program is chitosan modification to prepare materials with pH-sensitive properties for uses as drug delivery (Bloembergen & Pershan, 1967; Mahfouz et al., 1997; Schmitz et al., 2000). The free radical graft copolymerization polymethacrylic acid onto chitosan was carried out at 70 °C, bis-acrylamide as a cross-linking agent and persulfate as an initiator. Polymer-bonded drugs usually contain one solid drug bonded together in amatrix of a solid polymeric binder. They can be produced by polymerizing a monomer such as methacrylic acid (MAA), mixed with a particulate drug, by means of a chemical polymerization catalyst, such as AIBN or by means of high-energy radiation, such as X-ray or γ-rays. The modified hydrogel and satranidazole as a model drug were converted to nanoparticles by freeze-drying method. The equilibrium swelling studies and in vitro release profiles were carried out in enzyme-free simulated gastric and intestinal fluids [SGF (pH 1) and SIF (pH 7.4), respectively)]. The influences of different factors, such as content of MAA in the feed monomer and swelling, were studied (Saboktakin et al., 2008).

Materials and Methods

Derivatives of NOCCS, N-sulfonato-N,O-carboxymethylchitosan (NOCCS),containing sulfonato groups, SO_3 −Na+, were prepared by the reaction of sulfur trioxide–pyridine complex with NOCCS in alkaline medium at room temperature. Typically, 10 g of NOCCS (0.045 mol) dissolved in 0.60 L of water was treated with repeated additions of 12 g SO_3–pyridine. The SO_3–pyridine was slurried in 50–100mL of water and added dropwise, over 1 h. Both the NOCCS solution and the sulfur trioxide reagent slurry were maintained at a pH above 9 by the addition of sodium hydroxide (5 M). Following the last addition of the sulfating reagent, the sodium hydroxide solution was added until the pH stabilized (approx. 40 min.).

The pH of the mixture was adjusted to 9; the mixture was heated to 33 °C and held for 15 min. After filtration through a 110μm nylon screen, the filtered SNOCCS solution was poured into 6 L of 99% isopropanol. The resulting precipitate was collected and air-dried overnight. The dried precipitate was dissolved in 0.45 L boiling water, solution was poured into dialysis sacks (M_W = 12,000) and dialyzed for three to four days against dionized water. The contents of the sacks were lyophilized to yield the final product (6.2 g). N-sulfonato-N,O-carboxymethylchitosan with 1:1 molar ratios of methacrylic acid was polymerized at 60–70 °C in a thermostatic water bath, bis-acrylamide as a cross-linking agent (CA), using persulfate as an initiator ([I] = 0.02M) and water as the solvent (50 mL). The polymeric system was stirred by mechanical stirrer to sticky hydrogel, and it was separated from medium without solvent addition. All experiments were carried out in Pyrex glass ampoules. After the specific time (48 h), the precipitated network polymer was collected and dried in vacuum. Copolymer (50 mg) and satranidazole (10 mg) were dispersed with stirring in 25mL deionised water. After approximately 180 min, the sample was sprayed into a liquid nitrogen bath cooled down to 77K, resulting in frozen droplets. These frozen droplets were then put into the chamber of the freeze-dryer. In the freeze-drying process, the products are dried by a sublimation of the water component in an iced solution. The following procedure was used to assess the stability of satranidazole during the bead preparation process. The prepared beads were extracted twice with a solvent mixture of 1:1 acetonitrile and ethanol (v/v), the extract was evaporated, and the residue was injected onto HPLC column. Stability-indicating chromatographic method was adopted for this purpose. The method consisted of a symmetry C18 column (254mm×4.6mm; 5_m) run using a mobile phase of composition methanol: water (70:30, v/v) at a flow rate of 0.5 mL/min, a Waters pump (600E), and eluants monitored with Water photodiode array detector (996 PDA) at 227 nm. A definite weight range of 10–15mg of bead was cut and placed in a 1.5mL capacity microcentrifuge tube containing 1mL of release medium of the following composition at 37 ∘C: phosphate buffered saline (140mM, pH 7.4) with 0.1% sodium azide and 0.1% Tween 80. At predetermined time points, 100_L of release medium was sampled with replacement, to which 3mL of scintillation cocktail was added and vortexed before liquid scintillation counting. The cumulative amount of satranidazole released as a function of time was calculated. To study the molecular properties of satranidazole and Nsulfonato- N,O-carboxymethyl chitosan/PMAA, the solid-state characterization of samples were done by the application of thermal, X-ray diffraction, and microscopy technique. During this study, the characteristics of satranidazole and Nsulfonato-N,O-carboxymethylchitosan/PMAA were compared with the beads to reveal any changes occurring as a result of bead preparation. Differential scanning calorimetry (DSC) studies were performed with a Mettler Toledo 821 thermal analyzer (Greifensee, Switzerland) calibrated with indium as standard. For thermogram acquisition, sample sizes of 1–5mg were scanned with a heating rate of 5 ∘C/min over a temperature range of 25–300 ∘C. In order to check the reversibility of transition, samples were heated to a point just above the corresponding transition temperature, cooled to room temperature, and reheated up to 300 ∘C. Satranidazole samples and N-sulfonato-N,O-carboxymethylchitosan/ PMAA beads were viewed using a Philips XL-30 E SEM scanning electron microscope (SEM) at 30 kV (max.) for morphological examination. Powder samples of satranidazole and beads were mounted onto aluminum stubs using double-sided adhesive tape and then sputter coated with a thin layer of gold at 10 Torr vacuum before examination. The specimens were scanned with an electron beam of 1.2 kV acceleration potential, and images were collected in collected in secondary electron mode. Molecular arrangement of satranidazole and N-sulfonato-N,Ocarboxymethylchitosan/ PMAA in powder as well as in beads was compared by powder X-ray diffraction patterns acquired at room temperature on a Philips PW 1729 diffractometer (Eindhoven, Netherlands) using Cu-K α radiation. The data were collected over an angular range from 3∘ to 50° 2Θ in continuous mode using a step size of 0.02° 2Θ and step time of 5 s.

Results

In the present study, the beads were prepared by the classical method, which involves spreading a uniform layer of polymer dispersion followed by a drying step for removal of solvent system.

Since the methodology of bead preparation involved a heating step, it may have had a detrimental effect on the chemical stability of drug. Hence, the stability assessment of satranidazole impregnated in bead was done using stability-indicating method. For this purpose, satranidazole was extracted from bead and analyzed by HPLC. A single peak at 21.2m representing satranidazole (with no additional peaks) was detected in the chromatogram, suggesting that the molecule was stable during preparation of beads Satranidazole was extracted from different regions of Nsulfonato- N,O-carboxymethylchitosan/PMAA bead using acetonitrile: ETOH (1:1, v/v) solvent system. After normalization of satranidazole concentration on weight basis of bead, the variation in distribution of satranidazole in different regions of bead was <16%. The composition of the polymer defines its nature as a neutral or ionic network and furthermore, its hydrophilic/hydrophobic characteristics.

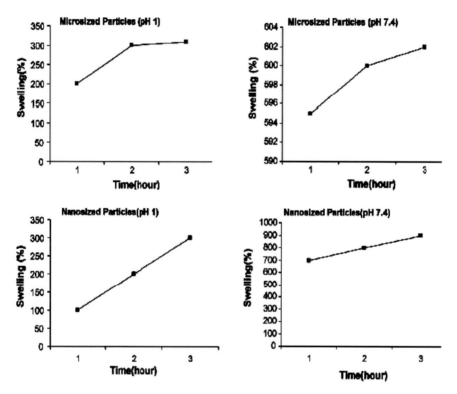

Figure 1. Time-dependent swelling of micro- and nano-carriers satranidazole drug model as a time at 37°C.

Ionic hydrogels, which could be cationic, containing basic functional groups or anionic, containing acidic functional groups, have been reported to be very sensitive to changes in the environmental pH. The swelling properties of the ionic hydrogels are unique due to the ionization of their pendent functional groups. Hydrogels containing basic functional groups is found increased swelling activity in acidic conditions and reduced in basic conditions. The pH-sensitive anionic hydrogels show low swelling activity in acidic medium and very high activity in basic medium. As shown in Fig. 1, an increase in the content of MAA in the feed monomer mixtures resulted in less swelling in simulated gastric fluid but greater swelling in and simulated intestinal fluids.

This is because the increase of MAA content in the hydrogels provides more hydrogen bonds at low pH and more electrostatic repulsion at high pH. Fig. 2 shows the scanning electron microscope (SEM) of graft NOCCS copolymer with polymethacrylic acid and nano-polymer-bonded drug, respectively.

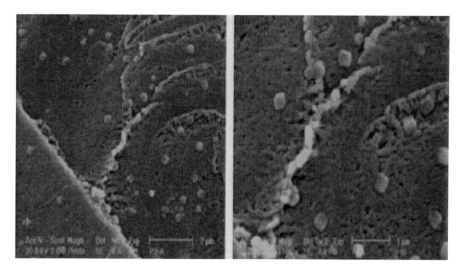

Figure 2. SEM of NOCCS-polumethacrylic acid hydrogel beads with satranidazole.

Nano- and micro-polymer-bonded drugs (50 mg) were poured into 3mL of aqueous buffer solution (SGF: pH 1 or SIF: pH 7.4) (Fig. 3). The mixture was introduced into a cellophane membrane dialysis bag. The bag was closed and transferred to a flask containing 20mL of the same solution maintained at 37 °C.

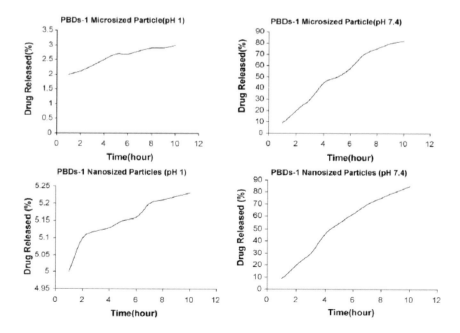

Figure 3. Release of satranidazole drug from micro- and nanocarriers as a functions of time at 37°C.

The external solution was continuously stirred, and 3mL samples were removed at selected intervals. The removed volume was replaced with SGF or SIF Fig. 3.

The triplicate samples were analyzed by UV spectrophotometer, and the quantity of satranidazole was determined using a standard calibration curve obtained under the same conditions. Nano- and micro-polymer-bonded drugs (50 mg) were poured into 3mL of aqueous buffer solution (SGF: pH 1 or SIF: pH 7.4) (Fig. 3). The mixture was introduced into a cellophane membrane dialysis bag. The bag was closed and transferred to a flask containing 20mL of the same solution maintained at 37 °C. The external solution was continuously stirred, and 3mL samples were removed at selected intervals. The removed volume was replaced with SGF or SIF Fig. 4.

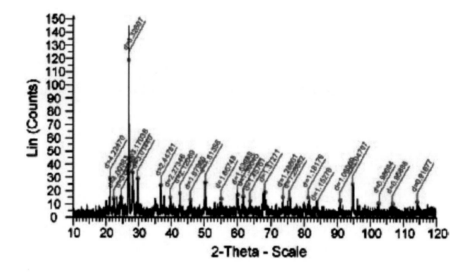

Figure 4. XRD pattern of NOCCS hydrogel beads.

The triplicate samples were analyzed by UV spectrophotometer, and the quantity of satranidazole was determined using a standard calibration curve obtained under the same conditions. It appears that the degree of swelling depends on their particle size; a decrease in the molecular size of carriers increased the swelling rate. The thermal behavior of a polymer is important for controlling the release rate in order to have a suitable drug dosage form. The glass transition temperature (Tg) was determined from the DSC thermograms. The values are given in Table 1.

The higher Tg values probably related to the introduction of cross-links, which would decrease the flexibility of the chains and the ability of the chains to undergo segmental motion, which would increase the Tg values. On the other hand, the introduction of a strongly polar carboxylic acid group can increase the Tg value because of the formation of internal hydrogen bonds between the polymer chains. X-ray diffraction is a proven tool to study crystal lattice arrangements, and it yields very useful information on degree of crystallinity. The X-ray diffraction patterns of hydrogel with satranidazole have several high-angle bead at the following 2Θ values: 21°, 22°, 25°, 26.5°, 28°, and 32°.The 26.5° 2Θ peak had the highest intensity, and the hump in the baseline occurred from 7° to 45° 2Θ, as observed for chitosan bead.

Table 1. DSC data and composition of copolymer

Polymer samples	Molar composition of monomers in the feed				Degree of substitution (DS)	T_g (°C)
	NOCCS (g)	MAA (g)	CA (g)	IN (g)		
P-1	1	3	0.05	0.05	0.52	135
P-2	1	2	0.05	0.05	0.49	142

Discussion

In order to develop a local delivery system for satranidazole, a biodegradable N-sulfonato-N,O-carboxymethylchitosan–PMAA bead was attempted, which to our knowledge was the first effort of its kind. The potential formulation problems were anticipated since chitosan is only soluble in aqueous acidic solutions, whereas satranidazole, being a hydrophobic drug, is insoluble under similar conditions.

In early stages of formula optimization studies, the satranidazole was incorporated into bead. The primary mechanisms for release of drugs from matrix systems in vitro are swelling, diffusion, and disintegration. In vitro degradation of chitosan beads were prepared by solution casing method occurred less rapidly as the degree 73% deacetylated showed slower biodegradation. Since the grade of chitosan used in the present study was of high molecular weight with a degree of deacetylation ≥85%, significant retardation of release of satranidazole from bead is attributed to the polymer characteristics.

In addition, diffusion of satranidazole may have been hindered by increased tortuosity of polymer accompanied by a swelling mechanism. As observed from SEM photomicrographs, the crystals of satranidazole have a different appearance than recrystallized satranidazole. These nanoparticles do not have clearly defined crystal morphological features in the SEM photomicrographs.

Hence, it appears that the irregularly shaped particles are surface deposited with poloxamer, which gives them an appearance resembling that of coated particles. X-ray diffraction technique is also used to study the degree of crystallinity of pharmaceutical drugs and excipients. A lower 2Θ value indicates larger d-spacings, while an increase in the number of high-angle reflections indicates higher molecular state order. In addition, broadness of reflections, high noise, and low peak intensities are characteristic of a poorly crystalline material. A broad hump in the diffraction pattern of chitosan extending over a large range of 2Θ suggests that chitosan is present in amorphous state in the bead.

Conclusion

The swelling and hydrolytic behavior of the hydrogels beads were dependent on the content of MAA groups and caused a decrease in gel swelling in SGF or an increase in gel swelling in SIF.

Modified chitosan with different contents of MAA and CA by graft copolymerization reactions were carried out under microwave radiation. The swelling of the hydrogels beads was dependent on the content of MAA groups and caused a decrease in gel swelling in SGF or an increase in gel swelling in SIF.

Incorporation of MAA made the hydrogels beads pH-dependent and the transition between the swollen and the collapsed states occurred at high and low pH. The swelling ratios of the hydrogels beads increased at pH 7.4 but decreased at pH 1 with increasing incorporation of MAA.

References

Bloembergen, N., & Pershan, P. S. (1967). Model catalysis of ammonia synthesis of iron–water interfaces – a sum frequency generation vibrational spectroscopic study of solid–gas interfaces and anion photoelectron spectroscopic study of selected anion clusters. *Physical Review,* 128(2), 606.

BrOndsted, H.,&Kope˘cek, J. (1990). Hydrogels for site-specific oral delivery. *Proceedings of the International Synposium on Controlled Release of Bioactive Materials,* 17, 128–129.

Chiu, H. C., Hsiue, G. H., Lee, Y. P., & Huang, L. W. (1999). Synthesis and characterization of pH-sensitive dextran hydrogels as a potential colon-specific drug delivery system. *Journal of Biomaterials Science, Polymer Edition,* 10, 591–608.

Fanta, G. F., & Doane, W. N. (1986). Grafted starches. In O. B. Wurzburg (Ed.), *Modified starches: properties and uses* (pp. 149–178). Boca Raton (FL): CRC.

Furda, I. (1983). Aminopolysaccharides – their potential as dietary fiber. In I. Furda (Ed.), *Unconventional sources of dietary fiber, physiological and in vitro functional properties* (pp. 105–122). Washington, DC: American Chemical Society.

Giammona, G., Pitarresi, G., Cavallora, G., & Spadaro, G. (1999). New biodegradable hydrogels based on an acryloylated polyaspartamide cross-linked by gamma irradiation. *Journal of Biomedical Science Polymer, Edition,* 10, 969–987.

Honghua, Xu., & Tiejing, Li. (2005). The analysis of boundary functions of CMS reaction factors. *Journal of Nature and Science,* 3(2), 25–28.

Jabbari, E., & Nozari, S. (2000). Swelling behavior of acrylic acid hydrogels prepared by _-radiation cross-linking of polyacrylic acid in aqueous solution. *European Polymer Journal,* 36, 2685–2692.

Krogars, K., Heinamaki, J., Vesalahti, J., Marvola, M., Antikainen, O., & Yliruusi, J. (2000). Extrusion-spheronization of pH-sensitive polymeric matrix pellets for possible colonic drug delivery. *International Journal of Pharmaceutics,* 187–194.

Mahfouz, A., Hamm, B., & Taupitz, M. (1997). Contrast agents for MR imaging of the liver: clinical overview. *European Radiology,* 7, 507.

Peppas, N. A. (1987). *Hydrogels in medicine and pharmacy.* Boca Raton, FL: CRC Press.

Ratner, B. D. (1989). S. K. Aggarwal (Ed.), *Comprehensive polymer science – the synthesis, characterisation, reactions & applications of polymers* (pp. 201–247). Oxford: Pergamon Press.

Saboktakin, M. R., Maharramov, A.,& Ramazanov, M. A. (2007). Synthesis and characterization of aromatic polyether dendrimer/mesalamine(5-ASA) nanocomposite as drug carrier system. *Journal of American Science,* 3(4), 45.

Saboktakin, M. R., Maharramov, A., & Ramazanov, M. A. (2008). Poly(amidoamine) (PAMAM)/CMS dendritic nanocomposite for controlled drug delivery. *Journal of American Science,* 4(1), 48.

Schmitz, S. A., Winterhalter, S., Schiffler, S., Gust, R., Wagner, S., Kresse, M., Coupland, S. E., Semmler, W., & Wolf, K. J. (2000). Superparamagnetic iron oxide nanoparticles functionalized polymers. *Investigative Radiology,* 35, 460.

Thierry, B., Winnik, F. M., Mehri, Y., & Tabrizian, M. (2003). A new Y3Al5O12 phase produced by liquid-feed flame spray. *Journal of American Chemical Society,* 125, 7494.

In: Handbook of Research on Nanomaterials, Nanochemistry ...   ISBN: 978-1-61942-525-5
Editors: A. K. Haghi and G. E. Zaikov   © 2013 Nova Science Publishers, Inc.

Chapter VIII

Molecular Dynamics Simulation in Modeling Carbon Nanotubes

R. Ansari and B. Motevalli
Department of Mechanical Engineering,
University of Guilan, Iran

1. Introduction

To investigate the properties of nanostructures, and in particular carbon nanotubes (CNTs), a strong tool is computational experiments. In this regard, computational nanomechanics have an indispensable role in the field of current nanotechnology and nanosciences. Different methods are applied in the literature in order to model carbon nanotubes. Primarily, Yakobson et al. [1] implemented molecular dynamics (MD) simulations to investigate the buckling behavior of these nanostructures under different loading conditions. They also tried to extract a fitted continuum model based on thin-shell theories. Indeed, MD simulations played a stronger tool in modeling carbon nanotubes, in comparison with continuum models, due to the fact that they can describe the atomistic level of these structures. As a result, an extensive study on carbon nanotubes has been conducted based on MD simulations, in recent years [1-30]. However, the important deficiencies of these simulations are their time expense and their limitations in the size scale of the structures under investigation. Subsequently, with the current computational computer abilities, MD simulations cannot be performed on very large systems. On the other hand, continuum models can overcome these problems, but as mentioned before, in some cases, they cannot appropriately model the real behavior of the system (e.g., properties of carbon nanotubes containing defects [3]). However, in the last two decades, continuum models have been numerously applied to various problems of nanostructured materials [31-40].

To overcome both limitations of continuum and molecular dynamic simulations, new approaches are based on multiscale models [41-46]. In these models, the critical parts of a material (e.g., defects, dislocation patterns occurring in fatigue, etc.), are modeled through atomistic methods, while the surrounding parts are modeled by continuum-based approaches using the finite element method, mesh-free techniques, etc. Through the simulation steps, both models are coupled and are solved simultaneously. However, some difficulties are encountered in coupling the different time scales of the two regions and the effect of wave reflections from the intermediate zone [46]. Due to the interesting features of multiscale models, recently, they have attracted much attention in the research area of nanocomputations, and it is expected that they can successfully and efficiently model various problems in this area of research.

As mentioned, molecular dynamics simulation is a strong tool to model and describe various problems in the field of nanocomputations. Even in the multiscale models, an MD simulation is a core part of it, whose information is feed to the continuum field of the model. Thus, this book chapter aims to describe the art of handling an MD simulation program, for the special case of carbon nanotubes. The text is organized in such manner that can be utilized as a self-learner, even for a beginner in the field. It should be noted that the general scheme and concepts presented in this book chapter can be also used for other case of nanostructures. It is expected that at the end of this book chapter, the reader will be able to conduct an MD code in order to investigate the properties of carbon nanotubes on his own. The organization of this book chapter is as follows: in Section 2, the important empirical potentials that are commonly used to model carbon nanotube structures are introduced. It is of interest to note that these potentials play a pivotal role in MD simulations so that the more accurate the potentials are, the more reliable the simulation is. Furthermore, the calculation of the interaction potential force from the potential energy function is illustrated in detail. Section 3 is devoted to the algorithms for conducting an MD code. As an example, the Nose-Hoover thermostating algorithm is explained in more detail in this section. In the end, some important factors in implementing a successful MD code are discussed.

2. Empirical Potentials

2.1. General Structure of Potential Function

Although quantum mechanics models are the most precise ones in modeling materials at atomistic scales, they can only handle very small nanoscale systems of about a few hundred atoms. Due to this fact, a more efficient way of describing the energetic of nanostructures is utilizing interatomic potential energy functions. These potentials are only a function of relative positions of the atoms, and through them, the total interaction energy of the whole atomic system is evaluated. Obviously, the more accurate these potentials are and the better they describe the bonding and energetic of the atomistic structure, the more reliable are the results obtained and the nearer is the model to reality. Indeed, there is not a specific potential function that is the best one for modeling all systems of particles. The appropriate potential function that can model all the bonding and dynamics of the nanostructure to the best possibility may vary with respect to the case of study and the system of particles under investigation. Therefore, the first step for implementing a successful MD code is to choose an appropriate potential energy function for the case study.

For a system of particles, a general structure of the total potential energy can be expressed by a potential function, which is in terms of the position vectors of the constituent particles, as follows:

$$V(\vec{r}_1, \vec{r}_2, ..., \vec{r}_N) = \sum_i v_1(\vec{r}_i) + \sum_i \sum_{j>i} v_2(\vec{r}_i, \vec{r}_j) + \sum_i \sum_{j>i} \sum_{k>j} v_3(\vec{r}_i, \vec{r}_j, \vec{r}_k) + ... \qquad (2.1)$$

Where $\vec{r}_i$ is the position vector of particle i, and v_n is the n-body interatomic potential function. The first term in Eq. (2.1) considers the effect of an external field; the second term includes the pair-wise interaction between particles i and j; and the third term computes for the three-body potential containing particles i, j, and k. Higher order interactions can be included in the potential energy function of Eq. (2.1), but in many cases of study, these terms have an insignificant effect. Additionally, these terms increase the computational time of simulation extensively.

2.2.1. Pair-Wise Potentials

There are number of potential functions to model the interaction between pairs of atoms or molecules efficiently. These potential energy functions are usually composed of two terms of attraction and repulsion. In long distances, the attraction term is dominant and the pair of particles tends to each other, while in short distances the repulsion term prevails and the pair of particles tends to repel. Actually, the pair-wise potentials that are applied for the case of carbon nanotubes are for the evaluation of the vdW interaction potential energies between the walls of multiwalled carbon nanotubes. Here, the two most popular pair-wise potentials, which are utilized extensively in the literature, are introduced. The first one is known as the Lennard-Jones (LJ) potential, which is used in the simulation and modeling of various nanoscale system of particles. This function is given as follows:

$$V_{LJ}(r_{ij}) = 4\varepsilon \left(\left(\frac{\sigma}{r_{ij}} \right)^{12} - \left(\frac{\sigma}{r_{ij}} \right)^{6} \right) \qquad (2.2)$$

The first term in Eq. (2.2) stands for the repulsion effect, while the second one is the attraction partition. Note that the pair-wise interaction potentials depend only on the magnitude of the separation distance between particles i and j (i.e. $r_{ij} = |r_i - r_j|$). The constant parameters in Eq. (2.2) are the collision diameter σ and the dislocation energy ε. These parameters depend upon the system of particles to be simulated. Subsequently, the corresponding LJ interaction force between a pair of particles is computed as

$$\vec{F}_i(r_{ij}) = -\vec{\nabla}_i V_{LJ} = 24\varepsilon \left(\frac{2\sigma^{12}}{r_{ij}^{13}} - \frac{\sigma^6}{r_{ij}^7} \right) \frac{\vec{r}_{ij}}{r_{ij}} \qquad (2.3)$$

Another commonly used potential function is the Morse pair-wise interaction potential, which is given by

$$V_M(r_{ij}) = \varepsilon\left(e^{2\beta(\rho - r_{ij})} - 2e^{\beta(\rho - r_{ij})}\right)$$
(2.4)

Similarly, the first term stands for the repulsion part, while the second term is the attraction one. The corresponding interaction force due to the Morse potential is obtained from

$$\vec{F}_M(r_{ij}) = 2\varepsilon\beta\left(e^{2\beta(\rho - r_{ij})} - e^{\beta(\rho - r_{ij})}\right)\frac{\vec{r}_{ij}}{r_{ij}}$$
(2.5)

Ref. [30] have applied different pair-wise potential functions for the systems of double-walled carbon nanotubes oscillators and have studied the behavior of such systems under these potentials. Based on their results, they concluded that applying different potentials may lead to a total different conclusion on the behaviors of carbon nanotube oscillators. Thus, in respect to the type of the system to be modeled, different pair-wise potentials may cause quite different results.

2.2.4. Many Body Potential Functions

As mentioned, the pair-wise potentials introduced in the previous subsection compute the vdW potential energy between two distinct carbon nanotubes. To accurately evaluate the interatomic potential energy within the atomistic structure of a carbon nanotube, an appropriate potential energy function must be applied. Herein, the most popular potential energy functions that are appropriate to model atomistic interaction within the structure of a carbon nanotube are presented.

One famous potential function is the Tersoff's potential, which was primarily suggested to model the bonding in Si [47, 48]. Actually, this potential function is a pair-wise potential, which takes the local atomistic environment into account. The total potential energy is given by

$$E = \sum_i \sum_{j>i} V^{Tr}(r_{ij})$$
(2.6)

where, $V^{Tr}(r_{ij})$ is the Tersoff potential energy function between atom i and its nearest neighbor atoms j, which is computed by

$$V^{Tr}(r_{ij}) = f_c(r_{ij})\left[V_R(r_{ij}) - B_{ij}V_A(r_{ij})\right]$$
(2.7)

The functions V_R and V_A represents the repulsive and attractive pair-wise interactions, and are given by

$$V_R(r_{ij}) = A_{CC}e^{-\lambda_R r_{ij}}$$
$$V_A(r_{ij}) = B_{CC}e^{-\lambda_A r_{ij}}$$

(2.8)

The local atomic environment is imposed through the B_{ij} coefficient in the potential energy function of Eq. (2.7). This coefficient is evaluated as follows

$$B_{ij} = (1+\beta^n \zeta^n)^{-\frac{1}{2n}},$$

$$\zeta = \sum_{k \neq i,j} f_C(r_{ij}) e^{\left[\lambda_A^3 (r_{ij}-r_{ik})^3\right]} g(\theta_{ijk})$$

(2.9)

$$g(\theta_{ijk}) = 1 + \frac{c^2}{d^2} - \frac{c^2}{d^2 + \left(h - \cos\theta_{ijk}\right)^2}$$

Table 2.1. Parameters in the Tersoff potential

A_{CC}	$1.3936 \times 10^3\ eV$
B_{CC}	
λ_A	$3.4674 \times 10^2\ eV$
μ	3.4879 Å^{-1}
	2.2119 Å^{-1}
β	1.5724×10^{-7}
n	
c	7.2751×10^{-1}
d	
h	3.8049×10^4
	4.3484
R_{CC}^1	-0.57058
	1.8 Å
R_{CC}^2	
	2.1 Å

The constant parameters in Eq. (2.9) are given in Table 2. 1 for the case of carbon-carbon interaction. θ_{ijk} is the angle between bonds i-j and i-k, and f_C is a cutoff function, which considers only the nearest neighbors of atom i in the calculation of its potential energy. This function is given by

$$f_c(r_{ij}) = \begin{cases} 1, & r_{ij} < R_{cc}^1 \\ \dfrac{1}{2} + \dfrac{1}{2}\cos\left(\dfrac{\pi(r_{ij} - R_{cc}^1)}{R_{cc}^2 - R_{cc}^1}\right), & R_{cc}^1 < r_{ij} < R_{cc}^2 \\ 0, & r_{ij} > R_{cc}^2 \end{cases}$$

(2.10)

Based on Tersoff potential, Brenner [49] proposed a potential that overcomes some deficiencies of the Tersoff potential and can model more precisely the bonding between the carbon atoms. This potential is as follows

$$V^{Br}(r_{ij}) = f_c(r_{ij}) \left[V_R(r_{ij}) + \overline{B}_{ij} V_A(r_{ij}) \right] \tag{2.11}$$

Similarly, V_R and V_A stand for the repulsive and attractive terms of the potential function of Eq. (2.11), respectively. These functions are expressed as

$$V_R(r_{ij}) = \frac{D_{cc}}{S_{cc}-1} e^{-\sqrt{2S_{cc}}\beta_{cc}\left(r_{ij}-R_{cc}^e\right)},$$

$$V_A(r_{ij}) = \frac{-D_{cc}S_{cc}}{S_{cc}-1} e^{-\sqrt{2/S_{cc}}\beta_{cc}\left(r_{ij}-R_{cc}^e\right)}, \tag{2.12}$$

Likewise, f_C is the cut-off function and its expression is the same as Eq. (2.10). Moreover, the coefficient $\overline{B}_{ij}$, which includes the local atomistic structure, is given by

$$\overline{B}_{ij} = \frac{B_{ij} + B_{ji}}{2} \tag{2.13}$$

$$B_{ij} = \left[1 + \sum_{k \neq i, j} f_c(r_{ik}) G(\theta_{ijk}) \right]^{-\delta_{cc}} \tag{2.14}$$

$$G(\theta_{ijk}) = a_0 \left[1 + \frac{c_0^2}{d_0^2} - \frac{c_0^2}{d_0^2 + \left(1 + \cos\theta_{ijk}\right)^2} \right] \tag{2.15}$$

The parameters in Eqs. (2,12) to (2.15) are given in Table 2.2. The Brenner potential has been used thoroughly in modeling of carbon nanotubes, and successful results have been obtained for various types of problems such as axial, torsional, and bending loading of these structures. Note that the principles of handling an MD code are similar for other types of potential functions also. Therefore, an MD code based on a specific potential energy function can be easily adapted to the other types of potentials. Since the Brenner potential function is applied successfully in modeling of carbon nanotubes, this potential is used in this book chapter in the loops of the MD code. Other potential functions can be also found in the literature (e.g., see Refs. [50, 51]).

The computation of the total force exerted on each individual atom is the most expensive and important part of a molecular dynamics simulation. The expression of this force is obtained by taking the gradient of the potential energy with respect to the position of an instant atom i. The computation of this gradient for the special case of Brenner potential energy function is presented as follows

$$\vec{F}_i = -\vec{\nabla}_i E = -\left(\frac{\partial}{\partial x_i}\hat{i} + \frac{\partial}{\partial y_i}\hat{j} + \frac{\partial}{\partial z_i}\hat{k} \right) \sum_{i=1}\sum_{j>i} V^{Br}(r_{ij}) \qquad (2.16)$$

Table 2.2. Parameters in the Brenner first generation potential function

Parameter	1st parameterization	2nd parameterization
R_{CC}^e	1.315 Å	1.39 Å
D_{CC}	6.325 eV	6 eV
β_{CC}	1.5 (1/ Å)	2.1 (1/ Å)
S_{CC}	1.29	1.22
δ_{CC}	0.80469	0.5
a_0	0.011304	0.00020813
c_0	19	330
d_0	2.5	3.5
R_{CC}^1	1.7 Å	1.7 Å
R_{CC}^1	2 Å	2 Å

Eq. (2.16) can be written in three scalar equations for each component of the interatomic force. As an example, the computation of the x component is presented; obviously, other components are obtained, similarly. The x component is obtained as

$$F_{i_x} = -\sum_{j\neq i} \frac{\partial V^{Br}(r_{ij})}{\partial x_i} - \sum_{j\neq i}\sum_{k\neq i,j} \frac{\partial V^{Br}(r_{jk})}{\partial x_i} \qquad (2.17)$$

The second term on the right hand side of Eq. (2.17) is the contribution, which is associated with $\bar{B}_{jk}$. Utilizing Eqs. (2.11) to (2.15), one can obtain

$$\frac{\partial V^{Br}(r_{ij})}{\partial x_i} = \frac{\partial f_c(r_{ij})}{\partial r_{ij}}\left[V_R(r_{ij}) + \bar{B}_{ij}V_A(r_{ij}) \right].\frac{x_i - x_j}{r_{ij}}$$
$$+ f_c(r_{ij})\left[\left(\frac{\partial V_R(r_{ij})}{\partial r_{ij}} + \bar{B}_{ij}\frac{\partial V_A(r_{ij})}{\partial r_{ij}} \right).\frac{x_i - x_j}{r_{ij}} + V_A(r_{ij})\frac{\partial \bar{B}_{ij}}{\partial x_i} \right] \qquad (2.18)$$

The derivatives of each term in Eq. (2.18) are obtained as follows

$$\frac{\partial V_R(r_{ij})}{\partial r_{ij}} - \frac{D_{cc}\beta_{cc}\sqrt{2S_{cc}}}{S_{cc}-1} e^{-\sqrt{2S_{cc}}\beta_{cc}(r_{ij}-R_{cc})}$$

$$(2.19a)$$

$$\frac{\partial V_A(r_{ij})}{\partial r_{ij}} = \frac{D_{cc}S_{cc}\beta_{cc}\sqrt{2/S_{cc}}}{S_{cc}-1}e^{-\sqrt{2/S_{cc}}\beta_{cc}\left(r_{ij}-R_{cc}^e\right)}$$

(2.19b)

$$\frac{\partial \bar{B}_{ij}}{\partial x_i} = \frac{1}{2}\left(\frac{\partial B_{ij}}{\partial x_i} + \frac{\partial B_{ji}}{\partial x_i}\right)$$

(2.19c)

$$\frac{\partial B_{ij}}{\partial x_i} = -\delta_{cc}\left[1+\sum_{k\neq i,j}G(\theta_{ijk})f_{ik}\right]^{-\delta_{cc}-1}\cdot\left[\sum_{k\neq i,j}\left\{f_{ik}\frac{\partial G(\theta_{ijk})}{\partial \cos(\theta_{ijk})}\cdot\frac{\partial \cos(\theta_{ijk})}{\partial x_i} + \frac{\partial f_{ik}}{\partial r_{ik}}G(\theta_{ijk})\frac{x_i-x_k}{r_{ik}}\right\}\right]$$

(2.19c)

$$\frac{\partial B_{ji}}{\partial x_i} = -\delta_{cc}\left[1+\sum_{k\neq i,j}G(\theta_{jik})f_{jk}\right]^{-\delta_{cc}-1}\cdot\left[\sum_{k\neq i,j}\left\{f_{jk}\frac{\partial G(\theta_{jik})}{\partial \cos(\theta_{jik})}\cdot\frac{\partial \cos(\theta_{jik})}{\partial x_i}\right\}\right]$$

(2.19d)

As mentioned, θ_{ijk} is the angle between bonds i-j and i-k; therefore,

$$\cos(\theta_{ijk}) = \frac{\vec{r}_{ij}\cdot\vec{r}_{ik}}{r_{ij}r_{ik}} = \frac{r_{ij}^2 - r_{ik}^2 + r_{jk}^2}{2r_{ij}r_{ik}}$$

(2.20)

$$\frac{\partial \cos(\theta_{ijk})}{\partial x_i} = \frac{r_{ij}^2 - r_{ik}^2 + r_{jk}^2}{2r_{ij}^2 r_{ik}}\cdot\frac{x_i-x_j}{r_{ij}} + \frac{r_{ik}^2 - r_{ij}^2 + r_{jk}^2}{2r_{ij}r_{ik}^2}\cdot\frac{x_i-x_k}{r_{ik}}$$

(2.21)

Similarly,

$$\frac{\partial \cos(\theta_{jik})}{\partial x_i} = \frac{r_{ij}^2 - r_{jk}^2 + r_{ik}^2}{2r_{ij}^2 r_{jk}}\cdot\frac{x_i-x_j}{r_{ij}} - \frac{r_{ik}}{r_{ij}r_{ik}}\cdot\frac{x_i-x_k}{r_{ik}}$$

(2.22)

Before introducing the second term in Eq. (2.17), a description in how to compute the series in $\dfrac{\partial B_{ij}}{\partial x_i}$ and $\dfrac{\partial B_{ji}}{\partial x_i}$ is presented. As mentioned, these potential functions evaluate the interaction of an atom with its nearest neighbors. Fig 2. 1 represents the local structure of atoms within a carbon nanotube.

Atoms j in all series are the first neighbors of atom i, and the bond i-j is the bond between these atoms. Since the function G_{ijk} includes triplet atoms where the atom i is the center, the atoms k in series of B_{ij} and $\dfrac{\partial B_{ij}}{\partial x_i}$ are also the first neighbors where when j specifies one of the neighbors, k is the other neighbors of atom i. These atoms are identified by k_1 in Fig 2. 1. In function G_{jik}, which appears in B_{ji} and $\dfrac{\partial B_{ji}}{\partial x_i}$, one of the first neighbors (atom j) is the center atom in the triplets of *jik*; therefore, in these triplets, atoms *k* are the first neighbors of atom *j*, which means they are in the second neighborhood of atom *i*. These atoms are represented by atoms k_2 in Fig 2. 1.

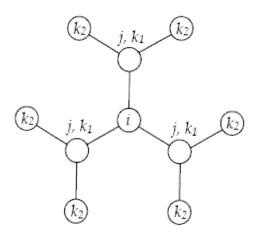

Figure 2.1. Adjacent atoms of an instant atom *i* within a carbon nanotube structure.

Note that in the total potential energy of atoms *j* (the first neighbors of atom *i*), the coefficient B_{jk} ($k \neq i$) includes bond *i-j* through $G(\theta_{jkl})\big|_{k\neq i,j}^{l\neq j,k}$, which must be taken into account in the computation of the force exerted on atom *i*, since this coefficient has derivative with respect to x_i. The second term in Eq. (2.17) is the contribution on the atom *i* by the potential energy of atom *j*. This term is extracted as follows

$$\frac{\partial V^{Br}(r_{jk})}{\partial x_i} = \frac{-D_{cc}S_{cc}}{S_{cc}-1} f_{jk} e^{-\sqrt{2/S_{cc}}\,\beta_{cc}\left(r_{jk}-R_{cc}^e\right)} \frac{\partial B_{jk}}{\partial x_i}$$

(2.23)

$$B_{jk} = \left[1 + \sum_{l\neq j,k} f_c(r_{jl})G(\theta_{jkl})\right]_{k\neq i,j}^{-\delta_{cc}}$$

(2.24)

In the expression of B_{jk}, *k* is the neighbor of *j*, which is not atom *i* (i.e., *k* is the second neighbor of atom *i*), while *l* is a neighbor of *j*, which is not atom *k*. Obviously, one of the neighbor atoms *l* can be atom *i*. Consequently, the terms including atoms *i* have derivatives with respect to x_i.

Finally, the last term is obtained as follows

$$\frac{\partial V^{Br}(r_{jk})}{\partial x_i} = \frac{1}{2}\frac{D_{cc}S_{cc}}{S_{cc}-1}\delta_{\dot{c}\dot{c}}f_{jk}e^{-\sqrt{2/S_{cc}}\beta_{cc}\left(r_{jk}-R_{cc}^e\right)}\left[\left(1+\sum_{l\neq j,k}G_{jkl}f(r_{jl})\right)^{-\delta_{cc}-1}\right.$$

$$\left.\left(\frac{\partial G_{jki}}{\partial\cos(\theta_{jki})}\frac{\partial\cos(\theta_{jki})}{\partial x_i}f_{ij}+G_{jki}\frac{\partial f_{ij}}{\partial r_{ij}}\frac{x_i-x_j}{r_{ij}}\right)\right]$$

(2.25)

3. Algorithms of Performing an MD code

3.1. Equations of Motions

In the classical dynamic simulations, the atoms are assumed to be mass point particles (ideal dimensionless particles). Due to the potential energy function and the position vectors of the atoms, the exerted forces on each individual atom are computed, as discussed in the previous section. Subsequently, the equations of motion of each particle can be solved through time steps. Obviously, the equations of motion of each particle are coupled to its nearest neighbor atoms due to the exerted force computed by the considered potential function. Actually, the most time-consuming part of a simulation code is in evaluating the exerted forces on the atoms in each time step. To derive the equations of motion, different methods can be applied. The most common ones in the classical dynamic simulations are the Newtonian, Lagrangian, and Hamiltonian equations of motions. An overview on these methods can be found elsewhere (e.g., see Refs [52, 53, 54]). In general, for a system of N particles without any constraints, 3N coordinates are needed for identification of their positions. Therefore, 3N equations of motion must be derived and be solved simultaneously. The equation of motion for an instant atom i can be written utilizing the Newton's second law as

$$m\ddot{\vec{r}}_i(t) = \vec{F}_i = -\nabla_i E, \qquad\qquad i = 1\,to\,N$$

(3.1)

or

$$\begin{cases} m\ddot{x}_i(t) = F_{i_x} = -\dfrac{\partial E}{\partial x_i} \\[2mm] m\ddot{y}_i(t) = F_{i_y} = -\dfrac{\partial E}{\partial y_i}, \qquad\qquad i = 1\,to\,N \\[2mm] m\ddot{z}_i(t) = F_{i_z} = -\dfrac{\partial E}{\partial z_i} \end{cases}$$

(3.2)

Different algorithms are suggested in the literature in order to solve these equations of motion [53, 54]. In most algorithms, the positions of atoms in the current time are used to compute the exerted forces on each individual atom. Then the current positions and velocities together with the exerted forces are utilized to update the new positions and velocities. To obtain the microscopic properties of a system, the molecular dynamics simulation must be conducted in an ensemble [53, 54]. The common ensemble used to model carbon nanotubes is the canonical ensemble or constant-NVT. This ensemble represents a system that is in contact with a heat bath; therefore, the temperature of the system is held constant through the simulation. Subsequently, a point of importance in applying algorithms to solve the equations of motion of a system of particles, in the canonical ensemble, is the treatment of the temperature, which must be held constant in the process. Different thermostat techniques are utilized to hold the temperature of the system constant [55-58]. Since the temperature of the system of particles is computed from the total kinetic energy of the system, most thermostat techniques are based on scaling the velocities of the particles. In this book chapter, the Nose-Hoover heat bath technique is illustrated as an example [56, 57, 58]. In this technique, the velocities of the particles involved in the simulation are scaled by introducing a dynamical friction coefficient defined as follows

$$\dot{\zeta}(t) = \frac{1}{Q}\left(\sum_{i=1}^{N} m_i \dot{r}_i^2 - (3N+1)k_B T_0\right) \tag{3.3}$$

Where, ζ is the dynamical frictional coefficient, Q is a parameterize parameter, and T_0 is the reference temperature. As revealed from Eq. (3.3), the derivative of the dynamical friction coefficient is in relation with the difference between the current kinetic energy and the reference one. Subsequently, this parameter is introduced in the equations of motion as

$$m\ddot{\vec{r}}_i = -\vec{\nabla}_i E - \zeta(t)m\dot{\vec{r}}_i \tag{3.4}$$

It can be seen from Eq (3.3) that if the current kinetic energy is higher than the reference kinetic energy corresponding to constant temperature T_0, ζ becomes positive; therefore, the frictional force imposed in Eq. (3.4) tends to decelerate the motion of the atoms. Obviously, when the current kinetic energy is low, ζ becomes negative, and the corresponding frictional force tends to accelerate the motion of the atoms. To solve the equation of motion with a Nose-Hoover thermostat technique, a typical algorithm is applied to Eqs. (3.3) and (3.4) as below

$$\vec{r}_i(t+\delta t) = \vec{r}_i(t) + \delta t\,\dot{\vec{r}}_i(t) + \frac{\delta t^2}{2}\left(\ddot{\vec{r}}_i(t) - \zeta(t)\dot{\vec{r}}_i(t)\right) \tag{3.5a}$$

$$\dot{\vec{r}}_i(t+\frac{\delta t}{2}) = \dot{\vec{r}}_i(t) + \frac{\delta t}{2}\left(\ddot{\vec{r}}_i(t) - \zeta(t)\dot{\vec{r}}_i(t)\right) \tag{3.5b}$$

$$\zeta(t+\frac{\delta t}{2}) = \zeta(t) + \frac{\delta t}{2Q}\left(\sum_{i=1}^{N} m_i \dot{r}_i^2(t) - 3(N-1)k_B T_0\right) \tag{3.5c}$$

$$\zeta(t+\delta t) = \zeta(t+\frac{\delta t}{2}) + \frac{\delta t}{2Q}\left(\sum_{i=1}^{N} m_i \dot{r}_i^2(t+\frac{\delta t}{2}) - 3(N-1)k_B T_0\right) \qquad (3.5d)$$

$$\dot{r}_i(t+\delta t) = \frac{2}{2+\delta t\,\zeta(t+\delta t)}\left(\dot{r}_i(t+\frac{\delta t}{2}) + \frac{\delta t}{2}\ddot{r}_i(t+\delta t)\right) \qquad (3.5e)$$

To demonstrate how to implement the above algorithm in simulation of a carbon nanotube, an example is illustrated in this section.

3.2. Implementation of the Nose-Hoover Algorithm in Simulation of Carbon Nanotubes

In this section, the way of handling a simulation code utilizing the Nose-Hoover algorithms of Eq. (3.5) is illustrated in more detail. Obviously, the general framework can be utilized in using other thermostat techniques. As an example, consider a carbon nanotube of (n, m), which is under axial loading. Under this loading, the Young modulus property of the CNT can be obtained, and its deformation can be studied. To perform this simulation, the following steps must be taken:

Step1. Generate the initial positions of the atoms of the carbon nanotube (n, m), and save them in the vectors X, Y, and Z. This is utilized by the concept of geometry of a carbon nanotube defined by (n, m) [59]. Utilize these positions to compute the initial acceleration vectors ax, ay, and az based on the Brenner potential energy function of Eqs. (2.17)-(2.25). It must be mentioned that each component of these vectors corresponds to a particular atom. Moreover, distinguish the boundary atoms (e.g., two rings of atoms from both ends). Note that to simulate the axial loading, one end of the nanotube is held permanent, while the other end is displaced incrementally after each relaxation step.

Step2. Generate the initial velocity of the atoms by a random distribution and save them in the vectors Vx, Vy, and VZ. Moreover, compute the initial kinetic energy based on these velocities.

Step3. Determine the time step δt, the increment axial displacement δL, an initial value for the dynamical friction coefficient ζ, and the reference temperature T0 (e.g., $\delta t = 1 ft$, $\delta L = 0.001L$, $\zeta = 0$, and $T_0 = 300\,°K$). Furthermore, define the parameter Q. A particular parameterization of Q is as

$$Q = (3N-1)k_B T_0 \tau^2 \qquad (3.6)$$

where τ is in the dimension of time (e.g. $\tau = 10\delta t$).

Step4. Determine the number of steps for initial relaxation K0 and relaxation after imposing each increment displacement K.

Step5. Let the nanotube be relaxed initially by K0 steps by the subroutine, which is based on the Nose-Hoover algorithm of Eqs. (3.5) as follows,
Subroutine for Nose-Hoover relaxation:

Step1. Compute the dynamical friction coefficient ζ in time $t+\frac{\delta t}{2}$ from Eq (3.5c).

Step2. Compute the positions and velocities of the N atoms in $t+\delta t$ and $t+\dfrac{\delta t}{2}$, from Eqs. (3.5a) and (3.5b), respectively. Moreover, compute the sum square of the velocities in $t+\dfrac{\delta t}{2}$ inside the loop of this step.

Step3. In the loop of Step2, update the positions of all atoms to the current position vectors else for the boundary ones.

Step4. Compute the kinetic energy in $t+\dfrac{\delta t}{2}$, then compute ζ in time $t+\delta t$ from Eq. (3.5d). Moreover, obtain the current acceleration vectors based on the updated position vectors.

Step5. Compute the velocities of the atoms in time $t+\delta t$ from Eq. (3.5e). Note that the acceleration in $t+\delta t$ is actually the current acceleration, which is updated in preceding step. Simultaneously, compute the sum square of the velocities, and update the velocities to the current velocities of the atoms.

Step6. Compute the current kinetic energy based on the updated velocities. Then computations begin again from Step2, until the loop of relaxation is finished.

Step 5. Displace the axial coordinates of one end of the boundary atoms in amount of δL. Then, let the nanotube be relaxed for K steps by the subroutine presented above.

Step 6. Compute the resultant axial force on the boundary atom, and save it in a vector. Additionally, compute the potential energy and the total mechanical energy, and save them also.

Step 7. Continue the simulation until the final axial strain is obtained.

Note that at the end of the simulation, the corresponding axial force to each axial strain is obtained. Therefore, the stress-strain curve can be depicted in order to investigate the properties of the CNTs. A similar procedure can be applied in order to impose torsional and bending loading. Fig 3. 1. represents the deformation of a (10, 10) carbon nanotube under torsional loading as an example of the illustrated algorithm in this section. The tube whose length is 9.72 nm contains 1,600 atoms (80 rings). Three rings from each end are held constant, while one of them is rotated with increment twisting angle of 1°. The time step is 1 fs, and the simulation is conducted at the constant temperature of 300 K. According to the conducted simulation, this instant armchair carbon nanotube buckles when it attains the twisting angle of 72°. Since this book chapter aims to note the important issues and the way of implementation of an MD simulation code for carbon nanotubes, presenting numerical results for investigation of the properties of these structures are out of its scope. One can conduct the procedure presented in this section in order to investigate these properties on his own. However, an extensive study on carbon nanotubes properties under different loadings is available in the literature, based on MD simulations (e.g., Refs. [1-27]). Fig 3.2. demonstrates the torsional buckling of a double-walled carbon nanotube [5]. To simulate multiwalled carbon nanotubes, a potential function such as Lennard-Jones potential must be utilized to model the vdW interaction force between the walls of the tubes, while each tube is treated by a suitable interatomic potential like the Brenner potential, individually. Therefore, in the presented procedure, only in the steps of computing the accelerations for each atom, the acceleration caused by the vdW interaction force must be added also.

To conduct a successive MD code, some important factors must be considered. These factors are listed in summary below

1) Choosing an appropriate potential energy function to model accurately the bonding in the system under investigation.

2) Choosing an appropriate thermostatic technique and algorithm scheme that can simulate the procedure correctly.

3) Choosing appropriate time steps, number of relaxation steps, and amount of displacement to the boundary atoms in each step. Obviously, smaller steps and displacements and higher number of relaxation steps result in a more accurate simulation, but, on the other hand, the time expense increases. Thus, in choosing these parameters, a balance must be taken between accuracy and the speed of the simulation.

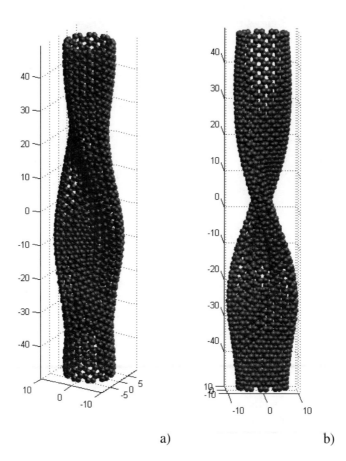

a)   b)

Figure 3.1. Deformation of a (10,10) carbon nanotube under torsional loading, a) at 100 twisting angle, b) at 120 twisting angle.

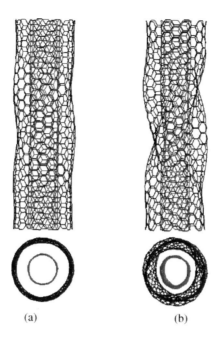

(a)   (b)

Figure 3.2. Deformation of a DWCNT under torsional loading, taken from Ref. [5].

Mylvaganam and Zhang [11] have discussed the above factors in their research article. They utilized two different potential functions of Tersoff and Tersoff-Brenner to model the binding between the atoms. Moreover, they applied different thermostat techniques and algorithm schemes in order to find the best for modeling of carbon nanotube structures. Fig 3. 3. displays the results taken from these different schemes [11]. As revealed from this Figure, in early stages of simulation, different schemes have almost similar results, but in higher stages, they follow quite different paths. These Figures demonstrate how critical the choice of the scheme utilized in MD simulation is when large deformations are examined. It is observed from Fig 3. 3b. that only two schemes could correctly predict the necking of the CNT and the formation of chain atoms in large strains, while the nanotube is suddenly separated into two parts after a certain strain in other schemes.

In the end, the procedure presented in this book chapter can be used in order to investigate different computational experiments on carbon nanotubes via MD simulations, such as:

1) Axial tension and compression of single-walled and multiwalled CNTs. The elastic properties, axial buckling, and large deformations can be investigated [1, 2, 4, 6-12, 14, 16, 17, 19-21, 23, 25, 26].
2) Imposing torsion and bending moment and studying buckling [1, 5, 12, 13, 22, 25].
3) Imposing different external pressure and examining the deformation of CNT [12].
4) Imposing the combination of foregoing loadings simultaneously.
5) Imposing the above-mentioned loadings to CNTs containing defects [3] and CNTs filled by other molecules such as fullerene.
6) Imposing axial tensile and compressive loads to bundle of CNTs [18].
7) Investigation of the thermal effects on the properties of CNTs [12, 15, 24], and so on.

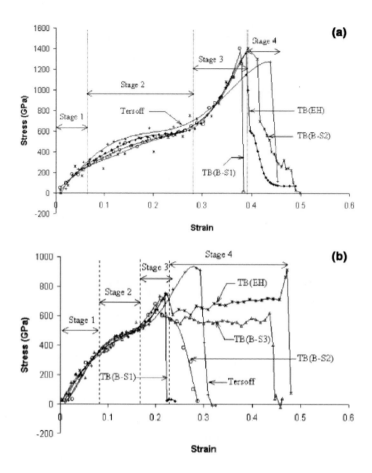

Figure 3. 3. The stress–strain curves of (a) a (10,10) armchair SWNT and (b) a (17,0) zigzag SWNT using Tersoff and Tersoff–Brenner potentials Ref. [11].

References

[1]   B.I. Yakobson, C.J. Brabec, J. Bernholc. Nanomechanics of carbon tubes: instabilities beyond the linear response. *Phys. Rev. Lett.* 76 (1996), 2511–4.

[2]   C. F. Cornwell, L. T. Wille. Simulations of the elastic response of single-walled carbon nanotubes. *Computational Materials Science,* Vol. 10, 42-45 (1998).

[3]   X. Hao, H. Qiang, Y. Xiaohu. Buckling of defective single-walled and double-walled carbon nanotubes under axial compression by molecular dynamics simulation. *Composites Science and Technology,* (2008).

[4]   C.L. Zhang, II. S. Shen. Predicating the elastic properties of double-walled carbon nanotubes by molecular dynamics simulation, *J. Phys. D: Appl. Phys.* 41 (2008) 055404.

[5]   Y. Y. Zhang and C. M. Wang. Torsional responses of double-walled carbon nanotubes via molecular dynamics simulations. *J. Phys.: Condens. Matter.* 20 (2008) 455214.

[6]   C.F. Cornwell, L.T. Wille. Elastic properties of single walled carbon nanotubes in compression. *Solid State Commun.* 1997;101:555–8.

[7]   Y. Wang, X. Wang, X. Ni, H. Wu. Simulation of the elastic response and the buckling modes of single-walled carbon nanotubes. *Computational Materials Science.* 32 (2005) 141-146.

[8]   B. W. Xing, Z. C. Chuna, C. W. Zhao. Simulation of Young's modulus of single-walled carbon nanotubes by molecular dynamics, *Physica.* B 352 (2004) 156–163.

[9]   Z. Yao, C. C. Zhu, M. Cheng, J. Liu. Mechanical properties of carbon nanotubes by molecular dynamics simulation, *Computational Materials Science.* (2001) 180-184.

[10] K. M. Liew, X. Q. He, C. H. Wong. On the study of elastic and plastic properties of multiwalled carbon nanotubes under axial tension using molecular dynamics simulation, *Acta Materialia.* 52 (2004) 2521–2527.

[11] K. Mylvaganam, L.C. Zhang. Important issues in a molecular dynamics simulation for characterising the mechanical properties of carbon nanotubes, *Carbon.* 42 (2004) 2025–2032.

[12] C. L. Zhang, H. S. Shen. Buckling and postbuckling analysis of single-walled carbon nanotubes in thermal environments via molecular dynamics simulation, *Carbon.* 44 (2006) 2608–2616.

[13] Zhao Wang, Michel Devel, Bernard Dulmet. *Twisting carbon nanotubes: A molecular dynamics study, Surface Science* (article in press).

[14] T. Kawaia, Y. Miyamotob, O. Suginob, Y. Koga. Fusion of ultra thin carbon nanotubes: tight-binding molecular dynamics simulations, *Physica B* 323 (2002) 190–192.

[15] K. Bi, Y. Chen, J. Yang, Y. Wang, M. Chen. Molecular dynamics simulation of thermal conductivity of single-wall carbon nanotubes, *Physics Letters A* 350 (2006) 150–153.

[16] P. M. Agrawal, B. S. Sudalayandi, L. M. Raff, R. Komanduri. A comparison of different methods of Young's modulus determination for single-wall carbon nanotubes (SWCNT) using molecular dynamics (MD) simulations, *Computational Materials Science* 38 (2006) 271–281.

[17] P. M. Agrawal, B.S. Sudalayandi, L.M. Raff, R. Komanduri. Molecular dynamics (MD) simulations of the dependence of C–C bond lengths and bond angles on the tensile strain in single-wall carbon nanotubes (SWCNT), *Computational Materials Science* 41 (2008) 450–456.

[18] K. M. Liew, C. H. Wong, M. J. Tan. Tensile and compressive properties of carbon nanotube bundles, *Acta Materialia.* 54 (2006) 225–231.

[19] H. W. Zhang, L. Wang, J .B. Wang. Computer simulation of buckling behavior of double-walled carbon nanotubes with abnormal interlayer distances, *Computational Materials Science.* 39 (2007) 664–672.

[20] J. M. Lua, C. C. Hwanga, Q. Y. Kuoc, Y. C. Wang. Mechanical buckling of multiwalled carbon nanotubes: The effects of slenderness ratio, *Physica E* 40 (2008) 1305–1308.

[21] G.A. Shen, S. Namilae, N. Chandra, Load transfer issues in the tensile and compressive behavior of multiwall carbon nanotubes, *Materials Science and Engineering A* 429 (2006) 66–73.

[22] H. W. Zhang, L. Wang, J. B. Wang, Z. Q. Zhang, Y. G. Zheng. Torsion induced by axial strain of double-walled carbon nanotubes, *Physics Letters A* 372 (2008) 3488–3492.

[23] C.C. Hwang a, Y.C.Wangb,n, Q.Y.Kuo b, J.M.Lu. Molecular dynamics study of multiwalled carbon nanotubes under uniaxial loading, *Physica E* 42 (2010) 775–778.

[24] C. L. Zhang, H. S. Shen. Thermal buckling of initially compressed single-walled carbon nanotubes by molecular dynamics simulation, *Carbon*. 45 (2007) 2614–2620.

[25] G. Cao, X. Chen. The effects of chirality and boundary conditions on the mechanical properties of single-walled carbon nanotubes, *International Journal of Solids and Structures*. 44 (2007) 5447–5465.

[26] Tarek Ragab, Cemal Basaran. A framework for stress computation in single-walled carbon nanotubes under uniaxial tension, *Computational Materials Science*. 46 (2009) 1135–1143.

[27] L.C. Zhang, On the mechanics of single-walled carbon nanotubes, *Journal of Materials Processing Technology*. 209 (2009) 4223–4228.

[28] S. B. Legoas, V. R. Coluci, S. F. Braga, P. Z. Coura, S.O. Dantas, and D. S. Galvao. Molecular-Dynamics Simulations of Carbon Nanotubes as Gigahertz Oscillators, *Phys. Rev. Let.* 90, 055504 (2003).

[29] H. Su, W. A Goddard III, Y. Zhao. Dynamic friction force in a carbon peapod oscillator, *Nanotechnology*. 17, 5691 (2006).

[30] P. Liu, Y.W. Zhang, C. Lu, Analysis of the oscillatory behavior of double–walled carbon nanotube–based oscillators, *Carbon*. 44 (2006) 27–36.

[31] R. Ansari, M. Hemmatnezhad, H. Ramezannezhad, Application of HPM to the nonlinear vibrations of multiwalled carbon nanotubes, *Numerical Methods for Partial Differential Equations*, Vol. 26, No. 2, Mar. 2010, pp. 490–500.

[32] B. Arash, R. Ansari, Evaluation of nonlocal parameter in the vibrations of single-walled carbon nanotubes with initial strain, 2010, *Physica E*, In Press. doi:10.1016/j.physe.2010.03.028.

[33] T. Murmu, S. C. Pradhan, Buckling analysis of a single-walled carbon nanotube embedded in an elastic medium based on nonlocal elasticity and Timoshenko beam theory and using DQM, *Physica E*, Vol. 41, No. 7, Jun. 2009, Pages 1232-1239.

[34] K. Y, Xu, E. C. Aifantis, Y. H. Yan, Vibrations of Double-Walled Carbon Nanotubes With Different Boundary Conditions Between Inner and Outer Tubes, *Journal of Applied Mechanics*, Vol. 75, No. 2, Mar. 2008, pp. 021013.

[35] W. H. Duan, C. M. Wang, Y. Y. Zhang, Calibration of nonlocal scaling effect parameter for free vibration of carbon nanotubes by molecular dynamics, *Journal of Applied Physics*, Vol. 101, No. 2, Jan. 2007, pp. 024305.

[36] W. H. Duan, C. M. Wang, Y. Y. Zhang, Calibration of nonlocal scaling effect parameter for free vibration of carbon nanotubes by molecular dynamics, *Journal of Applied Physics*, Vol. 101, No. 2, Jan. 2007, pp. 024305.

[37] Y. D. Kuang, X. Q. He, C. Y. Chen, G. Q. Li, Analysis of nonlinear vibrations of double-walled carbon nanotubes conveying fluid, *Computational Materials Science*, Vol. 45, No. 4, Jun. 2009, pp. 875–880.

[38] L. J. Sudak, Column buckling of multiwalled carbon nanotubes using nonlocal continuum mechanics, *Journal of Applied Physics*, Vol. 94, No. 11, Dec. 2003, pp. 7281–7287.

[39] P. Lu, H. P. Lee, C. Lu, P. Q. Zhang, Application of nonlocal beam models for carbon nanotubes, *International Journal of Solids and Structures*, Vol. 44, No. 16, Aug. 2007, pp. 5289–5300.

[40] Li, R., Kardomateas, G. A., Thermal Buckling of Multiwalled Carbon Nanotubes by Nonlocal Elasticity, *Journal of Applied Mechanics,* Vol. 74, No. 3, May 2007, pp. 399-405.

[41] W.K. Liu, E.G. Karpov, S. Zhang, H.S. Park. An introduction to computational nanomechanics and materials, *Comput. Methods Appl. Mech. Engrg.* 193 (2004) 1529–1578.

[42] S.P. Xiao, T. Belytschko. A bridging domain method for coupling continua with molecular dynamics, *Comput. Methods Appl. Mech. Engrg.* 193 (2004) 1645–1669.

[43] D. Qian, G. J. Wagner, W. K. Liu. A multiscale projection method for the analysis of carbon nanotubes, *Comput. Methods Appl. Mech. Engrg.* 193 (2004) 1603–1632.

[44] J. Knap, M. Ortiz. An analysis of the quasicontinuum method, *Journal of the Mechanics and Physics of Solids.* 49 (2001) 1899 – 1923.

[45] G. J. Wagner, W. K. Liu. Coupling of atomistic and continuum simulations using a bridging scale decomposition, *Journal of Computational Physics.* 190 (2003) 249–274.

[46] W. K. Liu, E. G. Karpov, H. S. Park. *Nano Mechanics and Materials, Theory, Multiscale Methods and Applications,* John Wiley & Sons (2006).

[47] J. Tersoff. New empirical model for the structural properties of silicon. *Physical Review Letters.* 56(6) (1986), 632–635.

[48] J. Tersoff. Empirical interatomic potential for carbon, with applications to amorphous carbon. *Physical Review Letters.* 61(25) (1988), 2879–2882.

[49] D.W. Brenner. Empirical potential for hydrocarbons for use in simulating the chemical vapor deposition of diamond films. *Physical Review B* 42(15) (1990), 9458–9471.

[50] D.W. Brenner, O.A. Shenderova, J.A. Harrison, S.J. Stuart, B. Ni and S.B. Sinnott. A second-generation reactive empirical bond order (rebo) potential energy expression for hydrocarbons. *Journal of Physics: Condensed Matter* 14 (2002), 783–802.

[51] S.L. Mayo, B.D. Olafson, W.A. Goddard III. Dreiding: a generic force field for molecular simulations. *J. Phys. Chem.* 94(26) (1990), 8897–909.

[52] H. Baruh. *Analytical Dynamics,* McGraw-Hill (1999).

[53] M.P. Allen and D. Tildesley. *Computer Simulation of Liquids.* Oxford University Press (1987).

[54] H. Rafii-Tabar. *Computational Physics of Carbon Nanotubes,* Cambridge University Press (2008).

[55] H.J.C. Berendsen, J.P.M. Postma, W.F. van Gunsteren, A. DiNola and J.R. Haak. Molecular dynamics with coupling to an external bath. *Journal of Chemical Physics.* 81(8) (1984), 3684–3690.

[56] S. Nose. A molecular dynamics method for simulations in the canonical ensemble. *Molecular Physics.* 53 (1984), 255–268.

[57] S. Nose. A unified formulation of the constant temperature molecular dynamics methods. *Journal of Chemical Physics.* 81(1) (1984), 511–519.

[58] W.G. Hoover. Canonical dynamics: Equilibrium phase-space distributions. *Physical Review. A* 31(3) (1985), 1695–1697.

[59] R. A. Jishi, M. S. Dresselhaus, G. Dresselhaus. *Symmetry properties of chiral carbon nanotubes,* 47 (24) (1993), 16671.

In: Handbook of Research on Nanomaterials, Nanochemistry ...   ISBN: 978-1-61942-525-5
Editors: A. K. Haghi and G. E. Zaikov   © 2013 Nova Science Publishers, Inc.

Chapter IX

Conductive Nanofiber

Z. M. Mahdieh, V. Mottaghitalab,
N. Piri and A. K. Haghi[*]
University of Guilan, Rasht, Iran

1. Introduction

Over the recent decades, scientists have become interested for creation of polymer nanofibers due to their promising potential in many engineering and medical applications [1]. According to various outstanding properties such as very small fiber diameters, large surface area- per-mass ratio, high porosity along with small pore sizes and flexibility, electrospun nanofiber mats have found numerous applications in diverse areas. For example, in the biomedical field, nanofibers play a substantial role in tissue engineering [2], drug delivery [3], and wound dressing [4]. Electrospinning is a sophisticated and efficient method by which fibers produces with diameters in nanometer scale entitled as nanofibers. In electrospinning process, a strong electric field applies on a droplet of polymer solution (or melt) held by its surface tension at the tip of a syringe needle (or a capillary tube). As a result, the pendent drop will become highly electrified, and the induced charges distribute over its surface. Increasing the intensity of electric field, the surface of the liquid drop will be distorted to a conical shape known as the Taylor cone [5]. Once the electric field strength exceeds a threshold value, the repulsive electric force dominates the surface tension of the liquid and a stable jet emerges from the cone tip. The charged jet then accelerates toward the target and rapidly thins and dries because of elongation and solvent evaporation. As the jet diameter decreases, the surface charge density increases and the resulting high repulsive forces split the jet to smaller jets. This phenomenon may take place several times, leading to many small jets. Ultimately, solidification is carried out and fibers deposits on the surface of the collector as a randomly oriented nonwoven mat [6-7]. Figure 1 shows a schematic illustration of electrospinning setup.

---

[*] Haghi@Guilan.ac.ir.

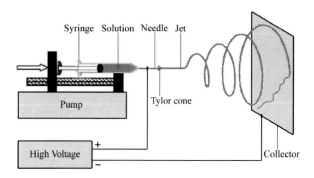

Figure 1. A typical image of Electrospinning process [8].

The physical characteristics of electrospun nanofibers such as fiber diameter depend on various parameters, which are mainly divided into three categories: solution properties (solution viscosity, solution concentration, polymer molecular weight, and surface tension), processing conditions (applied voltage, volume flow rate, spinning distance, and needle diameter), and ambient conditions (temperature, humidity, and atmosphere pressure) [9]. Numerous applications require nanofibers with desired properties, suggesting the importance of the process control. It does not come true unless having a comprehensive outlook of the process and quantitative study of the effects of governing parameters. In this context, Sukigara and his coworkers [10] assessed the effect of concentration on diameter of electrospun nanofibers.

Besides physical characteristics, medical scientists showed a remarkable attention to biocompatibility and biodegradability of nanofibers made of biopolymers such as collagen [11], fibrogen [12], gelatin [13], silk [14], chitin [15] and chitosan [16]. Chitin is the second most abundant natural polymer in the world, and Chitosan (poly-(1-4)-2-amino-2-deoxy-β-D-glucose) is the deacetylated product of chitin [17]. CHT is well known for its biocompatible and biodegradable properties [18].

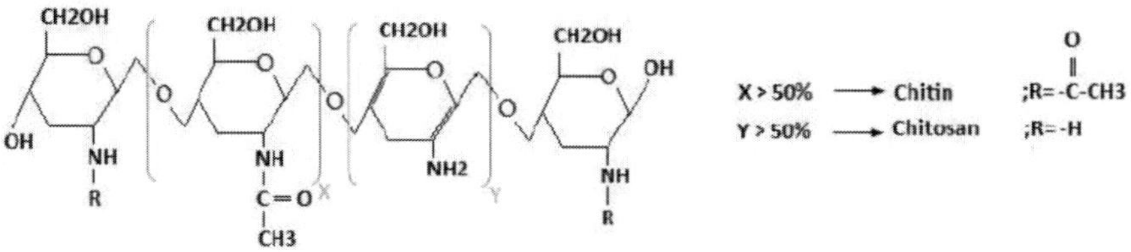

Scheme 1. Chemical structures of Chitin and Chitosan biopolymers.

Chitosan (CHT) is insoluble in water, alkali, and most mineral acidic systems. However, though its solubility in inorganic acids is quite limited, CHT is in fact soluble in organic acids, such as dilute aqueous acetic, formic, and lactic acids. CHT also has free amino groups, which make it a positively charged polyelectrolyte. This property makes CHT solutions highly viscous and complicates its electrospinning [19]. Furthermore, the formation of strong hydrogen bonds in a 3-D network prevents the movement of polymeric chains exposed to the electrical field [20].

Different strategies were used for bringing CHT in nanofiber form. The three top most abundant techniques include blending of favorite polymers for electrospinning process with

CHT matrix [21-22], alkali treatment of CHT backbone to improve electrospinability through reducing viscosity [23] and employment of concentrated organic acid solution to produce nanofibers by decreasing of surface tension [24]. Electrospinning of Polyethylene oxide (PEO)/CHT [21] and polyvinyl alcohol (PVA)/CHT [22] blended nanofiber are two recent studies based on first strategy. In the second protocol, the molecular weight of CHT decreases through alkali treatment. Solutions of the treated CHT in aqueous 70–90% acetic acid has been employed to produce nanofibers with appropriate quality and processing stability [23].

Using concentrated organic acids such as acetic acid [24] and triflouroacetic acid (TFA) with and without dichloromethane (DCM) [25-26] have been reported exclusively for producing neat CHT nanofibers. They similarly reported the decreasing of surface tension and at the same time enhancement of charge density of CHT solution without significant effect on viscosity. This new method suggests significant influence of the concentrated acid solution on the reducing of the applied field required for electrospinning.

The mechanical and electrical properties of neat CHT electrospun natural nanofiber mat can be improved by addition of the synthetic materials including carbon nanotubes(CNTs) [27]. CNTs are one of the key synthetic polymers that were discovered by Iijima in 1991 [28]. CNTs, either single-walled nanotubes (SWNTs) or multiwalled nanotubes (MWNTs), combine the physical properties of diamond and graphite. They are extremely thermally conductive like diamond and appreciably electrically conductive like graphite. Moreover, the flexibility and exceptional specific surface area-to-mass ratio can be considered as significant properties of CNTs [29]. The scientists are becoming more interested in CNTs for existence of exclusive properties such as superb conductivity [30-31] and mechanical strength for various applications. To the best of our knowledge, there has been no report on electrospinning of CHT/MWNTs blend, except those ones that use PVA to improve spinnability. Results showed uniform and porous morphology of the electrospun nanofibers. Despite adequate spinnability, total removing of PVA from nanofiber structure to form conductive substrate is not feasible. Moreover, thermal or alkali solution treatment of CHT/PVA/MWNTs nanofibers extremely influence on the structural morphology and mechanical stiffness [32]. The CHT/CNT composite can be produced by the hydrogen bonds due to hydrophilic positively charged polycation of CHT due to amino groups and hydrophobic negatively charged of CNT due to carboxyl and hydroxyl groups [33-34].

In current study, it has been attempted to produce a CHT/MWNTs nanofiber without association of processing agent to facile electrospinning process. In addition, a new approach explored to provide highly stable and homogenous composite spinning solution of CHT/MWNTs in concentrated organic acids. This in turn presents a homogenous conductive CHT scaffolds, which is extremely important for development of biomedical implants.

2. Experimental

2.1. Materials

CHT with degree of deacetylation of 85% and molecular weight of 5×10^5 was supplied by Sigma-Aldrich. The MWNTs, supplied by Nutrino, have an average diameter of four nm and the purity of about 98%. All of the other solvents and chemicals were commercially available and used as received without further purification.

2.2. Preparation of CHT-MWNTs Dispersions

A Branson Sonifier 250 operated at 30W used to prepare the MWNTs dispersions in CHT/organic acid (90%w/w acetic acid, 70/30 TFA/DCM) solution based on different protocols. In first approach, 3 mg of as received MWNTs was dispersed into deionized water or DCM using solution sonicating for 10 min (current work, sample 1). Different amount of CHT was then added to MWNTs dispersion for preparation of a 8-12 w/w % solution and then sonicated for another five min. Figure 2 shows two different protocols used in this study.

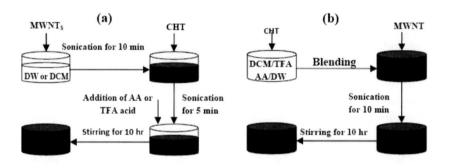

Figure 2. Two protocols used in this study for preparation of MWNTs/CHT dispersion (a) Current study(b) Ref. [35]. (*AA/DW abbreviated for acetic acid diluted in de-mineralized water*).

In the next step, the organic acid solution added to obtain a CHT/MWNT solution with total volume of 5 mL and finally the dispersion was stirred for another ten hours. The sample 2 was prepared using the second technique. Same amount of MWNTs were dispersed in CHT solution, and the blend with total volume of 5mL were sonicated for ten min and dispersion was stirred for ten hr [35].

2.3. Electrospinning of CHT/MWNTs Dispersion

After the preparation of spinning solution, it was transferred to a 5 ml syringe and became ready for spinning of nanofibers. The experiments were carried out on a horizontal electrospinning setup shown schematically in Figure 1. The syringe containing CHT/MWNTs solution was placed on a syringe pump (New Era NE-100) used to dispense the solution at a controlled rate. A high-voltage DC power supply (Gamma High Voltage ES-30) employed to generate the required electric field for electrospinning. The positive electrode and the grounding electrode of the high-voltage supplier attaches respectively to the syringe needle and flat collector wrapped with aluminum foil where electrospun nanofibers accumulates via an alligator clip to form a nonwoven mat. The voltage and the tip-to-collector distance fixed respectively on 18-24 kV and 4-10 cm. In addition, the electrospinning carried out at room temperature, and the aluminum foil was removed from the collector.

2.4. Measurements and Characterizations Method

A small piece of mat placed on the sample holder and gold sputter-coated (Bal-Tec). Thereafter, the micrograph of electrospun CHT/MWNTs nanofibers was obtained using scanning electron microscope (SEM, Phillips XL-30). Fourier transform infrared spectra (FT-IR) recorded using a Nicolet 560 spectrometer to investigate the interaction between CHT and MWNT in the range of 800- 4000 cm^{-1} under a transmission mode. The size distribution of the dispersed solution was evaluated using dynamic light scattering technique (Zetasizer, Malvern Instruments). The conductivity of nanofibre samples was measured using a homemade four-probe electrical conductivity cell operating at constant humidity. The electrodes were circular pins with separation distance of 0.33 cm, and fibres connected to pins by silver paint (SPI). Between the two outer electrodes, a constant DC current applied by Potentiostat/Galvanostat model 363 (Princeton Applied Research). The generated potential difference between the inner electrodes and the current flow between the outer electrodes was recorded by digital multimeter 34401A (Agilent). Figure 3 illustrates the experimental setup for conductivity measurement. The conductivity (δ: S/cm) of the nanofiber with rectangular surface can then be calculated according to equation 1, which parameters call for length (L:cm), width (W:cm), thickness (t:cm), DC current applied (mA) and the potential drop across the two inner electrodes (mV). All measuring was repeated at least five times for each set of samples.

$$\delta = \frac{I \times L}{V \times W \times t} \qquad\qquad (1)$$

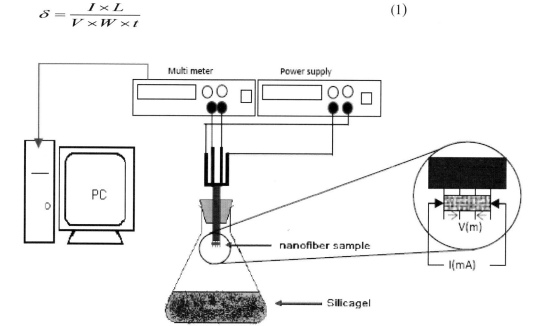

Figure 3. The experimental setup for four-probe electrical conductivity measurement of nanofiber thin film.

3. Results and Discussion

3.1. The Characteristics of CHT/MWNTs Dispersion

Utilisation of MWNTs in biopolymer matrix initially requires their homogenous dispersion in a solvent or polymer matrix. Dynamic light scattering (DLS) is a sophisticated technique used for evaluation of particle size distribution. DLS provides many advantages for particle size analysis to measures a large population of particles in a very short time period, with no manipulation of the surrounding medium. DLS of MWNTs dispersions indicate that the hydrodynamic diameter of the nanotube bundles is between 150 and 400 nm after ten min of sonication for sample 2. (Figure 4)

MWNTs bundled in sample 1 (different approach but same sonication time compared to sample 2) show a range of hydrodynamic diameter between 20-100 nm. (Figure 4).The lower range of hydrodynamic diameter for sample 1 correlates to more exfoliated and highly stable nanotubes strands in CHT solution. The higher stability of sample 1 compared to sample 2 over a long period of time was confirmed by solution stability test. The results presented in Figure 5 indicate that procedure employed for preparation of sample 1 (current work) was an effective method for dispersing MWNTs in CHT/acetic acid solution. However, MWNTs bundles in sample 2 showed re-agglomeration upon standing after sonication. Figure 5 indicates the sedimentation of large agglomerated particles.

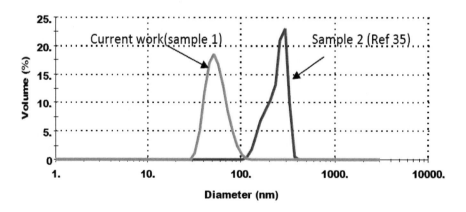

Figure 4. Hydrodynamic diameter distribution of MWNT bundles in CHT/acetic acid (1%) solution for different preparation technique.

Despite the method reported in ref. 35, neither sedimentation nor aggregation of the MWNTs bundles were observed in first sample. Presumably, this behavior in sample 1 can be attributed to contribution of CHT biopolymer to forms an effective barrier against re-agglomeration of MWNTs nanoparticles. In fact, using sonication energy, in first step without presence of solvent, makes very tiny exfoliated but unstable particle in water as dispersant. Instantaneous addition of acetic acid as solvent and long mixing most likely helps the wrapping of MWNTs strands with CHT polymer chain.

Figure 5. Stability of CHT-MWNT dispersions (a) Current work (sample1) (b) Ref. [35] (sample2).

Figure 6 shows the FT-IR spectra of neat CHT solution and CHT/MWNTs dispersions prepared using strategies explained in experimental part. The interaction between the functional group associated with MWNTs and CHT in dispersed form has been understood through recognition of functional groups. The enhanced peaks at ~1600 cm^{-1} can be attributed to (N-H) band and (C=O) band of amide functional group. However, the intensity of amide group for CHT/MWNTs dispersion increases presumably due to contribution of G band in MWNTs. More interestingly, in this region, the FT-IR spectra of MWNTs-CHT dispersion (sample 1) have been highly intensified compared to sample 2 [35]. It correlates to higher chemical interaction between acid functionalized C-C group of MWNTs and amide functional group in CHT.

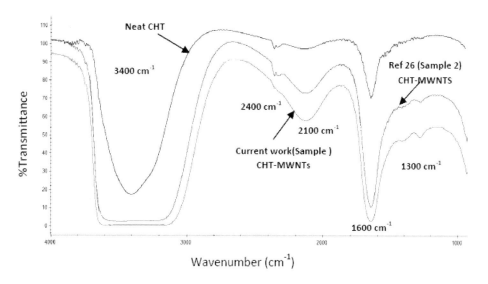

Figure 6. FT-IR spectra of CHT-MWCNT in 1% acetic acid with different techniques of dispersion.

This probably is the main reason for the higher stability and lower MWNTs dimension demonstrated in Figure 4 and Figure 5.

Moreover, the intensity of protonated secondary amine absorbance at 2400 cm^{-1} for sample 1 prepared by new technique is negligible compared to sample 2 and neat CHT.

Furthermore, the peak at 2123cm^{-1} is a characteristic band of the primary amine salt, which is associated with the interaction between positively charged hydrogen of acetic acid and amino residues of CHT.

In addition, the broad peaks at ~3410 cm^{-1} due to the stretching vibration of OH group superimposed on NH stretching bond and broaden due to hydrogen bonds of polysaccharides. The broadest peak of hydrogen bonds observed at 3137-3588 cm^{-1} for MWNTs/CHT dispersion prepared by new technique (sample 1).

3.2. The Physical and Morphological Characteristics of MWNTs/CHT Nanofiber

The different solvents, including acetic acid 1-90%, pure formic acid, and TFA/DCM tested for preparation of spinning solution-using protocol, are explained for sample 1. Upon applying the high voltage, even above 25 kV, no polymer jet forms using of acetic acid 1-30% and formic acid as the solvent for CHT/MWNT.

However, experimental observation shows bead formation when the acetic acid (30-90%) used as the solvent. Therefore, one does not expect the formation of electrospun fiber of CHT/MWNTs using prescribed solvents (data not shown).

Figure 7 shows scanning electronic micrographs of the CHT/MWNTs electrospun nanofibers in different concentration of CHT in TFA/DCM (70:30) solvent.

As presented in Figure 7a, at low concentrations of CHT, the beads deposited on the collector and thin fibers co-exited among the beads. When the concentration of CHT increases as shown in Figure 7a-c, the bead density decreases.

Figure 7c show the homogenous electrospun nanofibers with minimum beads, thin and interconnected fibers.

More increasing of concentration of CHT lead to increasing of interconnected fibers at Figure 7 d-e.

Figure 8 shows the effect of concentration on average diameter of MWNTs/CHT electrospun nanofibers. Our assessments indicate that the fiber diameter of CHT/MWNTs increases with the increasing of the CHT concentration. In this context, similar results were reported in previously published work [36-37]. Hence, CHT/MWNTs (10% wt) solution in TFA/DCM (70:30) were considered as resulted in optimized concentration. An average diameter of 275 nm (Figure 7c: diameter distribution, 148-385) was investigated for this conditions.

Table 1 lists the variation of nanofiber diameter and four-probe electrical conductivity based on the different loading of CHT. One can expect the lower conductivity, once the CHT content increases.

However, the higher the CHT concentration, the thinner fiber forms. Therefore, the decreasing of conductivity at higher CHT concentration damps by decreasing of nanofiber diameter. This led to a nearly constant conductivity over entire measurements.

Table 1. The variation of conductivity and
mean nanofiber diameter versus Chitosan loading

% CHT (%w/v)	% MWNT (%w/v)	Voltage (KV)	Tip to collector (cm)	Diameter (nm)	Conductivity (S/cm)
8	0.06	24	5	137 ± 58	NA
9	0.06	24	5	244 ± 61	9×10^{-5}
10	0.06	24	5	275 ± 70	9×10^{-5}
11	0.06	24	5	290 ± 87	8×10^{-5}
12	0.06	24	5	Non uniform	NA

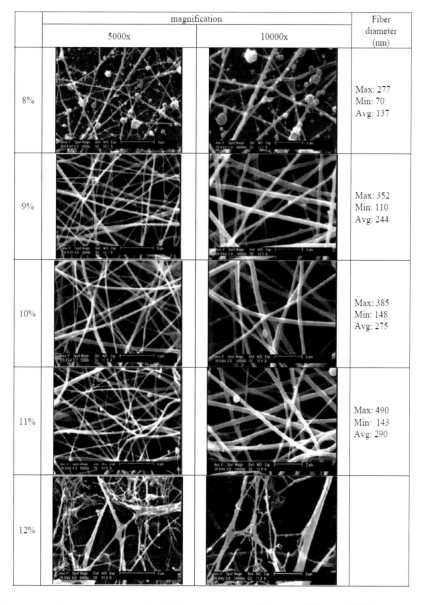

Figure 7. Scanning electron micrographs of electrospun nanofibers at different CHT concentration (wt %): (a) 8, (b) 9, (c) 10, (d) 11, (e) 12, 24 kV, 5 cm, TFA/DCM: 70/30, (0.06%wt MWNTs).

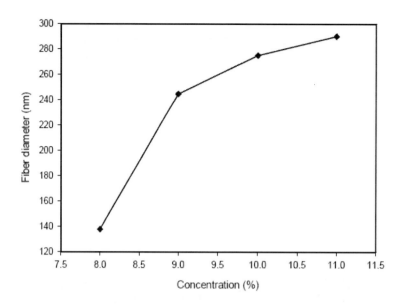

Figure 8. The effect of the CHT concentration in CHT/MWNT dispersion on nanofiber diameter.

Figure 9 shows the SEM image of CHT/MWNTs electrospun nanofibers produced in different voltage. In our experiments, 18 kV attains as threshold voltage, where fiber formation occurs.

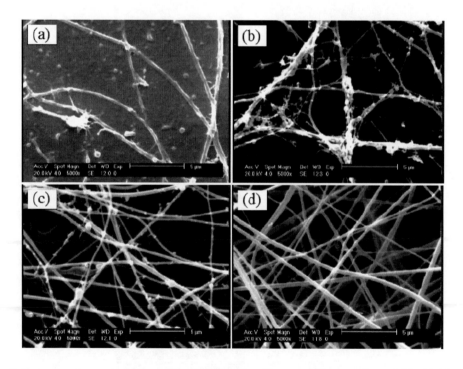

Figure 9. Scanning electronic micrographs of electrospun fibers at different voltage (kV): (a) 18, (b) 20, (c) 22, (d) 24, 5 cm, 10 wt%, TFA/DCM: 70/30. (0.06%wt MWNTs).

Table 2. The variation of conductivity and
mean nanofiber diameter versus applied voltage

% CHT (%w/v)	% MWNT (%w/v)	Voltage (KV)	Tip to collector (cm)	Diameter (nm)	Conductivity (S/cm)
10	0.06	18	5	Non uniform	NA
10	0.06	20	5	Non uniform	NA
10	0.06	22	5	201 ± 66	6×10^{-5}
10	0.06	24	5	275 ± 70	9×10^{-5}

For lower voltage, the beads and some little fiber were deposited on collector (Figure 9a). As shown in Figure 9a-d, the beads decreases while voltage increases from 18 kV to 24 kV. The collected nanofibers by applying 18 kV (9a) and 20 kV (9b) were not quite clear and uniform. The higher the applied voltage, the uniform nanofibers with narrow distribution starts to form. The average diameter of fibers, 22 kV (9c), and 24 kV (9d), respectively, were 204 (79-391), and 275 (148-385).

The conductivity measurement given in Table 2 confirms our observation in first set of conductivity data.

As can be seen from last row, the amount of electrical conductivity reaches a maximum level of 9×10^{-5} S/cm at prescribed setup.

The distance between tips to collector is another parameter that controls the fiber diameter and morphology.

Figure 10 shows the change in morphologies of CHT/MWNTs electrospun nanofibers at different distances. When the distance is not long enough, the solvent could not find opportunity for separation.

Hence, the interconnected thick nanofiber deposits on the collector (Figure 10a). However, the adjusting of the distance on 5 cm (Figure 10b) leads to homogenous nanofibers with negligible beads and interconnected areas.

However, the beads increases by increasing of distance of the tip-to-collector as represented from Figure 10b to Figure 10f. Similar results was observed for CHT nanofibers reported by Geng et al. [24].

Also, the results show that the diameter of electrospun fibers decreases by increasing of distance tip to collector in Figure 10b, 10c, 10d, respectively, 275 (148-385), 170 (98-283), 132 (71-224).

Similarly, the previous works report a decreasing trend in nanofiber diameter once distance increases [38-39].

A remarkable defect and non-homogeneity appears for those fibers prepared at a distance of 8 cm (Figure 10e) and 10 cm (Figure 10f). However, a 5 cm distance selected as proper amount for CHT/MWNTs electrospinning process. Conductivity results also are in agreement with those data obtained in previous parts (Table 3).

The non-homogeneity and huge bead densities plays as a barrier against electrical current and still a bead free and thin nanofiber mat shows higher conductivity compared to other samples. Experimental framework in this study was based on parameter adjusting for electrospinning of conductive CHT/MWNTs nanofiber.

It can be expected that the addition of nanotubes can boost conductivity and change morphological aspects, which is extremely important for biomedical applications.

**Table 3. The variation of conductivity and mean
nanofiber diameter versus applied voltage**

% CHT (%w/v)	% MWNT (%w/v)	Voltage (KV)	Tip to collector (cm)	Diameter (nm)	Conductivity (S/cm)
10	0.06	24	4	Non uniform	NA
10	0.06	24	5	275 ± 70	9×10^{-5}
10	0.06	24	6	170 ± 58	6×10^{-5}
10	0.06	24	7	132 ± 53	7×10^{-5}
10	0.06	24	8	Non uniform	NA
10	0.06	24	10	Non uniform	NA

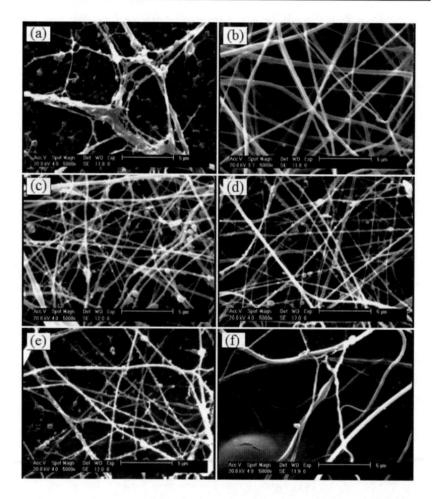

Figure 10. Scanning electronic micrographs of electrospun fibers of CHT/MWNT at different tip-to-collector distances (cm): (a) 4, (b) 5, (c) 6, (d) 7, (e) 8, (f) 10, 24 kV, 10 wt%, TFA/DCM: 70/30.

Conclusion

Conductive composite nanofibers of CHT/MWNTs were produced using conventional electrospinning technique. A new protocol was suggested for preparation of electrospinning solution, which shows much better stability and homogeneity compared to previous techniques. Several solvent including acetic acid 1-90%, formic acid and TFA/DCM (70:30) investigated in the electrospinning of CHT/MWNTs dispersion. Results of DLS and dispersion stability tests showed that the TFA/DCM (70:30) solvent is most preferred for nanofiber formation process with acceptable electrospinability. The formation of nanofibers with conductive pathways regarding to exfoliated and interconnected nanotube strands is a breakthrough in CHT nanocomposite area. This is a significant improvement in electrospinning of CHT/MWNTs dispersion. It has been also observed that the homogenous nanofibers with an average diameter of 275 nm could be prepared with a conductivity of 9 $\times 10^{-5}$ S/cm.

References

[1]   S. Agarwal, J.H. Wendorff, and A. Greiner, *Polymer*, 49, 5603 (2008)

[2]   M. Li, M.J. Mondrinos, M.R. Gandhi, F.K. Ko, A.S. Weiss, and P.I. Lelkes, *Biomaterials*, 26, 5999 (2005)

[3]   J. Zeng, L. Yang, Q. Liang, X. Zhang, H. Guan, X. Xu, X. Chen, and X. Jing, *J. Control. Release.*, 105, 43 (2005)

[4]   M. S. Khil, D.I. Cha, H.-Y. Kim, I.-S. Kim, and N. Bhattarai, *J. Biomed. Mat. Res. B.*, 67B, 675 (2003)

[5]   G. I. Taylor, *Proc Roy Soc London* 313, 453 (1969)

[6]   J. Doshi and D.H. Reneker, *J. Electrostat.*, 35, 151 (1995)

[7]   D. Li and Y. Xia, *Adv. Mater.* , 16, 1151 (2004)

[8]   M. Ziabari, V. Mottaghitalab, and A.K. Haghi, *Korean J. Chem. Eng.*, 25, 923 (2008)

[9]   S.H. Tan, R. Inai, M. Kotaki, and S. Ramakrishna, *Polymer*, 46, 6128 (2005)

[10] S. Sukigara, M. Gandhi, J. Ayutsede, M. Micklus, and F. Ko, *Polymer*, 44, 5721 (2003)

[11] J.A. Matthews, G.E. Wnek, D.G. Simpson, and G.L. Bowlin, *Biomacromolecules*, 3, 232 (2002)

[12] M.C. McManus, E.D. Boland, D.G. Simpson, C.P. Barnes, and G.L. Bowlin, *J. Biomed. Mater. Res. A.*, 81A, 299 (2007)

[13] Z.-M. Huang, Y.Z. Zhang, S. Ramakrishna, and C.T. Lim, *Polymer*, 45, 5361 (2004)

[14] X. Zhang, M.R. Reagan, and D.L. Kaplan, *Adv. Drug. Deliver. Rev.*, 61, 988 (2009)

[15] H.K. Noh, S.W. Lee, J.-M. Kim, J.-E. Oh, K.-H. Kim, C.-P. Chung, S.-C. Choi, W.H. Park, and B.-M. Min, *Biomaterials*, 27, 3934 (2006)

[16] K. Ohkawa, K.-I. Minato, G. Kumagai, S. Hayashi, and H. Yamamoto, *Biomacromolecules*, 7, 3291 (2006)

[17] O.C. Agboh and Y. Qin, *Polym. Adv. Technol.*, 8, 355 (1997)

[18] M. Rinaudo, *Prog. Polym. Sci.*, 31, 603 (2006)

[19] I. Aranaz, M. Mengíbar, R. Harris, I. Paños, B. Miralles, N. Acosta, G.Galed, and Á. Heras, *Curr. Chem. Biol.*, 3, 203 (2009)

[20] A. Neamnark, R. Rujiravanit, and P. Supaphol, *Carbohydr. Polym.*, 66, 298 (2006)

[21] B. Duan, C. Dong, X. Yuan, and K. Yao, *J. Biomater. Sci. Polymer. Ed.*, 15, 797 (2004)

[22] Y.-T. Jia, J. Gong, X.-H. Gu, H.-Y. Kim, J. Dong, and X.-Y. Shen, *Carbohydr. Polym.*, 67, 403 (2007)

[23] H. Homayoni, S.A.H. Ravandi, and M. Valizadeh, *Carbohydr. Polym.*, 77, 656 (2009)

[24] X. Geng, O.-H. Kwon, and J. Jang, *Biomaterials*, 26, 5427 (2005)

[25] S. Torres-Giner, M.J. Ocio, and J.M. Lagaron, *Angluis*, 8, 303 (2008)

[26] S.D. Vrieze, P. Westbroek, T.V. Camp, and L.V. Langenhove, *J. Mater. Sci.*, 42, 8029 (2007)

[27] K. Ohkawa, D. Cha, H. Kim, A. Nishida, and H. Yamamoto, *Macromol. Rapid Comm.*, 25, 1600 (2004)

[28] S. Iijima, *Nature*, 354, 56 (1991)

[29] A.M.K. Esawi and M.M. Farag, *Mater. Design.*, 28, 2394 (2007)

[30] W. Feng, Z. Wu, Y. Li, Y. Feng, and X. Yuan, *Nanotechnology*, 19, 105707 (2008)

[31] M.A. Shokrgozar, F. Mottaghitalab, V. Mottaghitalab and M. Farokhi , *J. Biomed. Nanotech.*, 7, 1–9(2011)

[32] S. H. Baek, B. Kim, and K.-D. Suh, *Colloid. Surface. A.*, 316, 292 (2008)

[33] Y. L. Liu, W.H. Chen, and Y.H. Chang, *Carbohydrate. Polym.*, 76, 232 (2009)

[34] J. Tkac, J.W. Whittaker, and T. Ruzgas, *Biosens. Bioelectron.*, 22, 1820 (2007)

[35] G.M. Spinks, M.Geoffrey, S.R. Shin, G.G. Wallace, P.G. Whitten, S.I. Kim, and S.J. Kim, *Sensor. Actuat. B-Chem.*, 115, 678 (2006)

[36] H. Zhang, Z. Wang, Z. Zhang, J. Wu, J. Zhang, and J. He, *Adv. Mater.*, 19, 698 (2007)

[37] J.M. Deitzel, J. Kleinmeyer, D. Harris, and N.C. Beck Tan, *Polymer*, 42, 261 (2001)

[38] S. Zhang, W.S. Shim, and J. Kim, *Mater. Design*, 30, 3659 (2009)

[39] Y. Li, Z. Huang, and Y. Lu, *Eur. Polym. J.*, 42, 1696 (2006)

In: Handbook of Research on Nanomaterials, Nanochemistry ...   ISBN: 978-1-61942-525-5
Editors: A. K. Haghi and G. E. Zaikov   © 2013 Nova Science Publishers, Inc.

Chapter X

Achievements in Production of PAN Monofilament in Nanoscale

M. Sadrjahani[1], S. A. Hoseini[1],
V. Mottaghitalab[2] and A. K. Haghi[2]

[1] Textile Eng. Faculty, Isfahan University of Technology, Isfahan, Iran
[2] Textile Eng. Dept, Faculty of Eng, University of Guilan, Rasht, Iran

1. Introduction

With potential applications ranging from protective clothing and filtration technology to reinforcement of composite materials and tissue engineering, nanofibers offer remarkable opportunity in the development of multifunctional material systems. The emergence of various applications for nanofibers is stimulated from their outstanding properties such as very small diameters, huge surface area-per-mass ratio and high porosity along with small pore size. Moreover, the high degree of orientation and flexibility beside superior mechanical properties are extremely important for diverse applications [1-3].

In this study, aligned and molecularly oriented PAN nanofibers were prepared using a novel technique comprised of two needles with opposite voltage and a rotating drum for applying take-up mechanism. The electrospinning process was optimized for increasing of productivity and improving the mechanical properties through controlling internal structure of the generated fibers.

2. Background

Electrospinning is a sophisticated technique that relies on electrostatic forces to produce fibers in the nano to micron range from polymer solutions or melts. In a typical process, an electrical potential is applied between droplet of a polymer solution, or melt, held at the end of a capillary tube and a grounded target.

When the applied electric field overcomes the surface tension of the droplet, a charged jet of polymer solution is ejected. The trajectory of the charged jet is controlled by the electric field.

The jet exhibits bending instabilities due to repulsive forces between the charges carried with the jet. The jet extends through spiraling loops, as the loops increase in diameter, the jet grows longer and thinner until it solidifies or collects on the target [3]. The fiber morphology is controlled by the experimental design and is dependent upon solution conductivity, solution concentration, polymer molecular weight, viscosity, applied voltage and distance between needle and collector [2,3]; due to initial instability of the jet, fibers are often collected as randomly oriented structures in the form of nonwoven mats, where the stationary target is used as a collector. These nanofibers are acceptable only for some applications such as filters, wound dressings, tissue scaffolds and drug delivery [4].

Aligned nanofibers are another form of collected nanofibers that can be obtained by using rotating collector or parallel plates [2,3]. Recent studies have shown that aligned nanofibers have better molecular orientation and as a result improved mechanical properties than randomly oriented nanofibers [3,5,6]. These nanofibers can be used in applications such as composite reinforcement and device manufacture [4].

Moreover, the aligned nanofibers are better suited for thermal or drawing treatment (the methods of collecting aligned nanofibers can be utilized for preparing of carbon nanofibers from electrospun PAN nanofibers precursor) [7]. Recently, PAN nanofibers have attracted a lot of interest as a precursor of carbon nanofibers. Fennessey and his coworkers prepared tows of unidirectional and molecularly oriented PAN nanofibers using a high speed, rotating take up wheel. A maximum chain orientation parameter of 0.23 was determined for fibers collected between 8.1 and 9.8 m/s. The aligned tows were twisted into yarns, and the mechanical properties of the yarns were determined as a function of twist angle. Their yarn with twist angle of 11° had initial modulus and ultimate strength of about 5.8 GPa and 163 MPa, respectively [3].

Zussman et al. have demonstrated the use of a wheel-like bobbin as the collector to position and align individual PAN nanofibers into parallel arrays. They are obtained from Herman's orientation factor of 0.34 for nanofibers collected at 5 m/s. [5]. Gu et al. collected aligned PAN nanofibers across the gap between the two grounded stripes of aluminum foil. They reported the increase of orientation factor from 0 to 0.127 and expressed improving mechanical properties in particular the modulus of the resultant carbon fibers [6].

3. Experimental

3.1. Materials

Industrial Polyacrylonitrile (PAN) was provided by Iran Polyacryle and dimethylformamide (DMF) was purchased from Merck. The weight average molecular weight (WM) and the number average molecular weight (nM) of PAN were WM =100,000 g/mol and nM =70000 g/mol. All solutions of PAN in DMF were prepared under constant mixing by magnetic mixer at room temperature.

3.2. Electrospinning Setup

The electrospinning apparatus consists of a high-voltage power supply, two syringe pumps, two stainless steel needles (0.7 mm OD) and a rotating collector with variable surface speed, which is controlled by an Inverter (Fig 1). In this setup, unlike the conventional technique, two needles were installed in opposite direction, and polymer solutions were pumped to needles by two syringe infusion pumps with same feed rate. The flow rate of solutions to the needle tip is maintained so that a pendant drop remains during electrospinning. The horizontal distance between the needles and the collector was several centimeters. When high voltages were applied to the needles with opposite voltage, jets were ejected simultaneously. Then the jets with opposite charges attracted each other, stuck together and a cluster of fibers formed. For collecting aligned nanofibers, the cluster of fibers formed between the two needles was towed manually to the drum and is rotated from 0 to 3,000 rpm.

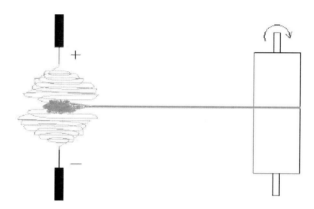

Figure 1. Schematic electrospinning setup for collection continuous aligned fibers.

3.3. Microscopy

Electrospun nanofibers were observed by scanning electron microscopy (SEM) and optical microscopy. Samples were mounted onto SEM plates; sputter coated with gold, and examined using a Philips XL-30 electron microscope to determine fiber diameters. A Motic optical microscope was used to capture images for alignment analysis. Fibers alignment was analyzed using image processing technique by Fourier power spectrum method as a function of take up speed. All images used in the process were obtained using the Motic optical microscopy at a resolution of 640×480 pixels in 1000× magnifications. The number of captured images was 30 at each of take up speeds.

3.4. FTIR

Dried, electrospun fiber bundles were examined using a Bomen MB-Series100 infrared spectrometer (FTIR) to measure crystallization index of PAN nanofibers as a function of collection speed. FTIR spectra were recorded over the range of 400-4000 cm^{-1} with

21scans/min. The crystallization index (A1730/A2240) for PAN fibers was acquired by rationing the absorbance peak areas of nitrile (2240 cm^{-1}) and carbonyl (1730 cm^{-1}) groups. The relation between degree of crystallinity and A1730/A2240 is indirect (i.e., the larger A1730/A2240, the less crystalline fibers are) [8].

3.5. Raman Spectroscopy

Raman spectroscopy was used for the survey of molecular orientation variability with take up speed. Raman spectra were obtained with the Thermo Nicolt Raman spectrometer model Almega Dispersive 5555. The spectra were collected with a spectral resolution of 2-4260 cm-1 in the backscattering mode, using the 532 nm line of a Helium/Neon laser. The nominal power of the laser was 30 mW. A Gaussian/Lorentzian fitting function was used to obtain band position and intensity. The incident laser beam was focused onto the specimen surface through a 100 × objective lens, forming a laser spot of approximately 1 μm in diameter, using a capture time of 50 s. The analysis of PAN nanofiber orientation using polarised Raman spectroscopy was carried out based on a coordination system defined in Fig 2. The nanofiber axis is defined as Z and molecular chain are oriented at angle θ° with respect to the Z axis. The nanofiber is mounted on the stage of the Raman microscope such that the incident laser comes in along the x' axis. The angle between the polarization plane and the nanofiber axis is ψ. The orientation studies were performed when fibers were at ψ = 0° and ψ = 90° to the plane of polarization of the incident laser. At each angle, the enhanced spectra at VV and VH configuration collected. In the VV configuration, the polarised laser and the analyzer are parallel to each other; and in the VH configuration, the polarized laser and the analyzer are perpendicular to each other.

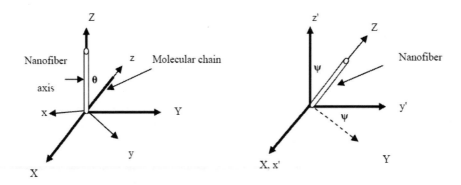

Figure 2. Demonstration of molecular chain coordination (xyz) in nanofiber (XYZ coordinates) and the nanofiber arrangement in reference frame of the Raman sample stage (x'y'z' coordinates).

3.6. Mechanical Testing

The mechanical behavior of bundles of aligned PAN nanofibers was examined using the Zwick 1446-60 with a crosshead speed of 60 mm/min and gauge length of 25mm under standard conditions. All samples were included in standard container during 24h under temperature of 25 oC and relative humidity of 65% before tested. The initial modulus, stress and strain of samples were determined.

3.7. Differential Scanning Calirometry (DSC)

DSC curves of electrospun fibers were obtained using a DSC 2010V4/4E by heating from 30 to 300 $^{\circ}$C at a heating rate of 10 $^{\circ}$C /min. The effect of take up speed on the Tg temperature and evolved heat per time was examined.

4. Results and Discussion

4.1. Productivity

In this study, fibers were electrospun in the aligned form by using a simple and novel method, which manufactures well-aligned polymer nanofibers with infinite length and over large collector area [4]. In this technique, the electric field only exists between two syringe needles and thus the movement of nanofibers in drawing area (distance between the needles and the collector) is due to mechanical force caused by rotating drum, and electric field doesn't have any portion at moving nanofibers from needles to collector. This is likely caused rupture of fibers in this technique. The concentration, the applied voltage, distance between two needles and distance between the syringe needles and the rotating drum were adjusted in order to obtain optimum and productive conditions. The conditions pertinent to minimum number of rupture for nanofibers prepared at different concentrations are shown in Table 1.

PAN nanofibers prepared at 14 and 15 wt% concentrations have lower rupture, which can be due to high chain entanglement in these concentrations. Fig 3 shows SEM images of PAN nanofibers electrospun in obtained optimum conditions. The generated nanofibers have uniform structure and beaded fiber don't observe. As expected, the average diameter of nanofibers increased by increasing solution concentration (Table 2).

Table 1. The conditions obtained to generate nanofibers with minimum amount of rupture

Solution concentration (Wt%)	Applied voltage (Kv)	Feed rate (mL/h)	Distance between two needles (cm)	Distance between needles and collector (cm)	Total number of rupture (at 15 miniutes)
13	10.5	0.293	13	20	12
14	11	0.293	13	20	6
15	11	0.293	15	20	6

**Table 2. Average diameter of electrospun
nanofibers at different solution concentrations**

Diameter (nm) / Concentration	Average	Coefficient Variation (CV%)	$\bar{x} \pm sd$ (nm)
13 wt%	323.45	9.59	323.45 ± 31.03
14 wt%	394.19	7.32	394.19 ± 28.84
15 wt%	404.67	10.82	404.67 ± 43.81

This is probably due to the greater resistance of the solution to be stretched by the charges on the jet with increasing the volume percent of solid in the solution and viscosity [2]. The prepared nanofibers at 14wt% concentration have better uniform structure since the percent of coefficient variation (cv%) of diameter at 14wt% is less than other concentrations. The viscosity is high at 15 wt% concentration and results in difficultly pumping the solution through syringe needle and solution dry at the tip of needle somewhat. By considering this thought and also the value of rupture, uniformity of diameter and suitable diameter in nanoscale range, the conditions acquired at the 14 wt% concentration were chosen as a desired option for producing PAN nanofibers.

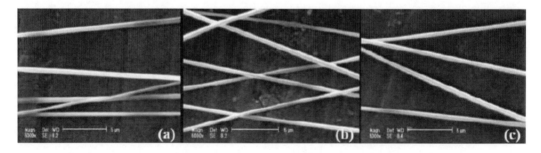

Figure 3. SEM images of PAN nanofibers at concentrations of (a) 13wt%; (b) 14wt%; (c) 15wt%.

4.2. Alignment

Analysis of fibers alignment was carried out by obtaining angular power spectrum (APS) of nanofibers collected at different take up speeds from 22.5 m/min to 67.7 m/min. The plot of normalized APS (ratio of intensity of the APS to the corresponding mean intensity of the Fourier power spectrum) versus angle was used for calculating degree of alignment (Fig 4). In this plot, area of peak at angle of 90o shows density of aligned nanofibers in the rotation direction of the collector. For relative comparing among samples, the ratio of Apeak/ATotal (area of peak at angle of 90o/ total area of APS plot) was utilized. The alignment of the collected fibers is induced by the rotation of the target and improves as the surface velocity of the target is increased (Table 3). As the rotation speed increases, the effective draw (difference between surface velocity of drum and final velocity of fiber) is increased, resulting in better alignment of the collected fiber and less deviation between the fiber and rotation direction. Also, the results show that more increase of take up speed causes no further increase of alignment. The maximum amount of degree of alignment obtained 37.5% at take up speed of 59.5 m/min.

Table 3. The degree of alignment of the collected fibers

Take-up speed (m/min)	22.5	31.6	40.6	49.6	59.5	67.7
Degree of alignment (%)	24.59±3.97	34.4±5.29	32.72±7.65	29.48±5.97	37.53±5.26	29.43±7.04

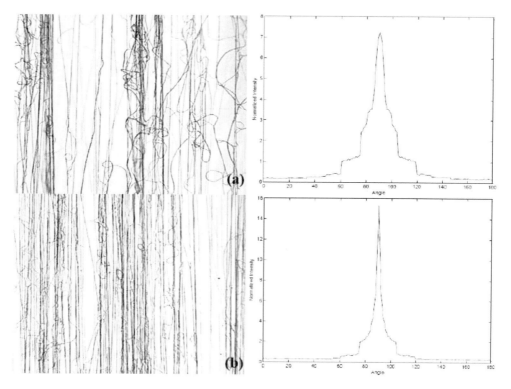

Figure 4. Optical micrograph of electrospun PAN nanofibers with corresponding normalized APS at take-up speeds of (a) 22.5 m/min; (b) 59.5 m/min.

4.3. Crystallization Index

The crystallization index (A1730/A2240) was calculated from FTIR spectra of PAN nanofibers collected at different take-up speeds from 22.5 m/min to 67.7 m/min in optimum conditions. The obtained results (Fig 5) and performance ANOVA statistic analysis over them show that the increase of surface velocity has no effect on the crystallization index, and as a result, the crystallinity of PAN nanofibers has not varied.

4.4. Molecular Orientation

The molecular orientation of the nanofibers was examined by Raman Spectroscopy. Raman spectra were collected from bundles of fibers electrospun at 11 kV from 14 wt% PAN in DMF solutions collected onto a drum rotating with a surface velocity between 22.5 m/min and 67.7 m/min. The main difference among different molecular structures of PAN fibers usually arise in the region of 500-1500 cm-1, which is called as finger point region [9]. In this region, the peaks over the ranges of 950-1090 cm-1 and 1100-1480 cm-1 are common [9,10], and can be observed at Raman spectra of generated PAN nanofibers (Fig 6).

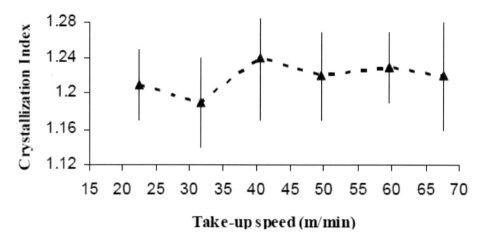

Figure 5. Crystallization Index of PAN nanofibers versus take up speed.

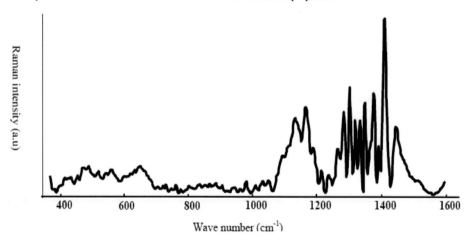

Figure 6. Raman spectra of PAN nanofiber.

Figure 7 shows the Raman spectra of different samples of the nanofibers using VV configuration for different amounts of ψ. Raman spectra were obtained in two directions, parallel (ψ=0⁵) and vertical direction (ψ=90⁵) with respect to the polarization plane.

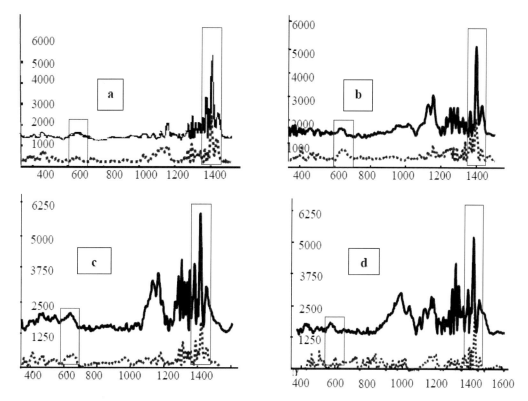

Figure 7. The Raman spectra under VV mode. (a) 22.5 m/min. (b) 49.6 m/min. (c) 59.5 m/min. (d) 67.7 m/min. From top to bottom, the angle between fiber axis and polarization plane is 0^0 and 90^0.

Compared to peak enhanced in 600 cm-1 with constant intensity in different polarization angle, the intensity at 1394 cm-1 monotonically decreases with increasing of ψ. The intensity dependence of peak enhanced in 1394 cm-1 to the angle of fiber and polarization plane can be considered as powerful tool for determination of nanofibers orientation.

Other peaks that enhanced between 1594 cm-1 and 1169 cm-1 did not decrease significantly, except the peak enhanced at 1190 cm-1 and 1454 cm-1. It is worth noting that trend in peak intensity observed for different samples shown in Figure 8 in VH configuration can also be attributed to different orientation magnitudes.

When the nanofibers were examined using the VH configuration, the intensity dependence of enhanced peak at 1394 cm-1 on ψ showed a different trend for different samples that cannot be directly correlated to degree of orientation. The higher intensity was obtained for sample d, however the intensity value decreases for samples c, b and a, respectively, as shown in Figure 8.

According to intensity ratios shown in Table 4, Raman spectra show a much stronger orientation effect in the sample c compared with the other spun nanofibers. It is clear that the higher drum take up speed is mostly responsible for the orientation of molecular chain in the fiber direction.

Therefore, as expected, the low orientation dependence of Raman modes was observed in lower take up speed.

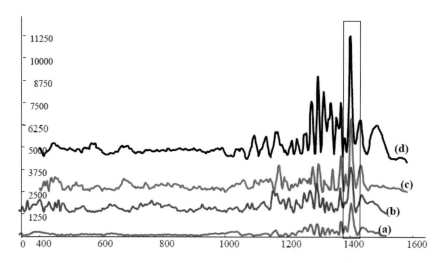

Figure 8. The Raman spectra under VH configuration at zero angle between fiber axis and polarization plane, (a) 22.5 m/min, (b) 49.6 m/min, (c) 59.5 m/min and (d) 67.7 m/min.

Table 4. The intensity ratios derived from Fig. 7 and Fig. 8 based on shown values of enhanced peak at 1394 cm^{-1}

Sample	I_{VV0}	I_{VV90}	I_{VH0}	I_{VV90}/I_{VH0}	I_{VV0}/I_{VH0}	I_{VV0}/I_{VV90}
A	4400	1900	1600	1.1875	2.75	2.31
B	3600	1500	2400	0.625	1.5	2.40
C	4000	1500	4277	0.35071	0.93523	2.67
D	3800	2200	6350	0.34646	0.59843	1.73

The mathematic formulation of Raman intensity in VV and VH modes are represented by the following expressions [11, 12]:

$$I^{VV}(\psi)\alpha\left(\cos^4\psi - \frac{6}{7}\cos^2\psi + \frac{3}{35}\right)\langle P_4(\cos\theta)\rangle$$
$$+\left(\frac{6}{7}\cos^2\psi - \frac{2}{7}\right)\langle P_2(\cos\theta)\rangle + \frac{1}{5}$$

(3.1.)

$$I^{VH}(\psi)\alpha\left(-\cos^4\psi + \cos^2\psi - \frac{4}{35}\right)\langle P_4(\cos\theta)\rangle$$
$$+\frac{1}{21}\langle P_2(\cos\theta)\rangle + \frac{1}{15}$$

(3.2.)

The orientation order parameters of <P2 (cosθ)> and <P4 (cosθ)>, which are, respectively, the average values of P2 (cosθ) and P4 (cosθ) for the SWNTs bulk product. The Pi(cosθ) is the Legendre polynomial of degree I, which is defined as P2(cosθ)= (3 cos 2 θ-1)/2 and P4(cosθ)= (35 cos 4 θ-30 cos2 θ+3)/8 for the second and fourth degree, respectively. More specifically, the <P2(cosθ)> is known as the Herman's orientation factor (f), which varies between values of 1 and 0 corresponding, respectively, to nanotubes fully oriented in the fiber direction and randomly distributed [11].

$$f = \frac{3\left\langle \cos^2 \theta \right\rangle - 1}{2}$$

(3.3.)

The orientation factors can be determined by solving of following simultaneous algebraic equations given in 3.4 and 3.5. These equations are obtained from equations 3.1 and 3.2 by dividing of both sides of these equations and substitution of angles of 0 and 90 degrees for ψ.

$$\frac{I_{G,RBM}^{VV}(\psi=0)}{I_{G,RBM}^{VH}(\psi=0)} = -\frac{24\langle P_4(\cos\theta)\rangle + 60\langle P_2(\cos\theta)\rangle + 21}{12\langle P_4(\cos\theta)\rangle - 5\langle P_2(\cos\theta)\rangle - 7}$$

(3.4.)

$$\frac{I_{G,RBM}^{VV}(\psi=90)}{I_{G,RBM}^{VH}(\psi=0)} = \frac{-9\langle P_4(\cos\theta)\rangle + 30\langle P_2(\cos\theta)\rangle - 21}{12\langle P_4(\cos\theta)\rangle - 5\langle P_2(\cos\theta)\rangle - 7}$$

(3.5.)

The left-hand side terms of equations of 3.4 and 3.5 are the depolarization ratios that can be experimentally determined. As it can be seen, only the intensity at 0ᵗ and 90ᵗ are required to determine the <P2 (cosθ)> and <P4 (cosθ)> for a uniaxially oriented nanofibers. Results calculated from Equations 3.4 and 3.5 for Herman orientation factor for different sample shows a range between 0.20 and 0.25 at take up speeds of 67.7 m/min and 59.5 m/min, respectively (Fig 9).

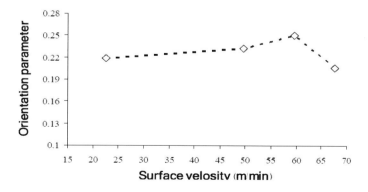

Figure 9. Orientation parameter versus take up speed of rotating drum determined by Raman Spectroscopy.

As results show, the maximum chain orientation parameter yields at speed of 59.5 m/min and further increase of it causes loss of molecular orientation, which corresponds to other studies [3]. As regards a distance between the needles and the collector is fixed, it may be due to decrease of drawing time of nanofibers with increasing take up speed and as a result, low sufficient opportunity for arranging molecular chains in draw direction. Therefore, it appears that applying high draw ratio at short time has no significant effect on molecular orientation. Comparing these results with orientation factor of 0.66 and 0.52, which have been observed for commercial wet-spun acrylic fibers and melt-spun acrylic fibers, respectively [13]; it can be stated that electrospun PAN nanofibers have lower molecular orientation than commercial fibers.

In general, based on the data of orientation factor for different sample, it can be clearly claimed that the increase of surface velocity of collector has a quite positive impact on molecular orientation of PAN nanofibers.

4.5. Mechanical Properties

Unidirectional bundles of aligned PAN nanofibers prepared from 14 wt% PAN in DMF solutions electrospun at 11 kV onto a drum rotating with a surface velocity ranging from 22.5 m/min to 67.7 m/min. The average linear density of the bundles was 176 den. The stress-strain behavior of the bundles was examined and the modulus, ultimate strength, and elongation at the ultimate strength were measured as a function of take up speed. The initial modulus and ultimate strength increase gradually with take up speed from 1.3 GPa (12.31 g/den) and 61.7 MPa (0.577 g/den) at a surface velocity of 22.5 m/min to 4.2 GPa (39.43 g/den) and 73.7 MPa (0.694 g/den) with a liner velocity of 59.5 m/min, respectively.

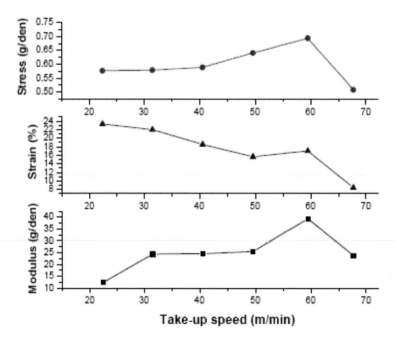

Figure 10. Stress, strain and modulus of PAN nanofibers versus take up speed.

The modulus and ultimate strength of the bundles decreased with take up speed greater than 59.5 m/min (Fig 10). The elongation at ultimate strength followed an inverse trend; it decreased from 23.3% to 8.4% with increasing take up speed from 22.5 m/min to 67.7 m/min, respectively.

4.6. Analysis of Mechanical Properties of PAN Nanofiber Bundles

In investigation of the results obtained for mechanical properties of PAN nanofibers bundles, the effect of parameters such as internal structure and arrangement of fibers with together (spatial orientation) were analyzed. As was shown, the increase of take up speed didn't affect on crystallization of PAN nanofibers; therefore, this parameter can't have main role in variation trend of fiber bundles strength. Figure 11-a, b show the plot of stress versus degree of alignment and molecular orientation parameter, respectively. Degree of alignment determines the number of fibers subjected in tension direction and increases stress of bundles of fiber. The positive correlation factor of 0.53 (acquired by SAS statistic software) between stress and alignment demonstrates direct relation of them, but the low amount of this coefficient offers low correlation. On the other hand, the positive and high correlation factor of 0.99 concludes liner and good correlation between stress and molecular orientation parameter. Thus, it can be stated that the molecular orientation performs a main and important task in response of mechanical properties and particularly stress of PAN nanofibers.

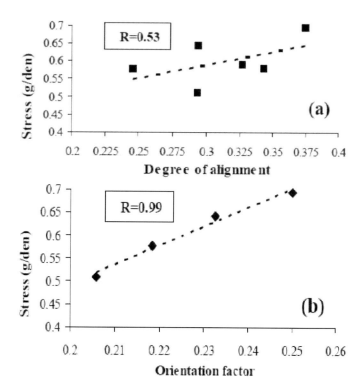

Figure 11. The stress of PAN nanofibers versus(a) degree of alignment(b) molecular orientation parameter.

4.7. Thermal Properties of PAN Nanofibers

The obtained results from DSC curves of PAN nanofibers (Fig 12) collected onto a rotating drum with surface velocity between 22.5 m/min and 67.7 m/min are summarized in Table 5.

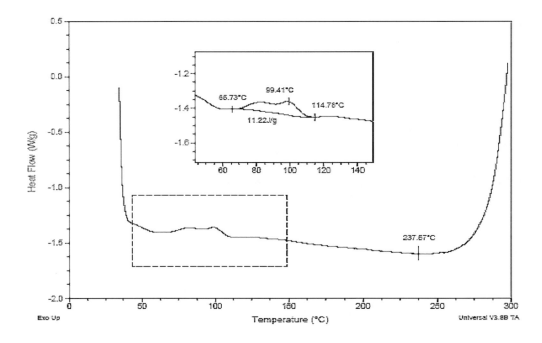

Figure 12. DSC curve of PAN nanofiber at take up speed of 59.5 m/min.

The glass transition temperature (Tg) of PAN nanofibers has no considerable change with increasing take up speed and approximately is equal with 100 oC. A high exothermic peak existed in DSC curve of PAN nanofibers is in relation to occurrence of dehydrogenation and cyclization reactions [14, 15, 16]. These chemical reactions must be conducted so to promote a slow release of heat, because a rapid release of heat can cause a loss of molecular orientation and melting of the fiber, which will have a detrimental effect on the mechanical properties of the final fiber [14, 15]. The minimum rates of released heat yielded for PAN nanofibers collected at liner speed of 67.7 m/min and 59.5 m/min. Therefore, the use of above take up speeds can be suitable for thermal treatment of PAN nanofibers.

**Table 5. Thermal properties of PAN nanofibers
collected at different take-up speeds**

Surface velocity (m/min)	22.5	31.6	40.6	49.6	59.5	67.7
T_g (°C)	100.9	101.5	102	100.6	99.4	100.5
$\Delta H/\Delta t$ (j/g.s)	0.258	0.415	0.524	0.342	0.242	0.205

Moreover, nanofibers with better initial molecular orientation must be employed to prevent fiber shrinkage during thermal treatment. It is due to relax stain acquired by stretching during spinning at over Tg temperature and also, chemical reaction at high temperature, which can disturb orientation of the molecular chains and causes poor mechanical properties of final fiber [14, 15].

Hence, the take up speed of 59.5 m/min by collecting nanofibers with better molecular orientation, low released heat caused by chemical reaction and suitable strength can be chosen as a desired speed for thermal treatment of PAN nanofibers. Work studying the effects of thermal treatment and stretching of electrospun PAN nanofiber simultaneously at elevated temperatures on the physical and mechanical properties is currently ongoing.

Conclusion

A simple and non-conventional electrospinning technique using two syringe pumps was employed for producing of highly oriented Polyacrylonitrile (PAN) multifilament nanofibers. The process was carried out using two needles in opposite positions and a rotating collector perpendicular to needle axis. Current procedure was optimized for increasing of orientation and productivity of nanofibers with diameters in the nanoscale range. PAN nanofibers were electrospun from 14 wt% solutions of PAN in Dimethylformamid (DMF) at 11 kv on a rotating drum with various linear speeds from 22.5 m/min to 67.7 m/min. The influence of take up velocity was also investigated on the degree of alignment, internal structure and mechanical properties of collected PAN nanofibers. Various characterization techniques were employed to find out the influence of operational parameters on degree of orientation. Based on micrographs captured by Optical microscope, the angular power spectrum (APS) was generated based on an image processing technique. The best degree of alignment was obtained for those nanofibers collected at take up velocity of 59.5 m/min. Moreover, polarized Raman spectroscopy under VV configuration at $\psi= 0°$ (parallel) and 90° (Vertical) versus polarization plane and also in VH configuration at $\psi=0$ was used as a new technique for measuring molecular orientation of PAN nanofibers. Similarly, a maximum chain orientation parameter of 0.25 was determined for nanofibers collected at take up velocity of 59.5 m/min.

References

[1] Z.M. Huang,Y.Z. Zhang, M. Kotaki, and S. Ramakrishna, "A review on polymer nanofibers by electrospinning and their applications in nanocomposites." *Composites Science and Technology*,Vol 63,2003,pp. 2223–2253.

[2] S. Ramakrishna, K. Fujihara, W.E. Teo, T.C. Lim, and Z. Ma, *An introduction to electrospinning and nanofibesr*. Singapore: World scientific, (2005).

[3] S.F. Fennessey and R.J. Farris, "Fabrication of aligned and molecularly oriented electrospun polyacrylonitrile nanofibers and the mechanical behavior of their twisted yarn." *Polymer*, Vol 45,2004,pp. 4217-4225.

[4] H. Pan, L. Li, L. Hu and X. Cui, "Continuous aligned polymer fibers produced by a modified electrospinning method." *Polymer*,Vol 47,2006,pp.4901-4904.

[5] E. Zussman, X. Chen, W. Ding, L.Calabri, D.A. Dikin, J.P. Quintana. and R.S. Ruoff, "Mechanical and structural characterization of electrospun PAN-derived carbon nanofibers." *Carbon*, Vol 43,2005 2175-2185.

[6] S.Y. Gu, J. Ren and Q.L. Wu, "Preparation and structures of electrospun PAN nanofibers as a precursor of carbon nanofibers." *Synthetic Metals* ,Vol 155,2005, pp. 157–161.

[7] R. Jalili, M. Morshed and S.A.Hosseini Ravandi, "Fundamental Parameters Affecting Electrospinning of PAN Nanofibers as Uniaxially Aligned Fibers." *Journal of Applied Polymer Science*, Vol 101,2006,pp4350-4357.

[8] V. Causin, C. Marega, S. Schiavone and A. Marigo, "A quantitative differentiation method for acrylic fibers by infrared spectroscopy." *Forensic Science International*, Vol 151,2006,pp125– 131.

[9] D. Mathieu and A. Grand, "Ab initio Hartree-Fock Raman spectra of Polyacrylonitrile." *Polymer, Vol* 39,1998,pp. 5011-5017.

[10] Y.S. Huang and J.L. Koenig, "Raman spectra of polyacrylonitrile." *Applied Spectroscopy*,Vol 25,1971,pp. 620-622.

[11] T. Liu, and S. Kumar, "Quantitative characterization of SWNT orientation by polarized Raman spectroscopy." *Chem. Phys. Lett*, Vol 378,2003,pp. 257-262.

[12] W.J. Jones, D.K. Thomas, D.W. Thomas and G. Williams, "On the determination of order parameters for homogeneous and twisted nematic liquid crystals from Raman spectroscopy." *Journal of Molecular Structure*, Vol 708, 2004, pp.145–163.

[13] J.A. Davidson, H.T. Jung, S.D. Hudson and S. Percec, "Investigation of molecular orientation in melt-spun high acrylonitrile fibers." *Polymer*,Vol 41,2000,pp. 3357–3364.

[14] S. Soulis, and J. Simitzis, "Thermomechanical behaviour of poly [acrylonitrile-*co*-(methyl acrylate)] fibres oxidatively treated at temperatures up to 180 ∘C." *Polymer International*,Vol 54, 2005, pp.1474–1483.

[15] M.S.A. Rahaman, A.F. Ismail and A. Mustafa, "A review of heat treatment on polyacrylonitrile fiber. Polymer Degradation and Stability," Vol 92,2007 pp. 1421-1432.

[16] K. Sen, P. Bajaj, and T.V. Sreekumar, "Thermal Behavior of Drawn Acrylic Fibers," *Journal of Polymer Science:Part B:Polymer Physics*, Vol 41,2003,pp.2949–2958.

In: Handbook of Research on Nanomaterials, Nanochemistry ...   ISBN: 978-1-61942-525-5
Editors: A. K. Haghi and G. E. Zaikov   © 2013 Nova Science Publishers, Inc.

Chapter XI

Supreme EMI Shielding Using Electro-less Plating of Metallic Nanoparticle on Cotton Fabric

A. Afzali [2], V. Mottaghitalab[1] and M. Saberi [2]
[1] Textile Eng. Dept, Faculty of Eng, University of Guilan, Rasht, Iran
[2] NanoRod Behan Co(LTD), Guilan Science and Technology Park(GSTP), Rasht, Iran

1. Introduction

Because of the high conductivity of copper, electro-less copper plating is currently used to manufacture conductive fabrics with high shielding effectiveness (SE). It can be performed at any step of the textile production, such as yarn, stock, fabric or clothing [1].

Electro-less copper plating as a non- electrolytic method of deposition from solution on fabrics has been studied by some researchers [1-9]. The early reported copper electro-less deposition method uses a catalytic redox reaction between metal ions and dissolved reduction agent of formaldehyde at high temperature and alkaline medium [1-2]. Despite technique advantages, such as low cost, excellent conductivity, easy formation of a continuous and uniform coating, experimental safety risks appear through formation of hazardous gaseous product during plating process, especially for industrial scale.

Further research has been conducted to substitute formaldehyde with other reducing agents coupled with oxidation accelerator such as sodium hypophosphite and nickel sulphate[4-8]. Incorporation of nickel and phosphorus particles provide good potential for creation of fabrics with a metallic appearance and good handling characteristics. These properties are practically viable if plating process followed by finishing process in optimized pH and in presence of ferrocyanide. Revealing the performance of electro-less plating of Cu-Ni-P alloy on cotton fabrics is an essential research area in textile finishing processing and for technological design [9-14].

The main aim of this chapter is to explore the possibility of applying electro-less plating of Cu-Ni-P alloy onto cotton fabric to obtain highest level of conductivity, washing and abrasion fastness, room condition durability and EMI shielding effectiveness. The fabrication and properties of Cu-Ni-P alloy plated cotton fabric are investigated in accordance with standard testing methods.

2. Experimental

Cotton fabrics (53×48 count/cm^2, 140 g/m^2, taffeta fabric) were used as substrate. The surface area of each specimen is 100 cm^2 .The electro-less copper plating process was conducted by multistep processes: pre-cleaning, sensitization, activation, electro-less Cu-Ni-P alloy deposition and post-treatment.

The fabric specimens (10cm × 10 cm) were cleaned with non-ionic detergent (0.5g/l) and NaHCO$_3$ (0.5g/l) solution for ten minutes prior to use. The samples then were rinsed in distilled water. Surface sensitization was conducted by immersion of the samples into an aqueous solution containing SnCl$_2$ and HCl. The specimens were then rinsed in deionized water and activated through immersion in an activator containing PdCl$_2$ and HCl. The substrates were then rinsed in a large volume of deionized water for ten min to prevent contamination of the plating bath. The electro-less plating process carried out immediately after activation. Then all samples immersed in the electro-less bath containing copper sulfate, nickel sulfate, sodium hypophosphite, sodium citrate, boric acid and potassium ferrocyanide. In the post-treatment stage, the Cu-Ni-P plated cotton fabric samples were rinsed with deionized water, ethylalcohol at home temperature for 20 min immediately after the metalizing reaction of electro-less Cu-Ni-P plating. Then the plated sample dried in oven at 70°C.

The weights (g) of fabric specimens with the size of 100 mm × 100 mm square before and after treatment were measured by a weight meter (HR200, AND Ltd., Japan). The percentage for the weight change of the fabric is calculated in equation (1).

$$I_W = \frac{W_f - W_0}{W_0} \times 100\%$$

(1)

where I_w is the percentage of increased weight, W_f is the final weight after treatment, W_o is the original weight.

The thickness of fabric before and after treatment was measured by a fabric thickness tester (M034A, SDL Ltd., England) with a pressure of 10 g/cm^2. The percentage of thickness increment were calculated in accordance to equation (2).

$$T_I = \frac{T_F - T_0}{T_0} \times 100\%$$

(2)

where T_I is the percentage of thickness increment, T_f is the final thickness after treatment, T_o is the original thickness.

A Bending Meter (003B, SDL Ltd., England) was employed to measure the degree of bending of the fabric in both warp and weft directions. The flexural rigidity of fabric samples expressed in N.cm is calculated in equation (3).

$$G = W \times C^3 \tag{3}$$

where G (N-cm) is the average flexural rigidity, W (N/cm^2) is the fabric mass per unit area, C(cm) is the fabric bending length.

The dimensional changes of the fabrics were conducted to assess shrinkage in length for both warp and weft directions and tested with (M003A, SDL Ltd., England) accordance with standard testing method (BS EN 22313:1992). The degree of shrinkage in length expressed in percentage for both warp and weft directions were calculated according in accordance to equation (4).

$$D_c = \frac{D_f - D_0}{D_0} \times 100 \tag{4}$$

where D_c is the average dimensional change of the treated swatch, D_o is the original dimension, D_f is the final dimension after laundering.

Tensile properties and elongation at break were measured with standard testing method ISO 13934-1:1999 using a Micro 250 tensile tester. Color changing under different application conditions for two standard testing methods, namely, (1) ISO 105-C06:1994 (color fastness to domestic and commercial laundering), (2) ISO 105-A02:1993 (color fastness to rubbing) were used for estimate. Scanning electron microscope (SEM, XL30 PHILIPS) was used to characterize the surface morphology of deposits. WDX analysis (3PC, Microspec Ltd., USA) was used to exist metallic particles over surface Cu-Ni-P alloy plated cotton fabrics. The chemical composition of the deposits was determined using X-ray energy dispersive spectrum (EDS) analysis attached to the SEM. The coaxial transmission line method as described in ASTM D 4935-99 was used to test the EMI shielding effectiveness of the conductive fabrics. The set-up consisted of a SE tester, which was connected to a spectrum analyzer. The frequency is scanned from 50 MHz to 2.7 GHz and is taken in transmission. The attenuation under transmission was measured equivalent to the SE.

3. Results and Discussion

3.1. Elemental Analysis

The composition of the deposits was investigated using X-ray energy dispersive spectrum (EDS) elemental analysis. The deposits consisted mainly of copper with small amounts of nickel and phosphorus. Table 1 shows the weight percent of all detected elemental analysis.

**Table 1. Elemental analysis of electro-less copper
plated using hypophosphite and nickel ions**

Element	Copper	Nickel	Phosphorous
~ wt%	96.5	3	0.5

The nickel and phosphorus atom s in the copper lattice possibly increase the crystal defects in the deposit. Moreover, as non-conductor, phosphorus will make the electrical resistivity of the deposits higher than pure copper. Electro-less plating of copper conductive layer on fabric surface employs hypophosphite ion to reduce copper ion to neural copper particle. However the reduction process extremely accelerates by addition of Ni^{2+}. Addition of Ni^{2+} also sediments tiny amount of nickel and phosphorus elements. Following formulations show the mechanism of copper electro-less plating using hypophosphite.

$$2H_2PO_2^- + Ni^{2+} + 2OH^- \rightleftharpoons Ni^0 + 2H_2PO_3^- + H_2$$

$$2H_2PO_2^- + 2OH^- \xrightarrow{\quad Ni \quad}_{surface} 2e^- + 2H_2PO_3^- + H_2$$

$$Cu^{2+} + 2e^- \rightleftharpoons Cu^0$$

$$Ni^0 + Cu^{2+} \rightleftharpoons Ni^{2+} + Cu^0$$

3.2. Fabric Weight and Thickness

The change in weight and thickness of the untreated cotton and Cu-Ni-P alloy plated cotton fabrics are shown in Table 2.

**Table 2. Weight and thickness of the untreated
and Cu-Ni-P-plated cotton fabrics**

Thickness(mm)	Weight (g)	Specimen (10 cm× 10 cm)
0.4378	2.76	Untreated cotton
0.696(↑5.7%)	3.72 (↑18.47%)	Cu-Ni-P plated cotton

The results presented that the weight of chemically induced Cu-Ni-P-plated cotton fabric was heavier than the untreated one. The measured increased percentage of weight was 18.47%. This confirmed that Cu-Ni-P alloy had clung on the surface of cotton fabric effectively. In the case of thickness measurement, the cotton fabric exhibited a 5.7% increase after being subjected to metallization.

3.3. Fabric Bending Rigidity

Fabric bending rigidity is a fabric flexural behavior that is important for evaluating the handling of the fabric. The bending rigidity of the untreated cotton and Cu-Ni-P-plated cotton fabrics is shown in Table 3.

The results proved that the chemical plating solutions had reacted with the original fabrics during the entire process of both acid sensitization and alkaline plating treatment. After electro-less Cu-Ni-P alloy plating, the increase in bending rigidity level of the Cu-Ni-P-plated cotton fabrics was estimated at 11.39% in warp direction and 30.95% in weft direction, respectively. The result of bending indicated that the Cu-Ni-P-plated cotton fabrics became stiffer to handle than the untreated cotton fabric.

**Table 3. Bending rigidity of the untreated
and Cu-Ni-P-plated cotton fabrics**

Specimen	Bending (N·cm)	
	warp	weft
Untreated cotton	1	0.51
Cu-Ni-P plated cotton	1.17(↑11.39%)	0.66(↑30.95%)

3.4. Fabric Shrinkage

The results for the fabric Shrinkage of the untreated cotton and Cu-Ni-P-plated cotton fabrics are shown in Table 4.

**Table 4. Dimensional change of the untreated
and Cu-Ni-P-plated cotton fabrics**

Specimen	Shrinkage (%)	
	warp	weft
Untreated cotton	0	0
Cu-Ni-P plated cotton	-8	-13.3

The measured results demonstrated that the shrinkage level of the Cu-Ni-P-plated cotton fabric was reduced by 8% in warp direction and 13.3% in weft direction, respectively.

After the Cu-Ni-P-plated, the copper particles could occupy the space between the fibres and hence more copper particles were adhered on the surface of fibre. Therefore, the surface friction in the yarns and fibres caused by the Cu-Ni-P particles could then be increased. When compared with the untreated cotton fabric, the Cu-Ni-P-plated cotton fabrics showed a stable structure.

4. Fabric Tensile Strength and Elongation

The tensile strength and elongation of cotton fabrics was enhanced by the electro-less Cu-Ni-P alloy plating process as shown in Table 5.

The metalized cotton fabrics had a higher breaking load with a 28.44% increase in warp direction and a 35.62% increase in weft direction than the untreated cotton fabric. This was due to the fact that more force was required to pull the additional metal-layer coating.

The results of elongation at break were 12.5% increase in warp direction and 7.8% increase in weft direction, indicating that the Cu-Ni-P-plated fabric encountered little change when compared with the untreated cotton fabric. This confirmed that with the metalizing treatment, the specimens plated with metal particles was demonstrated a higher frictional force of fibers. In addition, the deposited metal particles developed a linkage force to hamper the movement caused by the applied load.

Table 5. Tensile strength and percentage of elongation at break load of the untreated and Cu-Ni-P- Fabrics plated cotton fabrics

Specimen	Percentage of elongation (%)		Breaking load (N)	
	warp	weft	warp	weft
Untreated cotton	6.12	6.05	188.1	174.97
Cu-Ni-P plated cotton	6.98 (↑12.5%)	6.52 (↑7.8%)	241.5 (↑28.4%)	237.3 (↑35.62%)

5. Color Change Assessment

The results of evaluation of color change under different application conditions, washing, rubbing are shown in Table 6.

Table 6. Washing and rubbing fastness of the untreated and Cu-Ni-P Fabrics plated cotton fabrics

Specimen	Washing	Rubbing	
		Dry	Wet
Cu-Ni-P plated cotton	5	4-5	3-4

The results of the washing for the Cu-Ni-P-plated cotton fabric were grade 5 in color change. This confirms that the copper particles had good performance during washing. The result of the rubbing fastness is shown in Table 6. According to the test result, under dry rubbing condition, the degree of staining was recorded to be grade 4-5, and the wet rubbing fastness showed grade 3-4 in color change. This result showed that the dry rubbing fastness had a lower color change in comparison with the wet crocking fastness. In view of the overall results, the rubbing fastness of the Cu-Ni-P-plated cotton fabric was relatively good when compared with the commercial standard.

6. Surface Morphology

Scanning electron microscopy (SEM) of the untreated and Cu-Ni-P-plated cotton fabric is shown in Figures 1 with magnification of 250x. Microscopic evidence of copper-coated fabrics shows the formation of evenness copper particles on fabric surface and structure.

Figure 2 shows the SEM and WDX analysis copper-plated surfaces of cotton fiber. It was observed that the cotton fiber's surface was covered by Cu-Ni-P alloy particles composed of an evenly distributed mass. In addition, WDX analysis indicated that the deposits became more compact , uniform and smoother also existing homogenous metal particle distribution over coated fabric surface. These results indicate that the effect of chemical copper plating is sufficient and effective to provide highly conductive surface applicable for EMI shielding use.

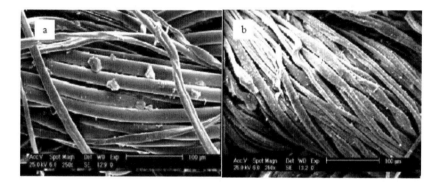

Figure 1. SEM photographs of the (a) untreated cotton fabric (b) Cu-Ni-P plated cotton fabric.

Figure 2. (a) SEM photograph of the Cu-Ni-P plated cotton fabric (b) WDX analysis of the Cu-Ni-P plated cotton fabric.

7. Shielding Effectiveness

Electromagnetic shielding means that the energy of electromagnetic radiation is attenuated by reflection or absorption of an electromagnetic shielding material, which is one of the effective methods to realize electromagnetic compatibility. The unit of EMISE is given in decibels (dB). The EMI shielding effectiveness value was calculated from the ratio of the incident to transmitted power of the electromagnetic wave in the following equation:

$$SE = 10\log\left|\frac{P_1}{P_2}\right| = 20\left|\frac{E_1}{E_2}\right| \tag{5}$$

where P_1 (E_1) and P_2 (E_2) are the incident power (incident electric field) and the transmitted power (transmitted electric field), respectively. Figure 3 indicates the shielding effectiveness (SE) of the copper-coated fabrics with 1 ppm $K_4Fe(CN)_6$ compared to copper foil and other sample after washing and rubbing fastness test. The shielding effectiveness test applied on five different conductive samples including copper foil, electro-less plated of Cu-Ni-P alloy particle on cotton fabric, electro-less-plated fabric after washing test, electro-less-plated fabric after dry and wet rubbing. As it can be expected, copper foil with completely metallic structure shows the best shielding effectiveness performance according to higher conductivity compared to other conductive fabric sample. However, SE of copper-coated cotton fabric was above 90 dB, and the tendency of SE kept similarity at the frequencies 50 MHz to 2.7 GHz.

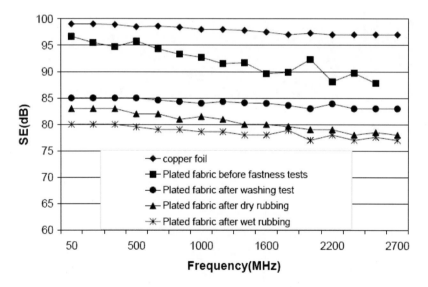

Figure 3. The shielding effectiveness of various conductive sample.

The acquired results for samples after washing fastness test shows nearly 10% decrease over frequency range, which is still applicable for practical EMI shielding use. Two other samples after rubbing show, respectively, 12% and 15% reduction in shielding effectiveness value, but the presented results still show an accepted level of shielding around 80 dB. The SE reduction after fastness tests is a quite normal behavior, which is likely due to removing of conductive particles from fabric surface. However, the compact and homogenous distribution of conductive particles provides a great conductive coating on fabric surface with high durability even after washing or rubbing tests. The copper-coated cotton fabric has a practical usage for many EMI shielding application requirements.

Conclusion

In this study, electro-less plating of Cu-Ni-P alloy process onto cotton fabrics was demonstrated. Both uncoated and Cu-Ni-P alloy coated cotton fabrics were evaluated with measurement weight change, fabric thickness, bending rigidity, fabric shrinkage, tensile strength , percentage of elongation at break load and color change assessment. The results showed significant increase in weight and thickness of chemically plated cotton fabric. Coated samples showed better properties and stable structure with uniformly distributed metal particles. The SE of copper-coated cotton fabric was above 90 dB, and the tendency of SE kept similarity at the frequencies 50 MHz to 2.7GHz. Also, the evaluation of SE after standard washing and abration confirms the supreme durable shielding behavior. The copper-coated cotton fabric has a practical usage for many EMI shielding application requirements.

References

[1] R.H Guo , S.Q Jiang , C.W.M Yuen , M.C.F Ng , "An alternative process for electroless copper plating on polyester fabric," *Journal of Material Science: Mater Electron,* DOI 10.1007/s10854-008-9594-4 (2008).

[2] Y.M Lin, S.H Yen, "Effects of additives and chelating agents on electroless copper plating," *Applied Surface Science,* Vol 178,2001, pp.116-126.

[3] G. Xueping ,W. Yating , L. Lei , Sh. Bin, H. Wenbin , "Electroless plating of Cu–Ni–P alloy on PET fabrics and effect of plating parameters on the properties of conductive fabrics," *Journal of Alloys and Compounds,* Vol 455,2008, pp. 308–313

[4] J.Li, H.Hayden, P.A. Kohl, "The influence of 2,2'-dipyridyl on non-formaldehyde electroless copper plating," *Electrochim Acta,* Vol 49 , 2004,pp.1789–1795

[5] H. Larhzil, M. Cisse, R. Touir, M. Ebn Touhami, M. Cherkaoui," Electrochemical and SEM investigations of the influence of gluconate on the electroless deposition of Ni–Cu–P alloys," *Electrochimica Acta* Vol 53, 2007,pp.622–628

[6] J.G. Gaudiello, G.L. Ballard," Mechanistic insights into metal-mediated electroless copper plating Employing hypophosphite as a reducing agent," *IBM Journal of Research and. Development,* Vol 37 1993.

[7] G.Xueping ,W.Yating , L. Lei , Sh.Bin, H. Wenbin, "Electroless copper plating on PET fabrics using hypophosphite as reducing agent," *Surface & Coatings Technology,* Vol 201,2007 pp 7018–702.

[8] E.G Han, E.A. Kim, K.W Oh, "Electromagnetic interference shilding effectiveness of electroless Cu-platted PET fabrics," *Synthetic Metals.* Vol 123 ,2001, pp 469–476

[9] S. Q Jiang, R. H. Guo, "Effect of Polyester Fabric through Electroless Ni-P Plating," *Fibers and Polymers,* Vol 9 2008,pp. 755-760.

[10] S.S. Djokic, *"Fundamental Aspects of Electrometallurgy: Chapter 10, Metal Deposition without an External Current, "* Eds., K. I. Popov, S. S. Djokic B. N. Grgur, pp. 249-270, Kluwer Academic Publishers, New York, 2002.

[11] O. Mallory, B. J Hajdu, Eds., *Eleolesctrs Plating: Fundamentals and Applications,* Noyes Publication, New York, 1990.

[12] R-C. Agarwala, V- Agarwala, "Electroless alloy/composite coatings: a review," *Sudhunu*, Vol 28,2003,pp. 475- 493.

[13] M. Paunovic, "Electrochemical Aspects of Electroless Deposition of Metals," *Plating,* Vol 51, 1968, pp 1161-1167.

[14] H.F. Chang, W. H. Lin, "TPR study of electroless plated copper catalysts," *Korean J. Chem. Eng,* Vol 15,1998, pp. 559-562.

In: Handbook of Research on Nanomaterials, Nanochemistry ...   ISBN: 978-1-61942-525-5
Editors: A. K. Haghi and G. E. Zaikov   © 2013 Nova Science Publishers, Inc.

Chapter XII

A Note on Electrochemical Synthesis of Cobalt Nanoparticles

A. F. Dresvyannikov, M. E. Kolpakov,
E. V. Pronina and M. V. Dudnik
Kazan State Technological University, Kazan, Russia

Introduction

At the present time, development of new methods and perfection of already known and approved methods of metal nanoparticles obtaining are in the center of attention. According to one of modern classification methods of nanoparticles, obtaining can be subdivided conditionally on physical and chemical [1]. In case of physical method, nanoparticles are formed of separate atoms as a result of metal sublimation and their subsequent condensation on various substrates or are formed owing to crushing of the big metal particles by means of corresponding devices (colloidal mills, ultrasonic generators, etc.). On the data of nanoparticle obtaining chemical methods process of chemical reduction in a solution of metal complex ions in the conditions favouring the subsequent formation small metal clusters or aggregates is a lie [1].

Depending on the reducing agent, natural chemical methods subdivide into classical, using chemical reducers (hydrazine, alkaline metal tetrahydrideborates, hydrogen, etc.), X-ray and electrochemical in which a reducer is solvated electron generated accordingly by ionising radiation in a solution or electrochemically (an external current source, contact exchange) on electrode surfaces [2].

Cobalt nanoparticles find a wide application at composite materials and alloys obtaining [3]. At the same time, they can be claimed in areas where their magnetic properties are important.

Studying of possibility and conditions of cobalt nanoparticles synthesis in water solutions by an electrochemical method is the purpose of the work.

Procedure

According to the theory of free volume, the most effective process of nanoparticle formation is the obtaining method in the liquid media, instead of in solid or gaseous environment. Metals nanoparticle electrochemical obtaining in the liquid media provides qualitative formation and process efficiency from the point of view of productivity and power inputs. The offered electrochemical method is based on reduction reactions of cobalt ions in water solutions. For cobalt nanoparticle obtaining, water solutions of cobalt salts are used; aluminum microparticles are used as fluidizated electrode.

The experiments have shown that cobalt aquacations are restored on an aluminum substrate from solutions with high speed (Fig. 1). Reaction of cobalt separation can be adequately described the equation (Eq. 1)

$$\alpha = \frac{A_0 - A_r}{1 + \left(\dfrac{t - t_0}{k_{-1}}\right)^D} + A_r, \tag{1}$$

A_0 – reaction index at time point t_0;
t_0 – reaction beginning time;
A_r - reaction index at the end of experiment;
k_{-1} – constant of reaction rate;
D – fractal dimension index.

Table 1. Values of equation parameters

Parameter	A_0	t_0	A_r	k_{-1}, c	D
Value	0	0	96	61	2,87

It is possible to assume with sufficient basis that cobalt discharge process on aluminium microparticles proceeds as follows. At the first stage, accompanied by surface oxide film transformation and simultaneous metal nucleation process begins and proceeds on an outer side of a particle surface. For short time, the aluminium surface becomes covered by a layer of metal nuclear. Across metal hydrogen is allocated. Hydrogen evolution in the process beginning occurs on cathode sites of an aluminium surface, but restoration of protons donors goes mainly on them in deposition process on cobalt surface. Further, after metal ions solution exhaustion, hydrogen deposition process proceeds only. High speed of cobalt sedimentation and intensive hydrogen deposition promote formation nucleation of nano-sized metal and interfere with formation of deposit dense layer.

For the purpose of reduction of sizes of particles obtained, metals deposit the solutions containing ions of a palladium (Pd^{2+}) or platinum ($[PtCl_6]^{2-}$) are added to the initial solution containing metal ions. Using foreign metal, which intensively absorbs hydrogen (platinide family elements) in an element condition, as an ion, hydride dispersion of system containing this metal occurs [4]. It is also known [3] that bringing in a solution of electropositive metal ion can lead to dispersion of metal deposit (in quantities, ten times smaller in comparison with quantity of deposited metal).

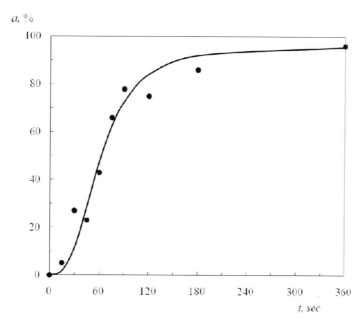

Figure 1. Kinetic curves of cobalt deposition from 1 M $CoCl_2$ water solution. Points - experiment result, a line - calculation result (Eq. 1).

At all investigated samples obtained from cobalt (II) solutions, the element cobalt presented by two crystal polymorphic modifications (α -Co - hexagonal and β -Co - cubic) is found out as the basic phase. On diffraction patterns of these samples (Fig. 2), reflexes 2.17; 2.04; 1.91; 1.254; 1.069 Å correspond hexagonal cobalt modifications; reflexes 2.04; 1.78; 1.253; 1.068 Å correspond cubic cobalt modifications (some reflexes of these crystal phases are blocked). The halo-liked background raising on diffraction pattern in the rate approximately 10-35 ° 2 θ allows assuming presence phases amorphous to X-rays.

On diffraction patterns of the sample 1 (Fig. 2 a), cobalt hydroxide and oxihydroxide reflexes are observed; they obviously formed at the set process mode owing to local cathode region alkalinization. Hydroxide and oxihydroxide colloidal particles formed in cathode region layer are deposited together with metal. In samples 2 and 3 (Fig. 2b, c) on diffraction patterns, palladium reflexes absence is possibly caused by a condition amorphous to X-ray. On diffraction patterns of samples 4 and 5 (Fig. 2 d, f), fine-dispersed platinum reflexes are shown.

According to electronic microscopy data, cobalt deposits contain nanoparticles in the size of 30-50 nm. The sample 1 (Fig. 3 a) represents nanoparticles in the size ~30-50 nm located separately. They tend to dark aggregates formation of the various size (to 0.5-1 μm). The sample 2 (Fig. 3b) also consists of nanoparticles in the size ~30-50 nm, located separately forming a continuous field. On this background (approximately to 5% from total of particles), larger dark quasi-hexagonal particles in the size to 100-150 nm are located at the sizes. Their sides are outlined by a light aura that is inherent in the objects incorporating a hydrated component (for example, OH $^-$ or H_2O). The sample 3 (Fig. 3c) differs from the sample 2 by presence of the big number of nanoparticle aggregates forming conglomerates of the various forms. The sample 4 (Fig. 3d) represents nanoparticles in the size ~30-50 nm. Hexagonal separate particles with the linear sizes 100-200 nm are accurately visible. They are absolutely opaque, their thickness more than 1 μm. They also form aggregates of any form. In the sample 5 (Fig. 3d), except the separate nanoparticles in the size ~30-50 nm, "rash" from sub-individuals in the size of ≤10 nm is noted on some sample fragments. The similar picture is characteristic for fine-dispersed iron hydroxide present at natural mineral objects. Here, dark hexagonal particles in the size of ~150 nm and aggregates to 1 μm are located, and the last have the needle shape.

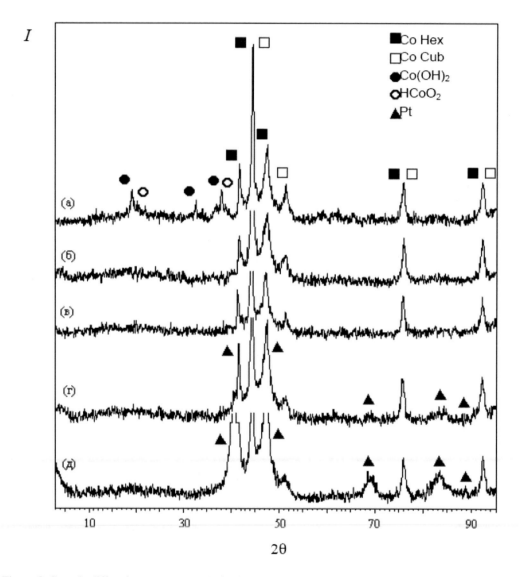

Figure 2. Sample diffraction patterns: a - 1; b - 2; c - 3; d - 4; e – 5.

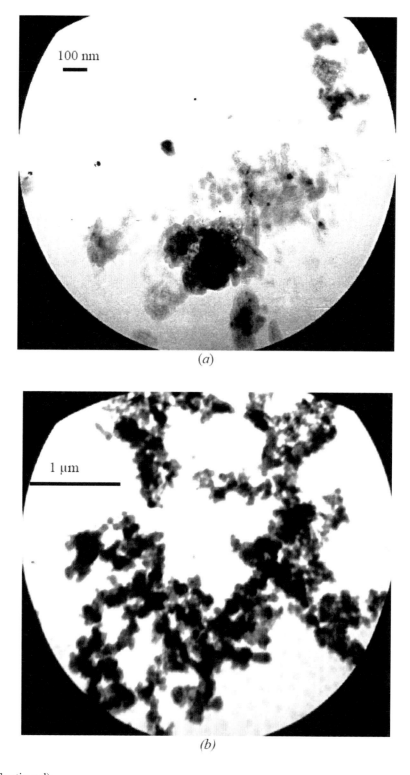

Figure 3. (Continued).

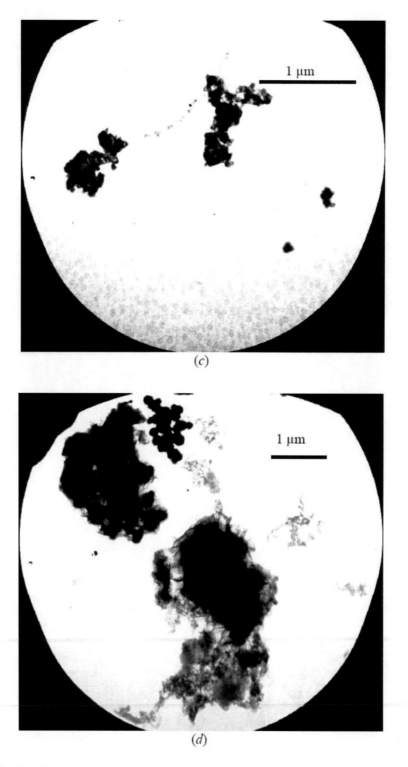

Figure 3. (Continued).

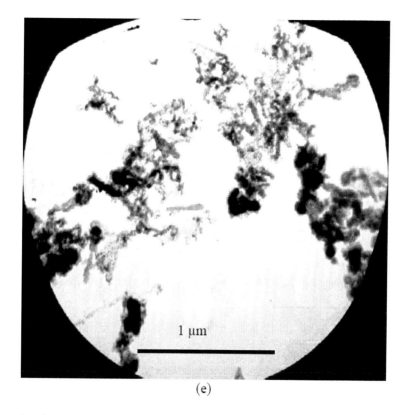

(e)

Figure 3. Sample microphotographies: *a* - 1, ×65000; *b* - 2, ×26000; *c* - 3, ×29000; *d* - 4, ×16000; *e* - 5, ×40000.

It is necessary to notice that using magnetic separation for branch of cobalt particles from mother solution leads to occurrence of residual magnetisation owing to this action particles aggregate in large conglomerates. It is also connected with aggregation as nanoparticles are always characterised by very high value of the relation "surface-volume," and aggregation process is thermodynamically favourable [1].

By results of X-ray diffractometry, it is established that in the investigated samples, the basic phase is cobalt. Management possibility by phase particle structure by using electrolyte of corresponding structure is established. It is revealed that the obtained particles practically do not contain oxide phase. It is possibly connected with the interfaced reaction of hydrogen separation during process of metal deposit formation. Saturation of electrolyte and deposit by hydrogen creates the reducing media, allowing performing synthesis without additional operation on removal of the dissolved oxygen from a solution.

Thus, it is possible to conclude that cobalt deposits obtained by electrochemical method consist basically from sphere- shaped or oval nanoparticles in the size of 50-100 nm, but it is observed the strong tendency to formation of large aggregates of the various sizes (to 0.5-6 μm), which it is possible to explain both the developed surface and residual magnetisation of powder samples. For the purpose of aggregate destruction, additional processing of the obtaining deposits by the HF-discharge or their crushing (mechanical, electro erosive) is necessary

Results

At the base of nano-sized metal synthesis electrochemical process of cobalt ion restoration on slurrying substrate in aluminium solution lays.

Experiments spent with aluminium samples (purity not less than 99.0%). $CoCl_2$ (qualifications «pure for analysis») without additional purification and also $PdCl_2$ (qualifications « pure for analysis») and H_2PtCl_6 (qualifications «pure») are used as the basic reactants.

Restoration kinetics is studied by sampling method through the fixed time intervals with the subsequent definition by X-ray fluorescence (VRA20L) and atomic absorption (AAS-1N) analyses.

The obtained deposit separated by magnetic separation from mother solution washed out bidistillated before neutral reaction and dried under vacuum at $60°$ C.

The X-ray analysis is spent by a powder method on D8 ADVANCE diffractometer (Bruker) using monochrome CuK α-radiations in a mode of step-by-step scanning. A scanning step is $0.02°$ 2θ, exposition time in a point is 1 sec, an interval of shooting 2θ is $3-95°$.

Samples prepared by press fitting of investigated material powder in standard disk ditch from quartz glass; during shooting, the preparation rotated in its own plane with a speed of 60 rpm. An operating mode of an X-ray tube is 40 kV, 30 mA.

Calculation of interplanar space values of diffraction reflexes was made automatically under EVA programmer entering into the complete set of device software.

Identification of crystal phases was carried out by standard way by comparison of the obtained experimental values of interplanar space and relative intensity with reference.

Investigations were spent by a method of transmission electronic microscopy (TEM) for detailed definition of dimension and morphological features of particles of studied samples according to methodical recommendations [5].

The synthesised sample particle size defined by the method of transmission electronic microscopy using of microscope-microanalyzer EMMA-4 at accelerating pressure of 75 kV. Preparations were prepared by suspension method, with preliminary preparation on ultrasonic dispergator UZDN-2 T and the subsequent drawing on collodion film-substrate or a carbon dusting on vacuum apparatus VUP-4. Microphotographies are obtained by means of digital camera OLYMPUS C-8080.

References

[1]   Foster L. Nanotechnologies. Science, innovation and possibilities. *M.: Technosphera,* 2008. 352 p.

[2]   Nanotechnology in nearest decade. Prognosis of investigation direction. / Edit. by *M.K. Roko, R.S. William, P. Alivisatosa.* Trans. From eng. M.: *Mir,* 2002. 292 p.

[3]   Dresvyannikov A.F., Grigor'eva I.O., Kolpakov M.E. Physical chemistry of nanostructured aluminium containing materials. Kazan: Pub. house «Fən» SA of TR, 2007. 358 p.

[4] Semenenko K.N., Yakovleva N.A., Burnasheva V.V. To a question on the reaction mechanism of hydride dispersion// *Russian journal of General chemistry,* 1994. v.64, № 4. P.529-534.

[5] Methodological recommendation № 137. Electronic microscopical analysis of minerals. M.*: NSOMMI,* 2000. 36 p.

In: Handbook of Research on Nanomaterials, Nanochemistry ...   ISBN: 978-1-61942-525-5
Editors: A. K. Haghi and G. E. Zaikov   © 2013 Nova Science Publishers, Inc.

Chapter XIII

Practical Hints on Energy-Saving Technology for Production of Nano-sized Particles

A. M. Lipanov[*] **and V. A. Denisov**[*]
The Institute of the Applied Mechanics UB RAS, Izhevsk, Russia

Introduction

In the practice of the breakage of various kinds of starting materials, there are two main mechanical-technological lines that have become widely used because they embrace coarseness of the starting material, requirements towards the end product, its physical-mechanical properties, etc. They are:

- Breakage during one or several intakes in one apparatus with the simultaneous classification of the mass flow or without it;
- Multistep breakage in several sequentially installed apparatus; here, the breakage process is accompanied by separation at the sizing screens.

The main criterion for choosing a mechanical-technological line for size-reduction is its efficiency, which means the yield of fractions with predetermined sizes for a certain period, energy inputs for receiving one unit of product, the number of personnel, laboriousness of maintenance and repair of the equipment, its reliability, and occupied production space.

It is not possible to obtain the product of milling with a predetermined granulometric composition in one apparatus by the breakage of starting material particles at one or several intakes with mass flow classification or without it due to the following:

[*] Lipanov Alexey Matveevich, academician RAS, director of Institute of Applied Mechanics UrD RAS, tel.: +7(3412)508200, e-mail: ipm@udman.ru.
[*] Denisov Valery Alexeevich, Dr. Tech. Sc., Professor, research fellow at IAM, UrD RAS, tel.: +7(3412)207658.

- All size-reduced materials have inhomogeneous structure, and as a result, separate particles are not uniform in their physical-mechanical properties;
- Diversity in physical-mechanical properties of the starting material excludes identity in the breakage of separate particles in the mill working chamber, which normally has constant working rate of the milling parts;
- The presence of one working chamber, where the working parts have a preset rate, determines the time of the starting material staying in it until the strongest particles are broken to the desired sizes, which is a preset condition;
- Introduction of the cycle ensuring size-reduction of the strongest particles inevitably leads to over-milling of the broken mass, which in its turn results in a high degree of non-uniformity of the granulometric composition, increased energy inputs, consumed by unnecessary over-milling, and quick wear of working parts;
- The non-uniform granulometric composition of the milling product leads to inefficiency of the use of the product (its energy of burning decreases, its activity in chemical reactions changes, its explosiveness increases, etc.)

Most of the mechanical-technological lines for the starting material size-reduction are based on the principle of multistep breakage and classification, which allows solving the main question, which is receiving the milled product with a predetermined granulometric composition, preventing over-milling and reducing the losses of a ready product. However, multistep lines for size-reduction have low effectiveness of starting material treatment and low performance factor as far as the energy consumption is concerned.

Multistep milling and classification process consists of separate stages, where breakage takes place at one intake in one mill. The number of stages is determined by the starting material nature and the desired degree of size-reduction.

In the multistep lines (Fig.1), the overall flow of starting material Q_{il} to be milled is fed at the first stage. Then, the material proceeds to the sizing-screen S_{il}, i.e.,

$$Q_{il} = S_{il} \tag{1}$$

As part of the material (the as-milled product), P^n_{il} is unloaded from the sizing-screen as a ready product, the flow P^c_{il} equal to

$$Q_{il} - P^n_{il} \tag{2}$$

proceeds to the second stage of milling; the indices (n) and (c) denote passage and exit of the product.

At the second stage, after the breakage of the product P^n_{il}, the ready product P^n_{i2} is formed, which is unloaded from the second sizing screen, and the remaining part proceeds to the following stage.

Theoretically, the process of re-milling and classification over the stages should be repeated until the magnitude $P^c_{i(m)}$ becomes zero. In terms of multistep milling and classification, the overall flow of a ready product can be written as follows

$$P^n_n = \sum P^n_{i(n)} \tag{3}$$

and re-milling load as $P^c_n = \sum P^c_{i(n)}$ (4)

where m is the number of stages.

To increase the performance factor of the technological line of the multistep milling and classification of the mass flow, it is necessary:

- that the selection of the apparatus for breakage and classification of the mass flow particles should be reasonable, taking into account the requirements towards the granulometric composition of the product of milling and the starting material physical-mechanical properties;
- that there is minimal necessary number of stages of breakage in a technological line or apparatus.

However, the analysis of the existing machinery for breakage of various solid-loose starting materials shows that there are no milling machines with high performance factor at present.

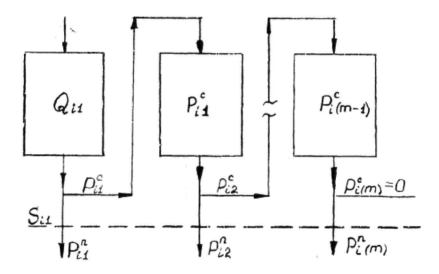

Figure 1. Technological line for multistep breakage of raw materials.

One of the main reasons of low performance factor is the design of the apparatus for breakage of materials, which is based on a simplified idea about the breakage process as a process containing a single act of loading. The simplified ideas about breakage, lying in the base of the energy theories of breakage and the theory of appearance and propagation of cracks, do not reflect the main process of reduction of coarseness. When the "index of breakage" of the starting material has different values and one desires to obtain the breakage product particles of the same size, it is necessary to consider the breakage process as the one consisting of the repeated stages of breakage, and each stage has three main characteristics [1]. They are a) screening the material particles for breakage, b) the breakage of the screened particles, and c) the classification of the broken particles with the purpose to sort out those of them that correspond to the ready product by their sizes. The quantitative characteristics of screening and breakage of the particles in the apparatus are determined by the functions:

- $P_n (y)$ – the probability of breakage of the particles, which have y size at the n-th stage of the breakage process,
- $F (x, y)$ – cumulative distribution over the coarseness range of $x < y$ mass of particles appearing as the result of the breakage of the mass unit of particles with y size.

Representation of the quantitative characteristics as the functions $P_n(y)$, $F(x, y)$ together with the assumption of the division of the process of breakage and classification into sequentially repeated stages creates the foundation for building the breakage machines, where the technology of multistep breakage can be used, and allows studying the change of the particle distribution in coarseness depending on the number of such stages. The function $P_n(y)$ describes the probability of the breakage of the particles of the size-reduced material as the result of one act or one stage of the breakage process. The $F(x, y)$ function characterizes the distribution of the mass of the particles in various ranges of coarseness occurring due to breakage of the mass unit of the initial particles.

Particular functions of screening, breakage and classification of the apparatus constitute one stage in the sequential process of the starting material reforming to the predetermined particle sizes. After each stage, the $F_n(x)$ function characterizing the distribution of the starting material mass in coarseness turns into a new $F_{n+1}(x)$ function, which characterizes the material after breakage.

The $F(x, y)$ function form determines the indicator of "grindability of a material." The above-mentioned conceptions of screening, breakage and classification lie in the base of the development of the models of multistep apparatus for milling various materials.

In the matrix models, the multistep milling process is considered as the sequence of breakage acts [2]. If there are n acts, and each of them is determined by the matrix X containing quantitative characteristics by the functions of screening and breakage, then the sequential breakage of particles of the mass flow can be described by the equations:

$$P_1=Xf;$$
$$P_2 =XP_1;$$
$$P_3 = XP_2; \tag{5}$$
$$P_n=XP_{n-1}$$

where P is the particle distribution in a ready product for n acts of breakage; f is the particle distribution in a starting material. The principle of multistep breakage can be used at modeling with the help of the parameter n. In this case, the process will be described by the equation:

$$P=(\Pi X_i)f, \tag{6}$$

where X_i is the matrix of the i-th act of breakage.

The number of breakage acts can be determined by the number of the chosen intervals of breakage by rates, necessary for complete breakage of a material, and features of the apparatus operation. In the presence of the identical breakage acts, the entire process is described by the equation:

$$P=X^n f. \tag{7}$$

The above equations show that the breakage of the starting material particles in apparatus of any type can be described by the equation $P = Xf$, where X can be a result both of a single milling process and a sequentially repeated milling process.

In case of a particular apparatus, the matrix X should be determined quantitatively so that it would determine the process and could be represented as a sole parameter influencing technological variables. Both a design and a mode of operation can influence the milling process in any apparatus.

The set of parameters determines quantitative value of the matrix X. The number of stages of milling, the frequency of the rotor rotation, the parameters of working parts, and the type of sizing screens belong to the design parameters. The technological parameters are mass consumption of feed, its granulometric composition and the requirements towards the product of milling.

In order to unambiguously specify the values of the elements of the functions of breakage and screening, constituting the matrix X, it is necessary to separate broken particles from unbroken ones using the sizing screen. The less is granulometric relation between the two sequential screen openings, the higher validity of the data will be. In this case, the assumption that the breakage of the particle leads to its removal from the class under consideration becomes more certain.

However, in order that the assumption of the transfer of the particles from one class to another at breakage will become acceptable in practice, the relation of the sizes of the screening openings should have a negligible value, infinitely small value. This is not possible experimentally.

In practice, it is possible to determine the breakage function regardless of the screening function. For this, it is necessary to make experiments using the apparatus model, which will define the screening function for this particular apparatus. Without the experiments, it is not possible to single out the breakage function as the characteristics of grindability of the starting material and the screening function as the characteristics of the apparatus from the matrix X.

Thus, to build up a useful model of the apparatus, it is necessary to make assumptions concerning its form, and the data on the functions of screening and breakage are needed as well. The studies on operation and technological peculiarities of the process of breakage and classification are to be conducted to substantiate the form of a model. These studies will allow making reasonable assumptions.

One of the main peculiarities of the operation of the multistep apparatus with centrifugal working parts is that each working part performs breakage with a constant rate. The breakage of particles by a working part takes place if its rate is more than the minimal one necessary for breaking. When the mass of the broken particle is constant and the rate is fixed, then the kinetic energy *(5 mV)** transferred to the particle via the working part will be equal to the work of breakage A_p. Therefore, $V \geq \sqrt{2A_p}\backslash m$.

As the analysis of the investigation data show (Table 1), the magnitudes A_p, m and V are variables. Thus, in order to attain the breakage of the particles of solid-loose material (Table 1) at one impact, it is necessary that the apparatus provides simultaneously the rate of movement-breakage within the limits of 8,9 ... 174,36 m/s. Attaining such a wide range of rates simultaneously is possible only in the apparatus of multistep type, where each stage of milling will impart its own average rate of breakage to the starting material.

**Table 1. The average boundary values of various parameters of the
breakage process of the solid-loose starting material at fine milling**

Parameters	Limestone	Sandstone	Quartzite	Rock salt	Steel-smelting slag	Grain
1. Average sizes of starting material grains, mm	3,86-5,9	3,92-5,88	4,14-6,94	4,09-7,22	3,24-7,11	2,37-5,58
2. Specific energy consumption, J\g	2,2-15,2	0,01-1,3	12,2-29	0,04-6,6	2,4-14,01	2,77-5,54
3. Rate range:						
3.1. minimal, m/s	66,33	4,47	156,2	8,9	69,28	74,43
3.2. maximal, m/s	174,35	50,99	240,83	114,89	167,39	105,26

At each stage, the particles are milled, and the indicator of their grindability corresponds to the conditions of breakage at the predetermined rate. The working breakage rates at the stages should increase from the first stage to the following ones. The rates at first and the last cycles of breakage correspond to the boundary operation rates.

The increase of the rates of movement-breakage from the first stage to the following ones is due to the increase of strength of a fraction of the material being milled and the decrease of the linear size of the particle. At the decrease of the linear size of the particles, the maximal performance factor of breakage decreases, tending to higher rates. After one-act breakage, which is the first stage of the process, the average size of the particles will be reduced. Thus, to maintain the maximal value of the performance factor at the following steps, it is necessary to increase the movement-breakage rate in the designed apparatus. The presence of the step-wise increase of the movement-breakage rates is due to the necessity of receiving the impact with the same force during the step-wise decrease of the mass of separate particles. This presupposition follows from the known theorem of momentum, according to which

$$Pt\Delta = \Delta(mV),$$

where m is the mass of milled material,

V is the rate of the particle movement at the impact,

P is the force of impact at the initial moment of impact,

Δt is the time of impact.

The basic scheme of the multistep milling of starting material in the apparatus with the centrifugal working parts is shown in Fig.2.

Taking into account the mass flow classification over the stages and ambiguity of the working parts' parameters determining the structure of the matrix X, equations (5) can be written in the form:

$$P_1 = Xf_{ini} ;$$
$$P_2 = X_1 m;$$
$$P_3 = X_2 m_1;$$
$$P_n = X_n m_{n-1},$$

(8)

m is the matrix of the distribution of the particles of the milling product, which is obtained at the n-th stage of breakage, and which did not pass through the sizing screen (undermilling).

Application of the substantiated models for multistep breakage and classification of the starting material to the particular designs of the working parts, the determination of the values of the functions of screening, breakage and classification for different starting materials will allow building and studying apparatus for various material treatments. The general view of these apparatus is given in Figs 3 – 6.

The technological parameters of the operation of the apparatus based on energy-saving technology for multistep milling and the classification of the mass flow are displayed in Table 2. The displayed data show that the economy of energy for the established power of the driver of the electric engines reaches 1,550%.

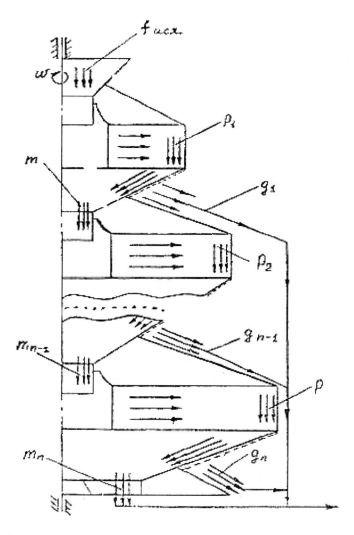

Figure 2. The basic scheme of the multistep milling of starting material in the apparatus with the centrifugal working parts.

Table 2. Performance specifications of the mills М Ц and М М Ц

Type	Productivity, ton/h	Coarseness of feed, mm	Electric engine power, kW	Fineness of milling, to μm	Energy-saving: МЦ \ ММЦ (times)	Sizes, LxWxH m	Mass, ton	Resources-economy: МЦ \ ММЦ (times)
МЦ-0.5	1.0	10	22	40	4,31	1.9x1.3x3.2	1.5	5,0
МЦ-3	3.0	40	110	40	-	5.5x3.0x7.6	8.0	-
МЦ-8	8.0	40	315	60	5,95	7.6x4.2x15.3	22.0	14,19
МЦ-25	25.0	80	1400	60	15,50	12.0x6.0x16.0	70.0	10,8
ММЦ-1	1.0	10	5.1	10	-	0.8x0.7x1.28	0.3	-
ММЦ-8	8.0	40	18.5	10	-	0.8x1.53x1.35	1.55	-
ММЦ-25	25.0	75*	90	10	-	1.53x1.53x2.5	6.5	-

Note: Centrifugal mills МЦ are developed by the research and production associations "Ural-Tsentr." /http://ume.sky.ru/main1.htm/. Multistep centrifugal mills ММЦ are developed by the Institute of Applied Mechanics UB RAS.
- the sizes of loaded pieces of the starting material are about 250 mm.

Figure 3. Multistep centrifugal mill, Productivity to 1,0 ton/h.

Figure 4. Multistep centrifugal mill, Productivity to 8,0 ton/h.

Figure 5. Multistep centrifugal mill, Productivity to 25,0 ton/h.

Figure 6. Impact mill, Productivity to 25,0 ton/h.

Conclusion

1) The increase of the performance factor of the technological lines for milling is possible due to building the apparatus with effective working parts, including also the operation of screening after each stage of breakage.

2) The offered milling matrix model includes the functions of screening, breakage and classification and allows describing the size-reduction process both at a single breakage and at multistep breakage.

3) To build up a useful model of the apparatus, it is necessary to have data on the functions of screening, breakage and classification, which characterize the grindability of a raw material and the design features of the apparatus.

4) The substantiated mathematical models of apparatus are of general character; they allow substantiating the technological lines for milling any solid-loose material.

5) The developed model for the process of milling mineral raw materials natural and technological origin allows one to analyse the mechanism of particles refinement down to nanosizes. The result obtained is about 300 nm.

References

[1]   Epstein B. Logarithmic – normal distributions in breakage of solids // *Ind. Eng. Chem.* 1940. 40. P. 2283-2291.

[2]   Denisov V.A. Thechnical report №8.4.4 on the theme IR-161-87 NIR join-stock company "Izhstal" – *"The study of elite-industrial samples of a multistep centrifugal crusher,"* Izhevsk, 1987, 242 p, illustrated.

In: Handbook of Research on Nanomaterials, Nanochemistry ...   ISBN: 978-1-61942-525-5
Editors: A. K. Haghi and G. E. Zaikov   © 2013 Nova Science Publishers, Inc.

Chapter XIV

Prediction of Polymer Composites Strength Properties

Yu. G. Bogdanova[1], V. D. Dolzhikova[1], I. M. Karzov[1],
A. Yu. Alentiev[1] and A. V. Shapagin[2]

[1] Chemistry Department of Moscow State University, Moscow, Russia
[2] A. N. Frumkin Institute of Physical Chemistry and Electrochemistry, Moscow, Russia

Introduction

The development of new polymer composites with predetermined properties is the urgent problem of modern material science. Adhesion of polymer binder to the filler material (fabrics, fibers, disperse particles) is one of the important factors influenced by the strength properties of polymer composites [1]. Thus, the prediction of polymer composite strength properties based on the polymer adhesion characteristics determination is the main task to develop the scientific principles of new polymer binder design. The work of adhesion (W_a) depends on the surface energy of components (σ_1 and σ_2, respectively) and the energy of polymer-filler interface (σ_{12}): $W_a = \sigma_1 + \sigma_2 - \sigma_{12}$ [2]. So, the development of express techniques permitted determining these energetic characteristics of interface, calculating the work of adhesion and optimizing the choice of polymer binder composition for the particular type of filler is the actual problem. The contact angle measurements permit to solve this problem.

Theory

The wetting is one of the few express techniques permitted to determine the surface energetic characteristics of solids being in contact with air or liquid. Such determination is possible due to Young's equation for equilibrium contact angle (θ):

$$\cos\theta = (\sigma_{SV} - \sigma_{SL})/\sigma_{LV},$$

where σ_{SV}, σ_{SL} и σ_{LV} are surface energies of solid/vapor, solid/liquid and liquid/vapor (surface tension) interfaces, respectively [2]. Experimental contact angle values of the test liquids with known surface tension dispersion and polar components (σ^d_{LV} and σ^p_{LV}) are usually applied to calculate the σ_{SV} and σ_{SL} values using the wetting molecular theory of Girifalco-Good-Fowkes-Young equations [3]. This approach allows determining the dispersion and polar forces contribution in surface energy values (σ_{SV} and σ_{SL}). So, it is possible to determine the polymer binder adhesive properties using contact angle measurements.

There are no systematic investigations connected the contact angle data with the polymer composites strength properties. But it is well known that good strength parameters of polymer composites are provided by high adhesion of the polymer matrix to the filler. So, the simplest way to predict the polymer composite strength properties using (W_a) values is to determine the filler and the polymer binder surface energies (σ_{FV} and σ_{BV}, respectively) and ones dispersion and polar components (σ^d_{FV}, σ^p_{FV} and σ^d_{BV}, σ^p_{BV}, respectively) and to calculate the work of adhesion value using eq.

$$W_a = 2(\sigma^d_{FV} \cdot \sigma^d_{BV})^{1/2} + 2(\sigma^p_{FV} \cdot \sigma^p_{BV})^{1/2} \qquad (1) \; [3, 4].$$

Often, the filler surface energy is not so simple to determine. So, another way of composite strength properties prediction is to determine the work of adhesion (W_a) of polymer binder to model liquids with different polarity.

Now we propose the new approach for polymer binder adhesive properties determination and polymer composite strength properties prediction. The work of adhesion values of polymer binder to model liquids are used for this prediction:

1) non-polar liquid $W_{dd} = \sigma_{SV} + \sigma_O - \sigma_{SO}$ (2),

2) polar liquid $W_{pp} = \sigma_{SV} + \sigma_W - \sigma_{S(W)W}$ (3),

3) polar and non-polar liquids $W_{dp} = \sigma_{SO} + \sigma_W - \sigma_{S(O)W}$ (4).

Water and octane were chosen as polar and non-polar liquids, respectively. The values σ_{SO}, $\sigma_{S(W)W}$ and $\sigma_{S(O)W}$ in eq. (2-4) are equilibrium surface energy values at the polymer-octane, polymer-water and polymer (equilibrated with octane)-water interfaces, respectively, σ_{SV} – polymer binder surface energy, σ_O и σ_W – octane and water surface tension values, respectively.

The molecular wetting theory allows calculating the σ_{SV}, σ_{SO}, $\sigma_{S(W)W}$ and $\sigma_{S(O)W}$. For example, σ_{SV} can be determined by the two-liquid method:

$$(1+\cos\theta_{L1}) \, \sigma_{LV(1)} = \left\{ 2(\sigma^d_{LV(1)} \, \sigma^d_{SV})^{1/2} + 2(\sigma^p_{LV(1)} \, \sigma^p_{SV})^{1/2} \right. \qquad (5)$$

$$(1+\cos\theta_{L2}) \, \sigma_{LV(2)} = \left. 2(\sigma^d_{LV(2)} \, \sigma^d_{SV})^{1/2} + 2(\sigma^p_{LV(2)} \, \sigma^p_{SV})^{1/2} \right. \qquad (6),$$

where θ_{L1} и θ_{L2} – test liquids advancing contact angles at the polymer surface (Fig.1a), $\sigma^p_{LV(1)}$, $\sigma^p_{LV(2)}$, $\sigma^d_{LV(1)}$, $\sigma^d_{LV(2)}$ are its polar and dispersion components, respectively; $\sigma_{SV} = \sigma^p_{SV} + \sigma^d_{SV}$, where σ^p_{SV} and σ^d_{SV} are polar and dispersion components of polymer surface free energy [4].

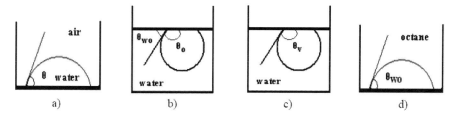

Figure 1. The scheme of contact angle measurements: (a) – water droplet at the polymer surface in the air environment; (b) - octane droplet at the polymer surface immersed in water (θ_O); (c) - the air bubble at the polymer surface immersed in water (θ_V); (d) θ_{WO} – the water droplet at the polymer surface immersed in octane. The contact angles were measured in accordance [7].

It is possible to calculate the polymer/liquid interfacial energy (σ_{SL}) using the molecular wetting theory equation:

$$\sigma_{SL} = \sigma_{SV} + \sigma_{LV} - 2(\sigma^d_{SV} \cdot \sigma^d_{LV})^{1/2} - 2(\sigma^P_{SV} \cdot \sigma^P_{LV})^{1/2} \text{ [2-4]}.$$

But such calculation way may be not quite correct because of the polymer chains mobility near the polymer/liquid interface. Priority scientific approach of equilibrium surface energy at the polymer/liquid interface (σ_{SL}) calculation has been proposed by E. Ruckenstein and co-workers [5]. We used technique developed by Ruckenstein to determine the σ_{SO}, $\sigma_{S(W)W}$ and $\sigma_{S(O)W}$ values. The polymer films were in contact with model liquids during 24 hours [5,6]. Then the air bubbles and octane or water drops contact angles at the polymer surfaces were measured. The following equations were used for calculations:

$$\sigma_{S(W)W} = \{(\sigma^P_{SW})^{1/2} - (\sigma^P_W)^{1/2}\}^2 + \{(\sigma^d_{SW})^{1/2} - (\sigma^d_W)^{1/2}\}^2 \tag{7},$$

where σ^P_{SW} and σ^d_{SW} are equilibrium interfacial energy polymer/water polar and dispersion components;

$$\sigma^P_{SW} = (\sigma_W - \sigma_O - \sigma_{OW} \cdot \cos\theta_O)^2 / 4\sigma^P_W \tag{8},$$

where $\sigma_{OW} = 50{,}8 \text{mJ/m}^2$ – octane/water interfacial tension, θ_O - octane droplet contact angle at the polymer surface immersed in water (Fig1b) [7];

$$\sigma^d_{SW} = (\sigma_{OW}\cos\theta_O - \sigma_W \cdot \cos\theta_V + \sigma_O)^2 / 4\sigma_O \tag{9},$$

where θ_V is the air bubble contact angle at the polymer surface immersed in water (Fig1c) [5,6].

$$\sigma_{S(O)W} = \sigma^P_{SW} + \{(\sigma^d_{SO})^{1/2} - (\sigma^d_W)^{1/2}\}^2 \tag{10},$$

where solid/octane equilibrium interfacial energy dispersive component value $\sigma^d_{SO} \approx \sigma^d_S$ [5,6];

$$_{SO} = \sigma^d_{SO} + \sigma^P_{SO} + \sigma_O - 2(\sigma^d_{SO} \cdot \sigma_O)^{1/2} \tag{11},$$

where σ^P_{SO} - solid/octane equilibrium interfacial energy polar component

$$\sigma^P_{SO} = (\sigma_{OW}\cdot\cos\theta_{WO} + \sigma_W - \sigma_O)^2/4\sigma^P_W \qquad (12),$$

where θ_{WO} – contact angle of the water droplet at the polymer surface immersed in octane (Fig1d) [5,6].

Experimental

The polymers of different type were used:

1) polyolefinketones – strictly olefin/carbon monoxide (CO) alternating copolymers (Fig.2).

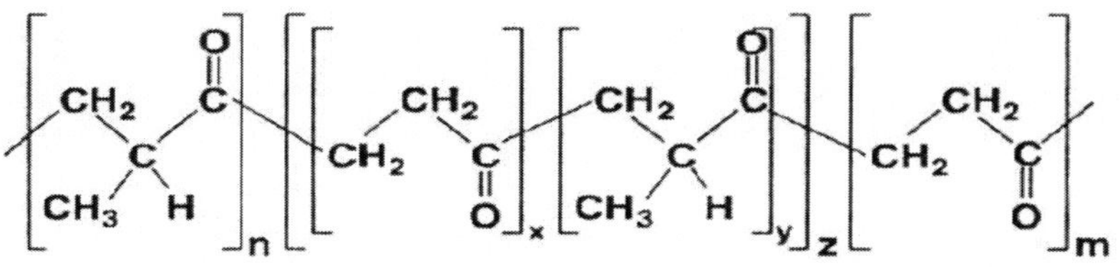

Figure 2. Structural formula of polyolefinketones monomer brunches (a) – PECO, R = CH₃; (b) – BECO, R = C₂H₅.

The butene-1/ethylene/CO (BECO) and propylene/ ethylene/CO (PECO) ternary copolymers were investigated. Physico-chemical characteristics of polymers are presented in Table 1.

Table 1. Physical-chemical characteristics of polyolefinketones

polymer	M	Tg, °C	Tm, °C	α,%
PECO	$M_w = 255000$	10	145,4	8,15
BECO	$M\eta = 38000^*$ [11]	-7,9 44,8	151,7	4,10

2) epoxy-novolac resin (ENR) (Fig.3a) modified by polyamide acid (PAA) based on resorcin diamine and oxydiphthalic dianhydride (Fig. 3b).

Films of BECO and PECO were prepared by coating to carrier surface (alumina plate) from 3% chloroform solution and drying in air atmosphere during 24 hours at the room temperature. ENR with various content of PAA was deposited from solutions (solvent – mixture of ethanol, acetone and dimethyl formamide), then hardened during six hours in 160°C [6].

Contact angles were measured at the closed chamber using horizontal microscope with measurement accuracy $\Delta\theta = \pm1\deg$. The droplet (or bubble) volume was (10÷20) µl. For the each point, the contact angles of (7÷10) droplets were measured. The accuracy of surface energy determination was $\Delta\sigma = \pm0,7$ mJ/m².

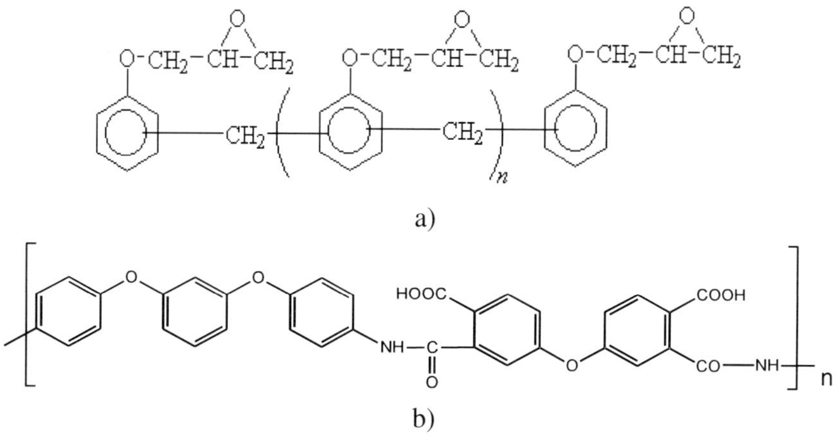

a)

b)

Figure 3. Structural formula of ENR (a) and PAA (b).

The polymers work of adhesion to the model liquids was calculated using eq. (2-4). The σ_{SV} values for polymers were determined by two-liquid method (eq. 4, 5) using water and ethylene glycol as the test liquids (Table 2).

Table 2. The test liquids surface tension (σ_{LV})

and its dispersive (σ_{LV}^d)and polar (σ_{LV}^p) components

liquid	σ_{LV}^p, mJ/m^2	σ_{LV}^d, mJ/m^2	σ_{LV}, mJ/m^2
water	50,8	21,8	72,6
ethylene glycol	19,0	29,3	48,3
octane	-	21,8	21,8

The test liquids surface tension ($\sigma_{LV} = \sigma_{SL}^p + \sigma_{SL}^d$) values were determined by Vilhelmy plate method; its polar (σ_{SL}^p) and dispersion (σ_{SL}^d) components were determined using experimental contact angle (θ) values of droplets at the Teflon-4 surface ($\sigma_{SV} = \sigma_{SV}^d = 18$ mJ/m^2[8]) and Girifalco-Good-Fowkes-Young equation:

$$\cos\theta = 2(\sigma_{SV}^d \cdot \sigma_{SL}^d)^{1/2}/\sigma_{LV} - 1 \ [2].$$

The data obtained appeared to be in accordance with literature data [9]. The Wdd, Wpp and Wdp obtained values were compared with tensile strength for composites obtained using those polymers matrix.

Tensile strength (P) of micro plastics (the simplest model of unidirectional tape composite) was investigated. The reinforcing element of micro plastics was carbon fiber «UKN» with linear density 380 tex. The polymer matrix was ENR modified by PAA. Tensile strength of micro plastics was determined by a standard method [10] using apparatus «Tinius Olsen H5KS.» Every sample was measured 80-10 times. The magnitudes of tensile strength were calculated on the cross-section area of micro plastics.

The carbon fibre free surface energy was determined using eq. (5, 6) and experimental contact angles of test liquids at the surface. The contact angles were determined using the meniscus photo imaging obtained with microscope «OLIMPUS BX51» at the 200 multiple extension.

The literature data of tensile strength of single-layer fiberglass plastics based on the glass fabric with BECO and PECO as coupling agents and polymer matrixes with different polarity (polyethylene, polyamide) also were analyzed [11].

The measurements were carried out at 20°C.

Results

Plots of the ENR work of adhesion to model liquids (Wpp and Wdd) and carbon fiber (Wa) versus the PAA content (eq. 2,3 and 1, respectively) are presented in the Fig. 4.

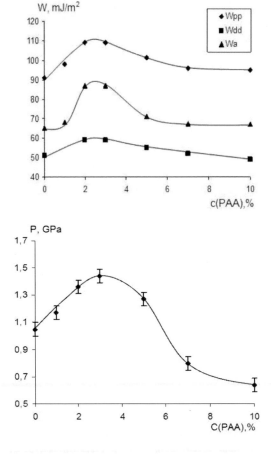

Figure 4. The work of adhesion Wdd (a), Wpp (b), Wa[1] (c) and of tensile strength P(d) dependences from PAA content.

---

[1] Wa was calculated using eq.5,6 and polar $\sigma^p_{FV} = 33,0 mJ/m^2$ and dispersion $\sigma^d_{FV} = 7,1 mJ/m^2$ experimental values for the carbon fibre.

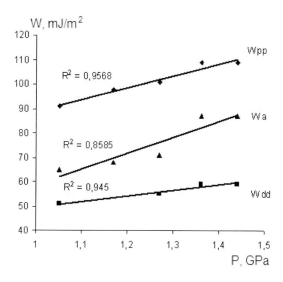

Figure 5. The correlation straight lines between the work of adhesion and tensile strength.

Each curve has a maximum with the value of PAA content 2-3%. The presented dependences correlate to ones of the micro plastics tensile strength (P) based on the carbon fiber from PAA content 2-3%. These results demonstrate that the adhesion interaction of carbon fiber with polymer matrix in micro plastics provided with dispersion and polar intermolecular forces. It might be connected with the presence of polar functional groups on the surface of carbon fiber as a consequence of preliminary oxidation or coupling agent treatment for the improvement its adhesion properties [12].

Scanning electron microscopy method showed[2] that phase separation in the curing system ENR-PAA occurs when the content of PAA more than 5% (Fig.6).

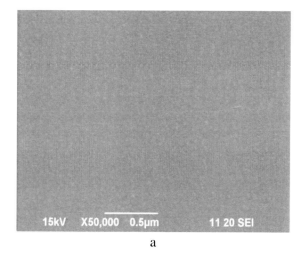

a

Figure 6. (Continued).

---

[2] Scanning electron microscopy investigation was provided on apparatus «Jeol.»

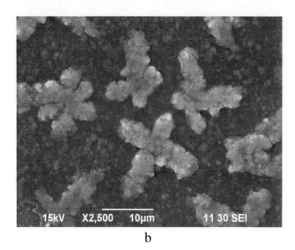

b

Figure 6. The SEM microphotography of curing ENR samples with a) 3% of PAA, b) 5% of PAA.

For lower contents of PAA correlation of work of adhesion to model liquids and tensile strength of micro plastics was set up (Fig5).

The work of adhesion values of polyolefinketones to model liquids and the tensile strength of single-layer fiberglass plastics are presented in tables 3 and 4.

Table 3. The work of adhesion of polyolefinketones to the model liquids

Wa, mJ/m^2	PCO	PECO	BECO
$Wa_{(1)} = \sigma_S + \sigma_W - \sigma_{S(W)W}$	90,9	105,6	108,8
$Wa_{(2)} = \sigma_S + \sigma_O - \sigma_{SO}$	47,0	48,4	57,2
$Wa_{(3)} = \sigma_{SO} + \sigma_W - \sigma_{S(O)W}$	6,1	54,8	40,5

Table 4. The tensile strength of single-layer fiberglass plastics based on the glass tissue with BECO and PECO coupling agents and polyethylene (PE) and polyamide (PA) matrixes: literature data [11]

object	tensile strength P, MPa	
	PE	PA
initial matrix	26±2	56±4
single-layer fiberglass plastic without precoat	60±8	123±6
single-layer plastic with PECO coupling agent	124±14	109±10
single-layer plastic with BECO coupling agent	113±10	156±14

The Wpp value is higher for BECO in comparison to PECO. So, the BECO is the better coupling agent for two polar phase superposition than PECO. The tensile strength of single-layer fiberglass plastics with BECO coupling agent and polar polyamide matrix does more than tensile strength with PECO one. The Wdp value is higher for PECO. This fact demonstrates the better PECO adhesion properties for polar and non-polar phase superposition in comparison with BECO coupling agent. The higher tensile strength values of fiberglass plastics with non-polar polyethylene matrix dressed with PECO in comparison to BECO coupling agent also prove the prediction using wetting method data.

Conclusion

The new approach to the polymer composite materials strength prediction using contact angle measurements was developed. It was founded that the work of adhesion of polymers to model liquids correlates with tensile strength of the different type polymer composites. It is necessary to note that the problem of adhesion is very complicated because the optimum of adhesion maybe exists for each kind of polymer matrix and fibers [1]. Nevertheless, the results obtained show the contact angle measurements are perspective to express analysis of the different compositions «connective-filler» and permit to find the optimal composition of polymer binders.

References

[1]   Gorbatkina Yu.A. Adhesive Strength in Fibre–Polymer System, *Khimiya*, Moscow, 1987; Ellis Horwood, New York, 1992.

[2]   Adamson A.W., Gast A.P. *Physical Chemistry of Surfaces*, A Wiley-Interscience Publication, 1997.

[3]   Van Oss C.J., Good R.J., Chaudhury M.K. The role of van der Waals forces and hydrogen bonds in "hydrophobic interactions" between biopolymers and low energy surfaces // *J. Colloid Interface. Sci.* 1986. v.111. p.378-392.

[4]   Vojtechovska J., Kvitek L. Surface energy – effects of physical and chemical surface properties//*Acta Univ. Palacki. Olomuc*, 2005. Chemica 44. P.25-48.

[5]   Ruckenstein E., Gourisankar S.V. Environmentally induced restructuring of polymer surfaces and its influence on their wetting characteristics in an aqueous environment//*J. of Colloid and Int. Sci.*, 1985, Vol.107. P.488-502.

[6]   Ruckenstein E., Gourisankar S.V. Preparation and characterization of thin film surface coatings for biological environments//*Biomaterials*, 1986. Vol.7. P.403-422.

[7]   Ruckenstein E., Lee S.H. Estimation of the equilibrium surface free energy components of restructuring solid surfaces//*J. of Colloid and Int. Sci.*, 1987. Vol. 120. P. 153-161.

[8]   Summ B.D., Goryunov Yu.V. Physico-Chemical Fundamentals of Wetting and Spreading, *Chemistry*, Moscow, 1976.

[9]   Lee L.H. Correlation between Lewis Acid–Base Surface Interaction Components and Linear Solvation Energy Relationship Solvatochromic α and β Parameters //*Langmuir*, 1996. Vol.12. P.1681-1687.

[10] ISO 10618:2044(E). Carbon fibre – Determination of tensile properties of resin-impregnated yarn.

[11] Smirnov Yu.N., Golodkov O.N., Olkhov Yu.A., Belov G.P. On the effect of polyketone structure on the strength properties of thermoplastic-matrix fiber glass composites // *Vysokomol. Soedin.*, 2008. Vol.50 P.199-207.

[12] Simamura S. Carbon Fibre, *Mir*, 1987.

In: Handbook of Research on Nanomaterials, Nanochemistry ...   ISBN: 978-1-61942-525-5
Editors: A. K. Haghi and G. E. Zaikov   © 2013 Nova Science Publishers, Inc.

Chapter XV

Priority Areas of Nanotechnologies Implementation: Hydroponic Crop Production

Yu. B. Chetyrkin[1], E. M. Basarygina[2] and P. M. Trushin[3]
[1] Rector, Candidate of Engineering Science, associate professor;
Chelyabinsk state agriengineering university, Russia
[2] Doctor of Engineering Science, professor;
Chelyabinsk state agriengineering university, Russia
[3] General director, Candidate of Engineering Science;
Public Limited Company "Teplichniy," Russia

Introduction

At the present stage, only active innovation should provide operational effectiveness of agro-based industries. The same is true of crops protected ground. Implementation of the latest advances in science technology and best practices can lead to improvement in innovative crop production, such as the developments in the fields of nanotechnology and nanomaterials.

The main purpose of the use of nanotechnology and nanomaterials in hydroponic crop production is to ensure high yield and product quality, coupled with rational use of energy and resources (Fig. 1).

In the context of a protected ground, where the main microclimatic parameters are regulated, the problems of root environment meeting the requirements of plants and allowing regulation of their mineral nutrition come to the forefront. In selecting priority areas for implementation of these technologies, it is also necessary to take into account the social value of output products, as food must not only meet physiological needs of human beings in the life-essential substances but also carry out environmental-preventive and therapeutic targets [4; 5].

The problems of protecting the population against the effects of various damaging factors and the search for means of prevention are very important. At present, it is common knowledge that a universal mechanism of various damaging effects is the intensification of oxidative radicals (PDS), which gains an unmanageable avalanche-like character. This process occurs with aging [3, 4]. In the first place, PDS damage cell membranes, their lipid structure, resulting in a violation of homeostasis. Uncontrolled formation of PDS is counteracted by the system for the protection of the organism, which includes a number of components of food. Food protectors are different substances (Table 1) [4].

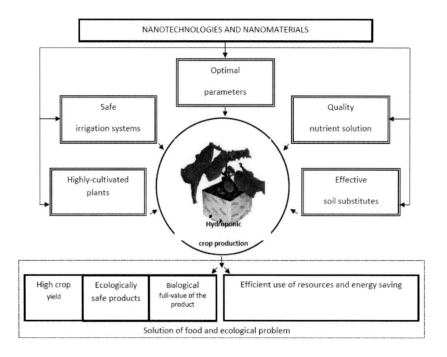

Figure 1. Usage of nanotechnologies and nanomaterials in hydroponic crop production.

Table 1. Protective effect of food protectors [4]

Protectors having this effect	Principal forms of protectors effect
Vitamins E, C, P, lipoic acid, sulfur, taurin, selenium	Restoration of FOR, interruption of reaction of peroxide oxidation
Se, ferrum, manganese, zinc, iodine, vitamins C, B_6	Biosynthesis of internal protectors, their activation
Folic acid, vitamins A, C, B_{15}, choline, inosite, carnitine-sitosterol, ballast substances	Improvement of detoxicating function of protective systems of organism, normalization of cholesterine exchange, other lipids
Vitamins P, C, A, PP, E, B_2	Providing the function of barrier physiological mechanisms
Carotenoids, vitamins A, E, C, selenium, kalium, calcium, iodine, ballast substances	Protection against cancer
Biologically valuable proteins, essential higher polyunsaturated fat, practically all vitamins, vitamin-like substances, essential mineral substances	Restoration of destroyed cell structures

Leafy vegetables: dill, cress, perennial onions, leaf lettuce, sorrel, parsley, beets, garlic, etc., are most effective for growing from the point of view of high content of ecologically protective elements and low energy costs [6]. A flow thin layer hydroponics is widely used when growing green vegetables. The advantage of this method is to create optimal conditions for the growth of root systems: a sufficient amount of moisture, nutrients and providing oxygen to the air. All this helps to obtain high yields of vegetable crops [1, 3]. However, this method of growing plants does not protect the root systems from pathogens that adversely affect the quantity and quality of products. Lack of soil with a high buffering leads to the fact that harmful microorganisms are transferred smoothly through the nutrient solution to the root system and affects the entire coenosis, causing crop losses. The capacity to protect plants during the growing season is limited, as the use of chemical protection is prohibited.

Results and Discussion

It seems appropriate to use nano-filters containing natural (zeolites) and specially produced nanomaterials (nanoparticles of silver) for clearing the nutrient solutions.

Research in this area was carried out in conjunction with Concern «nanotechnology» (Moscow), and an industrial method of obtaining silver nanoparticles in organic solvents, in water and in water-alcohol mixture under the trademarks AgBion-1, AgBion-2 and AgBion-3, was designed and developed. A special feature of these products is that the silver nanoparticles have spherical shape, and their size is in the range 9 - 15 nm. Studies conducted by the specialists of the group and outside organizations have shown that these particles are the most effective for the destruction of pathogens [7].

Chemical composition of zeolites obtained with X-Ray Fluorescence Spectrometry EDX, type 720, Shimadzu, is presented in Table 2.

Table 2. Chemical content of zeolites

Analyte	Resalt,%	Proc.-calc.	Line	Int. (cps/µA)
SiO_2	68, 394	Quan-FP	SiKa	0,1906
Al_2O_3	12, 134	Quan-FP	AlKa	0,0062
K_2O	6,220	Quan-FP	K Ka	0,4124
Fe_2O_3	5,388	Quan-FP	FeKa	1,4259
CaO	4,559	Quan-FP	CaKa	0,3758
SO_3	1,367	Quan-FP	S Ka	0,0188
TiO_2	0,878	Quan-FP	TiKa	0,1016
P_2O_5	0,719	Quan-FP	P Ka	0,0034
SrO	0,137	Quan-FP	SrKa	1,7636
MnO	0,070	Quan-FP	MnKa	0,0178
Rb_2O	0,039	Quan-FP	RbKa	0,4848
ZrO_2	0,038	Quan-FP	ZrKa	0,5010
ZnO	0,028	Quan-FP	ZnKa	0,1354
CuO	0,019	Quan-FP	CuKa	0,0765
Y_2O_3	0,007	Quan-FP	Y Ka	0,0872
NiO	0,005	Quan-FP	NiKa	0,0144

Schematic diagram of the developed technology hydroponic cultivation of green vegetables, including disinfection of nutrient solution, is shown in Fig. 2 [8, 9].

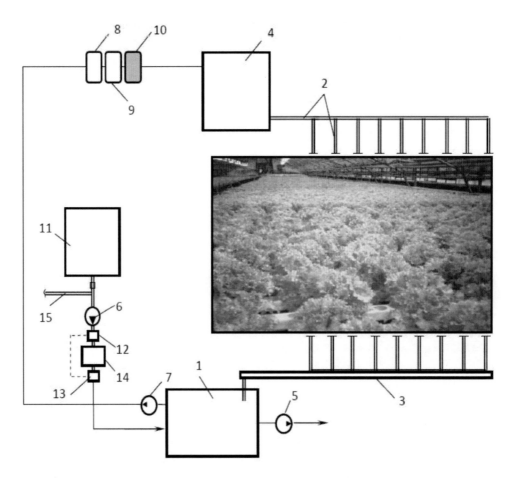

Figure 2. The scheme of the elaborated technology of leaf vegetable cultivation including disinfection of nutrient solution through nanofilter: 1 – storage tank for nutrient solution; 22 - pipeline; 3 – vegetation surface; 4 – receiving tank; 5, 6, 7 - pump; 8, 9 - filters; 10 – filter with nanoparticles of silver; 11-container with concentrated solution of fertilizers; 12- adjusted valve; 13- transducer of sensor for concentration measurement ; 14- mixing chamber; 15- pipeline for water supply.

Seedlings of green vegetables are placed in a tray at the vegetation surface 3. A nutrient solution is fed to tank 1, obtained from chamber 14 by adding water to the concentrated solution of fertilizer. In doing so, the composition of nutrient solution may be adjusted using the adjustable valve 12, connected to a concentration sensor 13. After filling the reservoir 1 to the desired level, the pump 5 is switched off. Nutrient solution pump 7 is fed into the receiving vessel 4, and thence by gravity comes to the vegetation surface 3. From the vegetation surface 3, the solution flows into tank 1 and then re-enters the receiving tank 4. Before entering growing surface 3, the nutrient solution is cleared of algae and impurities by using the filters 8, 9 and gets decontaminated by filter 10, containing nanoparticles of silver.

Among the priorities are also included studies related to the purification of natural water used in the preparation of nutrient solutions (Fig. 3). This is due to the fact that when growing without soil, physiological development and productivity of plants is largely determined by the quality of nutrient solutions.

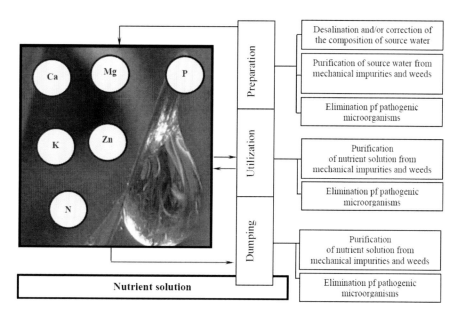

Figure 3. Technological operations related to preparation, utilization and dumping of nutrient solutions.

For small-scale culture on mineral substrate, an irrigation system was proposed, including water desalination and removal of pathogens by using a reverse osmosis membrane module (Fig. 4). The reverse osmosis membrane module was designed by OOO «NPP« TRIS» (Moscow) on the basis of technical specifications issued by Chelyabinsk State Agro-Engineering University (CSAU). Technical characteristics of the membrane clearing unit are presented in Table 3.

Table 3. Technical characteristics of the membrane clearing unit

System capacity:		
- source water, maximum	l/hour	120
- clarified water, minimum	l/hour	70
Degree of water unitilaztion, maximum	%	60,0
Installed capacity, maximum	kW	1,5
Output pressure, maximum	MPaG	1,6
Overall dimensions of membrane module	mm	680*260*1500
Guaranteed service period of membrane elements	years	1
Operating life of membrane elements	years	3
Reagent concentration for washing of membrane elements:		
- sodium	kg/year	2,0
- hydroxy tricarballylic (hydrochrolic) acid	kg/year	2,0(0,2)

Because of the severe pollution of entire agricultural areas and their unsuitability for production of products, a problem occurs with growing crops in artificial conditions. These studies were conducted with CSAU together with St. Petersburg State University of Technology and Design (SPSUTD). On the basis of previously performed works, the staff of SPSUTD proposed technology of gas-filled composite material based on the PAN-fibers and polivinilformal.

Ionite soil can be regarded as a highly effective type of artificial growth media for growing plants. They have a high margin of nutrients and allow plant cultivation for long periods without fertilization.

Ionite soils have a lot of valuable properties: high fertility, excellent agrochemical characteristics, ease of regeneration and sterilization are clearly defined and easily controlled chemical composition. Growing plants in ionite soil is just as easy as in natural soils, and in this respect, the first choice over the majority of hydroponic technology.

The advantage of ionite soil is that it can take any shape, does not require containers and may be in the form of fabrics, rugs or non-woven material [10].

Fine fibrous ionite served as an active filler in gas-based composite materials based on polivinilformal. The resulting composite is a porous elastic material, which through its capillary-penetrating structure and sorption capacity filler can be saturated with micro- and macroelements necessary for growth and development of plants (Fig. 5) [10].

The fundamental principle of producing ionite soils consists of the following: the mixture of cationite and anionite is nutrient medium for plants if its composition corresponds to the balance with nutrient solution. All the ions contained in the nutrient solution are accessible for a plant to such an extent that they are able to satisfy its needs. The quantity of the elements required for normal vital activity of plant − 14.

Figure 4. Osmotic-reverse membrane module designed by «NPP «TRIS» Ltd. For Public Limited Company "Teplichniy" (Chelyabinsk).

Figure 5. Gas-filled composite materials on the basis of polyvinylformal and PAN-fibers. To the left–support medium with growing plants, to the right – new support medium [10].

Uptake of ions of biogenic elements by the plant roots basically takes place by ion-exchange mechanism. In case of using ion-exchangeable mediums, this is the only opportunity because osmotically free ions are practically absent in them [10 - 12].

Conclusion

Laboratory and field tests on the presented innovative developments: technologies of leaf vegetable cultivation allowing disinfecting nutrient solution with the use of nanofilter; irrigation systems including water desalination and elimination of pathogenic microorganisms by means of osmotic-reverse membrane module, artificial soil mediums produced with the use of ion-exchange nanosystems are planned to be carried out on the basis of Public Limited Company "Teplichniy" (Chelyabinsk city).

Directions for further research and developments are related to the usage of bactericidal nanosurfacing in green crop cultivation, as well as nanosurfaces possessing enhanced adhesive strength in droplet irrigation of vegetable crops. This research is expected to be conducted jointly with the specialists of Izhevsk state technical university.

References

[1]   Autko A.A et al. *Vegeculture of protected.* M., 2006.

[2]   Kiselev V.N. et al. *Vegetable farming abroad.* M., 1990.

[3]   Krug G. Vegetable growing / translated from German. M.: Kolos, 2000.

[4]   Dudenko N.V., Pavlotskaya L.F., Chernykh N.F. et al. On the characteristics of biological activity of food ration // *Food issues.* 1996, № 3, p. 17 – 21.

[5]   Revich B.A. *Environmental pollution and population health.* M., 2001.

[6]   Blednykh V.V., Tsitzer O.Yu., Basarygina E.M. et al. Global agroecological problems: safety of agricultural products. M.: *Eco-Consent,* 2003.

[7]   Chmutin I.A. Nanotechnological developments of the Concern «Nanoindusrty» in the interests of agroindustrial complex // *Nanotechnics,* 2009, № 2 (special edition), p. 9 – 10.

[8]   Patent №69375. Hydroponic installation // *CSAU.* Authors Blednykh V.V., Basarygina E.M., Trushin P.M. et al. 2008.

[9]   Patent №67467. Filter for water purification //*CSAU.* Authors Blednykh V.V., Basarygina E.M., Trushin P.M. et al. 2008.

[10] Shirshova E.P., Kuznetsov A.Yu., Ananieva T.A. Nanostructural materials on the ground of ultra-high molecular polyethylene for hydroponic technologies // *Nanotechnics,* 2009, №2 (special edition), p. 72 – 77.

[11] Ananieva T.A., Kuznetsov A.Yu. Development of sorption action material for filtration of water and nutrient solutions and artificial soil for ornamental plant growing // *Research report to the agreement* № 1/18 dated 18.01.2008. St-Petersburg, 2008.

[12] Ananieva T.A., Kuznetsov A.Yu. Development of artificial soil mediums on the basis of ion-exchange nanosystems // *Research report to the agreement* № 9/08 dated 01.11.2008. St-Petersburg, 2009.

In: Handbook of Research on Nanomaterials, Nanochemistry ...   ISBN: 978-1-61942-525-5
Editors: A. K. Haghi and G. E. Zaikov   © 2013 Nova Science Publishers, Inc.

Chapter XVI

The Effect of Chemical Modification of Polymeric Films on the Structural-Power Characteristics of Their Surfaces

Anna K. D'yakova, Sergey A. Trifonov,
Eugene A. Sosnov and Anatoly A. Malygin
St.Petersburg State Institute of Technology
Technical University, Russia

1. Introduction

Among using polymers, the polyethylene and poly(vinyl chloride) (PVC) are very widespread materials. That is connected with their good dielectric performances, high chemical stability, strength, etc. These materials fall into vinyl type polymers with repeating monomeric unit $[-CH_2-CH_2-]_n$, where in the case of PVC, one of hydrogen atoms is replaced by atom of chlorine. However, in many cases, application of these materials is essentially restricted because of low values of surface energy that causes their poor adhesion performances and surface wettability. Noted properties largely depend on chemical composition and structure of surface layer and can be regulated by two basic expedients: removal of feeble boundary polymer layers and forming of new active centers changing their surface energy [1, 2].

One of methods of formation of new functional structures on the polymeric materials surface is gas-phase modification of solid matrixes, based on the molecular layering principles (ML) [3]. The essence of this method is that on a solid surface, the chemical reactions are carried out in requirements of maximal removal from equilibrium between reagents brought from the outside and functional groups of matrix. Molecular layering has appeared rather effective for regulating macroscopic properties of phenolphormaldehyde, polyamide, epoxy and other polymeric materials. According to IR-spectroscopy results, during chemical modification, grafting of element containing structures to the superficial reactive centers of polymers (olefinic bonds, oxygen-containing groups, hydrogen at tertiary atom of carbon, etc.) takes place. That results in changing of thermal-oxidative, electret, diffusive and other properties of materials [4-7].

The purpose of the present work was the examination of effect of chemical composition of poly(vinyl chloride) and high-pressure polyethylene (HPPE) films, modified by vapors of volatile halogenides (PCl_3, $TiCl_4$, $VOCl_3$ and $Si(CH_3)_2Cl_2$) on energy performances, wettability and topography of polymers surface.

2. Experimental Details

The influence of chemical modification was studied on HPPE films type No 158-03-020 and PVC films «Pentaprint,» which functional use, in many respects, is determined by energy performances, including wettability, of their surfaces. The volatile halogenides of phosphorus, vanadium, titanium and silicon have been chosen as modifying reagents.

Modified specimens carried out in flowing-type reactor in the described requirements [4, 6]. According to chemical analysis data, the content of element-modifiers in polymeric films is about 10^{-3} mmol·cm^{-2}. Thus it is possible to assume that as a result of chemical modification in superficial layer of polymeric materials, the element-hydroxo-, element-oxide-, element-methyl- groups of following structure:

$$
\begin{array}{cccc}
\text{HO}\diagdown\text{P}\diagup\text{OH} & \overset{\text{O}}{\underset{|}{\overset{\|}{\text{P}}}}\diagup\text{} & \text{HO}\diagdown\text{Ti}\diagup\text{OH} \\
 & & \\
\text{O}\!\!=\!\!\overset{\text{OH}}{\underset{|}{\text{V}}}\!\!\diagup\text{OH} & \text{O}\!\!=\!\!\underset{|}{\text{V}}\!\!\diagup\text{OH} & \text{HO}\diagdown\overset{\text{CH}_3}{\underset{|}{\text{Si}}}\diagup\text{CH}_3
\end{array}
$$

are formed.

Surface energetic characteristics of materials determined on the basis of measuring of wetting interfacial angles of the initial and modified films by test liquids (water and glycerin), certain in leak-in requirements. The volume of droplet is equal ~ 10 µl. The equilibrium value of an interfacial angle (θ) is determined from kinetic curves of droplet wetting and flowing as a section of plateau with stationary value of parameter during the fixed time. The average value of an interfacial angle received on the basis of ten measuring was used in calculations. The bias did not exceed 3%. The calculation of surface energy (γ_s), its polar (γ_s^p) and dispersive (γ_s^d) components has been executed by Kaelble-Dan-Fowkes method [10, 11] with use of two wetting liquids:

$$
\begin{cases}
\gamma_{l1} \dfrac{(1 + \cos\theta_{l1})}{2} = \sqrt{\gamma_{l1}^d \cdot \gamma_s^d} + \sqrt{\gamma_{l1}^p \cdot \gamma_s^p}, \\[2mm]
\gamma_{l2} \dfrac{(1 + \cos\theta_{l2})}{2} = \sqrt{\gamma_{l2}^d \cdot \gamma_s^d} + \sqrt{\gamma_{l2}^p \cdot \gamma_s^p},
\end{cases}
$$

where θ_l - leak-in angles of test liquids, and γ_l, γ_l^d, γ_l^p - a surface energy, its dispersive and polar components for test fluids, accordingly.

The polymer film surface morphology was studied by atomic force microscopy (AFM) with Solver P47 Pro (NT-MDT, Russia) scanning probe microscope in tapping mode. Specimens scanned in two regimes: topographic survey and phase contras, allowing receipt of the information about material's surface properties. At phase contrast regime, the analysis of phase shift changes ($\Delta\varphi$) of cantilever amplitude-frequency characteristics, caused by adhesion bonds, allows revealing differences in structure of specimen surface sections [12].

3. Results and Discussion

The values of interfacial wetting angles measured for water and glycerin on initial and modified specimens are presented in the Table.

Table. Interfacial wetting angles (θ) for PVC and HPPE films

Element-modifier	PVC		HPPE	
	Water	Glycerin	Water	Glycerin
-	88	72	98	82
phosphorus	36	43	66	73
vanadium	64	68	56	65
silicon	96	85	113	99
titanium	94	83	93	83

As follows from data of the Table, the surface of PVC initial samples is more polar in comparison with HPPE. Lower values of poly(vinyl chloride) wetting contact angles possibly are caused by presence of chlorine-containing groupings at polymer macromolecules. The treatment of the films by phosphorus chloride and vanadium oxychloride vapors promotes increase of surface hydrophily. After modifying polymeric matrixes by dimethyldichlorosilane, their surface becomes more hydrophobic. It is necessary to note that after the treatment of PVC and HPPE films by TiCl$_4$ pairs, wetting contact angles have the identical values (see Table). According to AFM investigations presented in [7], synthesis of titanium-oxide structures on the surface of composite material based on polyethylene leads to formation of uniform covering. Similar changes of surface topography of modified films, apparently, promote alignment of power condition of polymeric matrixes.

On the basis of the measured wetting contact angles, the surface energy and its components for the initial and modified samples have been calculated. The results of calculation are presented in Figures 1 and 2.

As follows from the data shown in Figures 1 and 2, the surface of initial polymers has non-polar character. The basic contribution to the surface energy (95% for PVC and 99% for HPPE) brings in dispersive component, caused by Van der Waals' forces. It is possible to explain presence of an insignificant polar component presence in polymers of defect structures. So, for example, the content of olefinic bonds in HPPE makes 0.3-0.5 on 1,000 carbon atoms [13], in PVC - 0.4-6 on 1,000 monomer units' [14]. Besides at polymers macromolecules, there are oxygen-containing groups of various types, which concentration makes 0.01-0.1 groups and 0.5-1 groups on 1,000 monomer units for HPPE and PVC accordingly [13, 14]. Apparently, not only presence of chlorine in polymer, but also higher content of chemical defects in the polymer structure results in higher surface energy of PVC films.

After the treatment of the films by phosphorus chloride, not only increment of surface energy in 1.8 times for PVC and in 1.4 times for HPPE but also disproportionation of values between its components (figs. 1, 2) was observed. The basic contribution in the total energetic characteristics brings by polar component, which increases more than 30 times for PVC and more than 100 times for HPHP. Magnitude of dispersive component of surface energy decreases in ~7 times and more than in 30 times for PVC and HPPE, accordingly. Such change of modified specimens' energy characteristics values is caused, apparently, by introduction to surface layer polar phosphorus-oxygen-containing structures, which actively interact with the test liquids due to electrostatic forces and formations of H-bonds and donor-acceptor bonds [5, 15, 16].

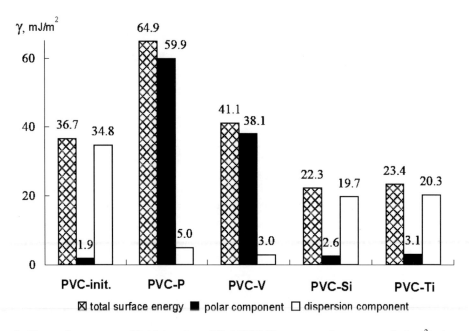

Figure 1. The surface energy of initial and modified PVC films. γ - surface energy (mJ·m^{-2}); the same on Fig. 2.

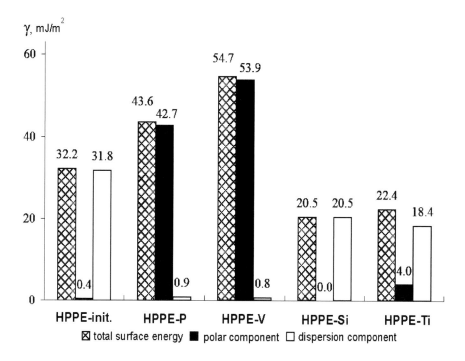

Figure 2. The surface energy of initial and modified HPPE films.

Analogous trends in surface energy and its components changes are shown for the polymer specimens modified by vanadium-containing structures. However, it is necessary to note that in case of HPPE, the magnification of energy parameter in 1.7 times appears more essential in comparison with the specimens handled by phosphorus chloride. For PVC films with vanadium-containing structures, the increasing of surface energy in comparison with an initial film is insignificant and makes only 10% (figs. 1, 2). According to literature data [5], the chemical grafting of vanadium-containing structures is carried out not only due to a leakage exchange but also red-ox interacting of $VOCl_3$ with the surface reactionary-capable censers of polymeric matrix. Apparently, unlike PVC, the processing by vanadium oxychloride vapors of polyethylene films, which is more inclined to oxidizing processes, results in raising of oxygen-containing groups in surface layer, which brings the additional contribution in the growth of polar component.

The gas-phase treatment of poly(vinyl chloride) and polyethylene films by dimethyldichlorosilane vapors results in decrease of surface energy values on 40% in comparison with initial matrixes. Introduction to polymers surface layer of non-polar silicon-methyl-containing structures ($=Si(CH_3)_2$) promotes decrease of dispersive component almost on 30%. But the magnitude of polar component for both polymers varies slightly (Figs. 1, 2). The calculation of surface energy for the polymers modified by titanium chloride vapors has shown that their surface also possesses lower energy than surfaces of initial polymers. The basic contribution (near 90 and 98% for PVC and HPPE, accordingly) in the total surface energy magnitude brings in dispersive component (Figs. 1, 2).

Studying of films with AFM using has allowed establishing that the surface of initial polymers is covered by the crystallites alloyed among themselves, with following typical dimensions: lateral 0.5-1.0 μm both 0.2-0.5 μm and altitude 10-20 nm and up to 40 nm for

PVC and HPPE, respectively (Figs. 3A, 4A). Scanning of materials surface of in phase contrast regime (figs. 3B, 4B) has not shown any areas that essentially differ in adhesion characteristics. The roughness of the HPPE surface layer, determined on section 10×10 μm^2 according to DIN 4768, makes ~32 nm. That exceeds magnitude of this parameter for PVC film (8 nm) and is caused by presence of topographical defects on the surface of polyethylene matrix (Fig. 4B). Last condition, probably, increased phase shift value ($\Delta\varphi$), which is ~30° and 1° for HPPE and PVC, accordingly.

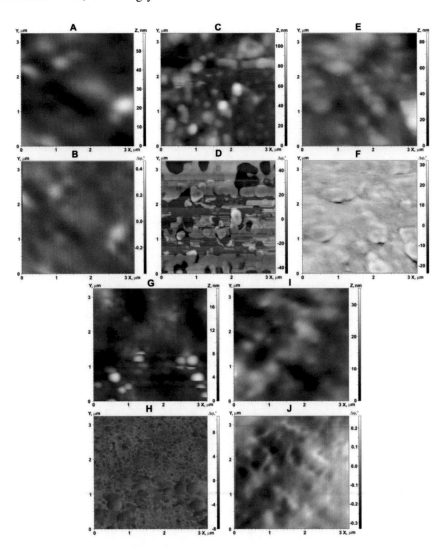

Figure 3. AFM images of PVC films surface.

A, B - PVC-init.; C, D - PVC-P; E, F - PVC-V; G, H - PVC-Si; I, J - PVC-Ti

A, C, E, G, I - topographical representation;
B, D, F, H, J - scanning in phase contrast regime; the same on Fig. 4.

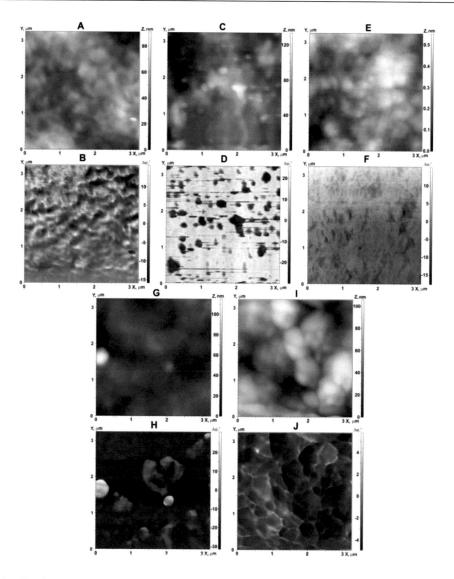

Figure 4. AFM images of HPPE films surface.

A, B - HPPE-init.; C, D - HPPE-P; E, F - HPPE-V; G, H - HPPE-Si; I, J - HPPE-Ti.

The modification of polymers by PCl_3 vapors results in changing of specimen's surface structure. On AFM-images in topography regime (Figs. 3C, 4C) appear the separate sections with lateral sizes of 200-300 nm, which adhesion characteristics essentially ($\Delta\varphi = 80°$ and $70°$ for PVC and HPPE, accordingly) differ from other surface. Thus in the phase contrast regime, it is not possible to fix precisely the boundary line of these structures, because at cantilever moving the partial conduction of material from these sections on matrix's surface to the bordering on sections (Figs. 3D, 4D) is observed. The specified phenomenon can be caused by that phosphor-oxide-structures formed during chemical modification on the film surface possess tall hydrolytic activity [17], which results in formation of hydrated shell around the grafted groups due to water vapor adsorption from air. The roughness of modified polymers surface changes slightly and makes ~15 nm (PVC), ~29 nm (HPPE).

According to [5], at the treatment of polyethylene by vanadium oxychloride vapor, the partial oxidation of reactionary-capable centers of polymer is observed. The leakage of the specified transformations on polymer's surface results in emersion of intercrystalline boundary lines and separate deep cavities with lateral sizes up to 200-250 nm in crystallites contact bands (Figs. 3E, 4E). Examination of the modified specimens in phase contrast regime confirms the change of surface chemical composition ($\Delta\varphi$~40-50°) and allows localizing areas of preferential grafting by element-containing structures exactly in the zones with morphology broken by red-ox reactions (Figs. 3F, 4F). Apparently, the consequence of the specified affecting is raise of surface roughness of vanadium-containing specimens in 1.2-2.0 times in comparison with initial films.

The treatment of the polymers by dimethyldichlorosilane vapor results in formation on surface of silicon-containing structures with lateral sizes from 0.3-0.5 μm (Fig. 3G) up to 1.0 μm (Fig. 4G) for PVC and HPPE, accordingly, and height up to ~10 nm. Scanning in the phase contrast regime allows revealing precise boundary lines of the synthesized structures (Figs. 3H, 4H). Thus the adhesion characteristics of the specified formations slightly differ from parameters of an initial surface of polymers, and magnitude of phase shift makes 6.5° (for PVC) and 30° (for HPPE).

As well as in case of modified by dimethyldichlorosilane specimens, the chemical grafting of titan-containing groups promotes burnishing of polymeric films landform, and that results in decreasing of surface roughness up to ~3 nm and ~29 nm for PVC and HPPE accordingly (Figs. 3I, 4I). Scanning in phase contrast regime does not reveal essential differences in the adhesion characteristics of surface to what decrease of phase shift value to two to three times in comparison with initial polymers and makes ~0.5° (for PVC) and ~10° (for HPPE) (Figs. 3J, 4J). Noted changes can be caused by uniform overlapping of polymeric matrixes surface layers by titanium-oxide groups, which results in equal values of wetting contact angle of polymeric matrixes (see Table).

Thus, AFM investigation of modified films have allowed showing not only differences of surface morphology of initial and modified polymers but also determining correlation between the chemical nature of synthesized element-containing groups and structure of surface layer with energy characteristics of obtained products. It must be mentioned that as a result, gas-phase treatment by volatile halogenides of different elements the topography of polymeric films surface does not undergo significant changes (with the exception for a case of treatment by $VOCl_3$ vapors). The synthesis on surface vanadium- and phosphorus-containing structures results in an increase of phase shift values (which characterize the adhesion properties of surface layer) in 1.3-1.8 times for HPPE and in 50-80 times for PVC. The specified changes testify to emersion on a surface of structures with other chemical nature that (see Table) results in significant decrease of wetting contact angle of the modified specimens, to magnification of superficial energy due to a polar component (Figs. 1, 2) and, in the final, increase wetting ability of polymeric films. The chemical grafting of the silicon-containing structures possessing, due to presence in their composition methyl-groups hydrophobic properties, promotes decrease of phase shift in two and three times for PVC and HPPE accordingly in comparison with initial matrixes. The basic contribution to decrease of modified specimen's surface energy brings its dispersive component (Figs. 1, 2). According to AFM results, the presence at polymers surface layer of titanium-containing groups results in formation of uniform coating that (see Table) promotes leveling of an energy state of polymeric matrixes.

Conclusion

1. The results of directional regulating of surface energy performances and wettability of PVC and HPPE films as a result of chemical modification of polymeric matrixes by vapors of halogenides of phosphorus, vanadium, titanium and silicon are presented. It was established that the synthesis vanadium- and phosphorus-containing groups increase wetting ability of polymer's superficial layer, silicon- and titanium-containing structures promote magnification of surface hydrophobicity.

2. Use of AFM has allowed revealing differences of surface morphology of initial and modified polymers and establishing interconnection of the additive's chemical nature and superficial layer's structure with their wettability. It's shown that the change of polymers surface energy depends on the chemical nature synthesized element-containing groups.

Acknowledgments

Financial support from Russian Foundation for Basic Research (grant No 07-03-00330) and Ministry of Education and Science of the Russian Federation (Program «Evolution of scientific potential of the higher school,» project No 2.1.1/2665) is gratefully acknowledged.

References

[1] V.I. Povstugar, V.I. Kodolov, and S.S. Mikhajlova *Structure and Properties of Polymeric Materials Surface* (rus.) (Khimia, Moscow, 1988).

[2] A.V. *Pocius Adhesion and Adhesives Technology: An Introduction.* (Carl Hanser Verlag, München, 2002).

[3] A.A. Malygin *Rus. J. Appl. Chem.* 69 (10) (1996) 1419.

[4] S.A. Trifonov, E.Yu. Semenova, and A.A. Malygin *Rus. J. Appl. Chem.* 69 (11) (1996) 1735.

[5] S.A. Trifonov, E.A. Sosnov, and A.A. Malygin *Rus. J. Appl. Chem.* 77 (11) (2004) 1854.

[6] A.A. Rychkov, S.A. Trifonov, A.E. Kuznetsov, E.A. Sosnov, D.A. Rychkov, and A.A. Malygin *Rus. J. Appl. Chem.* 80 (3) (2007) 461.

[7] S.A. Trifonov, E.A. Sosnov, Yu.S. Belova, A.A. Malygin, N.G. Razinkova, and G.G. Savkin *Rus. J. Appl. Chem.* 80 (8) (2007) 1413.

[8] V.A. Balandina, D.B.Gurvich, S.M.Kleshcheva, A.P. Nikolaeva, V.A. Nikitina, and E.M. Novikova *Analysis of polymeric plastics* (Rus.) (Khimia, Moscow-St.Petersburg, 1965)

[9] V.N. Muzgin, L.V. Halezina, V.I. Zolotavin, and I.Ya. Bezrukov *Analytical Chemistry of Vanadium* (Rus.) (Nauka, Moscow, 1981).

[10] F.M. Fowkes *Ind. Eng. Chem.* 56 (12) (1964) 40.

[11] N.B. Mel'nikova, V.I. Ignatov, V.D. Dolzhnikova, and B.D. Summ *Bull. Moscow State Univ., Series 2, Chemistry* (Rus.). 39 (6) (1998) 413.

[12] S.N. Magonov, V. Elings, and M.-H. Whangbo *Surf. Sci.* 375 (2-3) (1997) 385.

[13] A.V. Polyakov, F.I. Duntov, A.E. Sofiev, N.Ya. Tumarkin, Yu.N. Kondrat'ev, N.M. Domareva, A.L. Goldenberg, V.M. Kobyakov, and V.S. Zernov *High Pressure Polyethylene. Scientific and Technical Bases of the Production Synthesis* (Rus.), (Khimia, Leningrad, 1988).

[14] K.S. Minsker, and G.T. Fedoseeva *Destruction and Stabilization of Polyvinylchloride* (Rus.), (Khimia, Moscow, 1979).

[15] C.J. Van Oss, M.K. Chaudhury, and R.J. Good *Chem. Rev.* 88 (6) (1988) 927.

[16] C.J. Van Oss, R.J. Good, and M.K. Chaudhury *Langmuir* 4 (4) (1988) 884.

[17] J.R. Van Weser *Phosphorus and Its Compounds* (St. Louis, 1958).

In: Handbook of Research on Nanomaterials, Nanochemistry ...   ISBN: 978-1-61942-525-5
Editors: A. K. Haghi and G. E. Zaikov   © 2013 Nova Science Publishers, Inc.

Chapter XVII

Nanoparticle Finishes Influence on Color Matching of Cotton Fabrics

G. Rosace[1,], V. Migani[1], C. Colleoni[1], M. R. Massafra[2] and E. Sancaktaroglu[3]

[1] Dipartimento di Ingegneria Industriale - Università degli Studi di Bergamo, Italy
[2] Stazione Sperimentale per la Seta, Milano, Italy
[3] Department of Textile Engineering – Uludag University, Bursa, Turkey

1. Introduction

Cotton is one of the major textile fibers, and it has a unique combination of properties, including softness, durability, high strength, good dyeability and biodegradability, and for many centuries, it has found use in textile production [1]. Reactive dyes with vinylsulphone groups are widely used to dye cotton fibers because of the simplicity of application, the great choice of commercial products and their cheapness.

Even though a long tradition has given a solid and an in-depth knowledge of cotton textile fibers and of dyeing processes, the new research borders are moving to a development of the inherent textile materials properties; for this purpose, chemical finishing procedures are widely used. Textile materials can be treated with different functional finishes, such as water and oil repellent, durable press, soil-release, flame retardant, antistatic, and antimicrobial [2,3].

Water repellent finishing on fabrics is mostly imparted by the incorporation of low surface energy compounds, accompanied by the increase of the contact angle of liquids on its surface. The most recent approaches to improve repellency are based on the use of nanoparticles, such as highly branched 3D surface functional macromolecules called dendrimers, whose effect mechanism depends on being in a position to build-up crystal structures in nano-range, which produce wash-permanent, water-repellent and highly abrasion-resistant effects. When combined with fluoropolymers, dendrimers force them to co-crystallize leading to a self-organization of the whole system and to an enrichment of the fluoro polymers on the most outer layer of the textile [3].

Functionality and properties of dendrimers can be changed by filling their cavities or modifying the core and chain-ends [4]. Conformational flexibility of branches is capable of placing dendrimers hydrophilic interior in contact with aqueous subphase and extending their chains into the air above the air–water interface [5].

Particle size of these repellent finishes plays a vital role because when the inorganic particle size is reduced, the surface area is increased; this leads to good interaction with the matrix polymer, and a highest performance is achieved [6]. Alteration of surface properties by textile finishing applications and creation of a smoother reflection surface by a reduced superficial particle size could also give a color change [7].

In this study, the effect of particle sizes on surface roughness and color assessment after finishing of cotton fabrics was evaluated by surface reflectance, absorbance of finishes and color coordinates measurement. For this purpose, three types of commercially available dendrimer water repellent (DWR), fluorocarbon included dendrimer water-oil repellent (DWOR) and fluorocarbon water-oil repellent (FWOR) reagents were impregnated to dyed cotton fabrics and polymerized under optimum conditions. The dyeing of the fabric samples was carried out by three different color commercial dyes.

The reflectance and color coordinates of treated and untreated samples were measured by reflectance spectrophotometer according to the CIELAB, a CIE defined color space, which supports the accepted theory of color perception based on three separate color receptors in the eye (red, green and blue) and is currently one of the most popular color spaces [8].

Finally, water and oil repellency performances, treated-substrate characterization and fabrics mechanical properties were extensively investigated to estimate if finishing agent application gives rise to other changes besides color alterations [9].

2. Experimental

2.1. Fabrics

Scoured and bleached woven cotton fabrics (68 g/m^2) were employed in this study. Each sample (10 g) was immersed for 30 minutes at 60°C in 5 g/L ECE solution (a detergent free from fluorescent brightening agent, used in the ISO 105 series of color fastness test) with 0.5 g/L soaking agent, in a conical flask with continuous shaking. The samples were then thoroughly rinsed and dried at room temperature and stored at laboratory conditions (25 $\pm$ 2 °C and 65 $\pm$ 2% relative humidity).

2.2. Dyes

Three different colors of commercial Remazol (Dystar) reactive dyes with vinylsulphone groups (Table1) were used. The dye selection has been made according to the behaviour of finishing chemicals in their maximum reflectance intervals.

Table 1. Reactive dyes used

Commercial name	λ_{max}	C.I. generic name
Remazol Yellow Gold RNL	413	Reactive Orange 107
Remazol Red F3B	541	Reactive Red 180
Remazol Black B	600	Reactive Black 5

The molecular structures are reported in Fig. 1. All dyes were of commercial grade and were used as received.

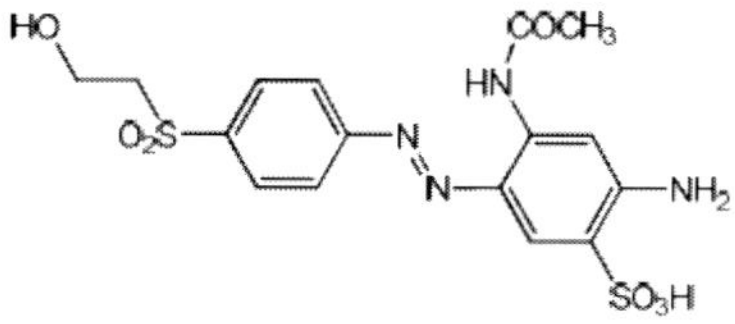

A

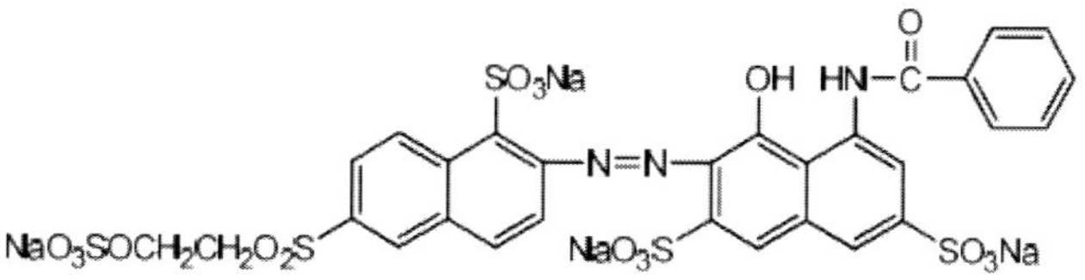

B

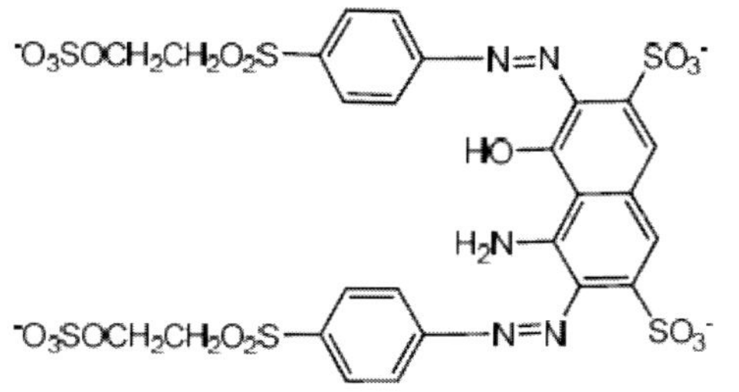

C

Figure 1. A: Reactive Orange 107 (Remazol Yellow Gold RNL). B: Reactive Red 180 (Remazol Red F3B). C: Reactive Black 5 (Remazol Black B).

2.3. Dyeing Agents

For level and consistent dyeing of cotton fabrics, Bersol CM (1g/L), Berdet WF(1g/L) (sequestering and wetting agents, respectively, supplied by Europizzi - Urgnano, Italy), and Depsolube ACA (1g/L) (anti-creasing agent supplied by Basf Italy) were used. All chemicals were commercial products.

2.4. Repellents

Patented dendrimer water repellent (density = 1.1 g/cm^3), fluorocarbon/dendrimer water-oil repellent (density = 1.03 g/cm^3) and also fluorocarbon water-oil repellent (density = 1.03 g/cm^3) provided by Rudolf Chemie (Turkey and Italy) were used as finishing agents and coded as DWR, DWOR and FWOR, respectively (Table 2).

Table 2. Finishing agent codes

Textile finish	Code
Dendrimer Water Repellent	DWR
Fluorocarbon/Dendrimer Water-Oil Repellent	DWOR
Fluorocarbon Water-Oil Repellent	FWOR

2.5. Buffer

The pH 5.5 buffer used for neutralization after dyeing and finishing application comprised acetic acid (9.60 g dm^{-3}) and sodium acetate (3.56 g dm^{-3}). Chemicals were obtained from Sigma-Aldrich, in analytical grade.

2.6. Dyeing

Untreated fabrics were dyed using method depicted in Fig. 2, according to selected dyes. Reactive dyeing was carried out using a liquor ratio of 10:1. Samples were then washed, rinsed and neutralized at proper temperature levels, to be finally left to dry under laboratory conditions.

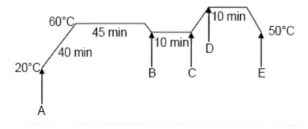

A= LR 10:1; 1% dye shade; 10 g fabric; 1 g/L Bersol CM;
50 g/L NaCl; 12 g/L Na$_2$CO$_3$; 1 g/L Berdet WE; 1 g/L
Depsolube ACA
B= rinsing
C= 0,7 ml buffer solution (pH= 5,5)
D= 1 g/L soaping
E= rinsing

Figure 2. Dyeing profile method for reactive dye used.

2.7. Application of Dendrimers

Dyed cotton samples were padded separately in aqueous baths with finishing chemicals (DWR, DWOR and FWOR). All products were applied to the fabrics with the pad-cure method, including full immersion into the bath to a wet pick-up of 60-80% at 20°C and were cured at 140-150 °C for 120 seconds. The application conditions for the three tested finishes are reported in Table 3; for all products, two different concentrations were investigated, taking into account the concentration ranges recommended by supplier.

Table 3. Dendrimer and fluorocarbon application conditions

Conditions	DWR		DWOR		FWOR	
Concentrations (g/l)	100	125	30	50	30	50
Pick up (%)	60-80		60-80		60-80	
pH	5-5.5		5-5.5		5-5.5	
Fixation temp (°C)	150		150		150	
Fixation time (min)	2		2		2	

2.8. Particle Size Measurement

The finishes particle size measurements were performed with a Malvern Instruments Mastersizer 2000. During the laser diffraction measurement, particles were passed through a focused laser beam and scattered light at an angle that was inversely proportional to their size. The angular intensity of the scattered light was then measured by a series of photosensitive detectors. The number and positioning of these detectors in the Mastersizer 2000 have been optimized to achieve maximum resolution across a broad range of sizes.

2.9. Textile Testing

GSM (Grams per square: the standard measurement of the gravitational force acting on fabrics), TS (Tensile strength: the resistance of a material to rupture when subject to tension) and ELB (Elongation at break: the increase in length when the last component of the specimen breaks) properties of cotton fabrics were also measured for all applications. GSM is a standard measurement of fabric weight per unit area carried out for all samples as follows: the weight of samples with the size of 20 square cm was measured by Mettler balance ($\pm 10^{-4}$g). The percentage change of fabric weight was calculated with (1).

$$Weight\ change\ (\%) = \frac{W_1 - W_0}{W_0} X\ 100$$

(1)

where W_1 is the weight of the substrate after treatment, and W_0 is the initial weight of the untreated substrate. Tensile strength and breaking elongation were measured according to ISO 13934-1, using standard test methods on dynamometer Instron Tensile Tester model 4501 (CRE). Ten specimens were prepared for testing, five for warp direction and five for weft direction. The standard deviations of all experimental results were lower than $\pm 2,5\%$.To thoroughly evaluate both water and oil repellency, above-mentioned performances were measured after abrasion, washing and iron treatments. The samples were conditioned at 20 ⊥ 2 °C and 65 ± 4% RH for 24 hours before textile testing, according to standard ISO 139:2005. All measurements were repeated for the three equally treated samples and then averaged. Water and oil repellency tests were reformulated after 5 and 15 repeated home launderings carried out at 40° C, as described in the ISO 6330, followed by tumbler drying at 80° C for 20 min. Half of the samples were also ironed at 150° C for 60 seconds for observation of heat effect on recovery of repellency. The abrading cycles of the treated and control fabric samples were performed on a Nu-Martindale abrasion tester (Mesdan Italy) according to ISO 12947-3 with two different abrasion cycles (150 and 300). Water resistance test measures the resistance to surface wetting, water absorption and penetration. To determine the extent of surface wetting, the ISO 4920 spray test was used. The substrate was held taut within a 15 cm diameter ring at a 45° angle, and 250 mL of water at a temperature of 23 ± 1°C was lightly dropped onto the substrate from a distance of 15.2 cm to remove excess water. A rating of 80 or higher is desirable, because it indicates better water repellency. Oil repellency was measured according to ISO 14419. In this test, oily drops were deposited onto substrate, allowed to remain for 30 seconds, and removed by wicking or wiping with a paper tissue. Eight different challenge liquids with different liquid tensions are numbered by a rating increasing from 1 in the case of vaseline, the easiest to repel, to 8 in the case of n-heptane, the most difficult to repel (Table 4). Oil repellency is reported as the maximum rating of the liquid that does not wet the substrate. For detailed comparisons, multiple drops of each liquid were tested. In general, an oil repellency rating of 5 or higher is desirable.

Table 4. Definition of oil repellency

Oil repellency rating	Test liquid	γ_L, mN/M
8	n-heptane	19.8
7	n-octane	21.4
6	n-decane	23.5
5	n-dodecane	24.7
4	n-tetradecane	26.4
3	n-hexadecane	27.3
2	35/65 mix n-hexadecane/vaseline	29.6
1	Vaseline	31.5

2.10. Surface Reflectance and Roughness Measurement

The reflectance of non-dyed samples (untreated and treated) was measured with a Lambda 950 Perkin Elmer apparatus, equipped with an RSA-PE-150 Labsphere accessory for reflectance measurements. Each reflectance value R (%) was determined as the average of four measurements, with an experimental error of about 1-2%. For a randomly rough surface having a Gaussian type surface heights distribution, the reflectance can be expressed as follows:

$$Rr = Rs \exp[-(4\sigma\cos i/\lambda)^2] \qquad (2)$$

where Rs and Rr are specular reflectances of perfectly smooth and rough surfaces, respectively; σ is the standard deviation of the surface from its mean level, function of the surface roughness; i and λ are the incident angle and wavelength of light, respectively. Changes in surface roughness of treated fabrics were expressed as differences in the root-mean-square of the vertical Z dimension values within the examined areas, which were calculated using the following equation:

$$RMS_{xy} = \sqrt{\sum_{xy=1}^{N} \frac{(Z_{xy} - Z_{average})^2}{N^2}}$$

(3)

2.11. Absorbance Measurement

The ultraviolet–visible absorption spectra of finishes were recorded on a Thermo Nicolet Evolution UV–Vis 500 spectrophotometer Vision 32 software, using a 1 cm path length cell. All spectra were recorded at room temperature in the wavelength range from 350 to 750 nm at the rate of 300 nm min^{-1}. Finish/dye solutions were evaluated from the change of absorbance at the λ_{max} of the dye in the UV-Vis spectra of the sample solution.

2.12. Color Difference Assessment

Color difference evaluation was based on the surface reflectance in the visible waveband: indeed, any effects that change the reflectance cause a color difference. The color coordinates of dyed cotton (untreated and treated) samples dyestuffs selected for dyeing cotton fabrics were measured on the same spectrophotometer used for absorbance measurements, equipped with an integrating sphere with a 10mm opening, under a D65/10^0 illuminant according to ISO 105 J01. The spectrophotometer measurements were made using the L*a*b* system. The evaluation of the overall color difference (ΔE), obtained before and after finishing, in the 3D color space is given by:

$$\Delta E^* = \sqrt{(\Delta L^*)^2 + (\Delta a^*)^2 + (\Delta b^*)^2}$$

(4)

where ΔL* represents the lightness difference; Δa and Δb, the differences in a and b values, wherein a* is a measure of redness/greenness, and b* is a measure of yellowness/blueness. The a* and b* coordinates approach to zero for neutral colors (white, greys) and increase in magnitude for more saturated or intense colors. The advantage of the CIE-lab system is that color differences can be expressed in units that can be related to visual perception and clinical significance [8].

Four reflectance measurements were made on each sample; the samples were rotated 90° before each measurement, and the averages of the reflectance values (%) at wavelengths between 400 and 700 nm were recorded. A white Thermo Nicolet Electronic calibration plate was used.

3. Results and Discussion

3.1. Particle Size and Surface Roughness Measurements

The results of nanoparticle sizes for DWR, DWOR and FWOR finishing agents are depicted in Fig. 3. DWR presents the greatest particle size, and volume (%) values are about the half of the ones observed for the two other finishes. Besides, DWR shows a second low concentration peak in the particles size range of 10-100 nm, probably due to a self-assembling of dendrimers [10]. From these results, specific surface areas of nanoparticles were determined: 6.6 m^2/g, 38.0 m^2/g and 40.1 m^2/g for DWR, DWOR and FWOR, respectively.

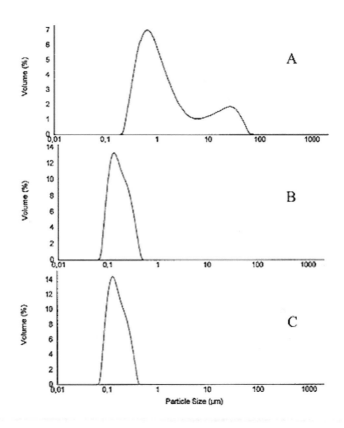

Figure 3. Particle size of DWR (A), DWOR (B) and FWOR (C) finishes.

Table 5 compares treated to untreated samples concentrations and RMS$_{xy}$ values, the latter computed using (3). For the selected chemicals concentrations, the smaller the particle size, the smoother the polymer over the textile surface, as expected. Particle surface area, in fact, increases as the particle size decreases: the highest particle surface area is achieved when there is a high concentration of nanometer or sub-nanometer sized particles. This is the reason why a higher concentration of DWR product is required to obtain the same repellency effects of DWOR and FWOR. Surface roughness reduces the specular reflectance of a wavelength of light, especially at high angle of incidence, and increases the scattering of light while decreasing the surface reflectance.

Textile samples treated with small particle sized chemicals, having a higher surface area, give a more uniform dispersion, a smoother reflecting surface and thus higher reflectance values (%) [8].With the application of finishing agent, it was believed that such treatment increased the inter-yarn and inter-fiber frictional force, as confirmed by the roughening effect on the textile surface.

Table 5. Surface characterization of untreated and treated samples

Samples	Concentrations (g/l)	RMS (Root-mean-square) nm
Untreated	0	25.41
DWR	125	13.7
DWOR	50	6.9
FWOR	50	6.9

3.2. Repellency Measurements

It was observed (Fig. 4) that high water repellency degrees of dendrimers have a tendency to improve when they are combined with fluorocarbon polymers. It was also observed that untreated fabrics showed no repellency behaviour and that, for all different treatments, water repellency was initially the same.

After abrasion and washing processes, a performance decrease was observed, particularly for lower finishing concentrations. In addition, the binding properties of dendrimers to textile surface had acceptable fastnesses.

After oil repellency testing of DWR finishing, no repellency effect was observed. The possible reason is related to the high DWR finishing surface energy in comparison with surface tension of oil (20-35 mN/m). For fluorocarbon groups supplied DWOR and FWOR, a very low critical energy of 6 mN/m can be achieved [11]. The change in oil repellency was shown in Fig. 5. The effect of concentration is evident in FWOR products in comparison with DWOR products, probably due to the orientation of fluorocarbon chains achieved by dendrimers. For this reason, lower concentrations of DWOR are more efficient than FWOR products, which do not include dendrimers [12].

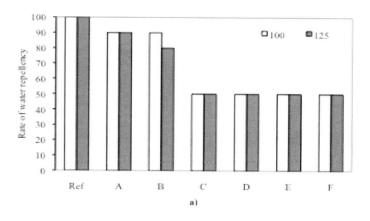

Figure 4. (Continued).

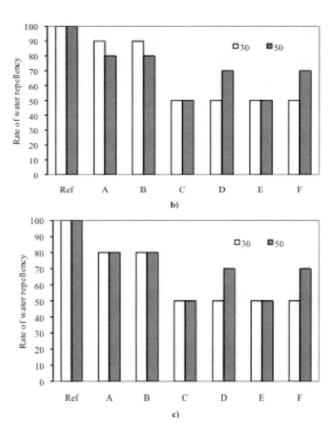

Figure 4. Water repellency a) after DWR finish at 80 g/L and 125 g/L concentrations, b) DWOR finish at 30 g/L and 50 g/L concentrations and c) FWOR finish at 30 g/L and 50 g/L concentrations after: application alone (ref); 150 cycles of abrasion (A); 300 cycles of abrasion (B); laundering (5 cycles) (C); laundering (5 cycles) and ironing (D); laundering (15 cycles) (E); laundering (15 cycles) and ironing (F).

Despite sufficient fixation, textiles finished with fluorocarbon suffer from a clear loss in effect after washing, caused by the unfavourable conditions prevailing in the washing machine. Surface-active surfactants, the extremely polar medium (water), the elevated temperature (at least 40°C) in combination with the mechanical influence lead the orientated fluorocarbon chains "diving away" from the edge or being very much disturbed. After drying, the polymer is still on the substrate, however, but the side chains are in disorder, and for this reason, no oil and water repellent effects are apparent. Since the fluorocarbon chains melting temperature is 80-90°C, the fluorocarbon finish is not able to spontaneously regenerate at room or drying temperature. A simple ironing and thus the heating up of the textile over the melting temperature of the fluorocarbon chains would, however, be sufficient to reorganize the finishing and to reproduce the performance level [11]. In our opinion, the change in water repellency after washing and abrasion seen on DWR products is also probably related to this reason. Anyway, samples before and after dendrimer and fluorocarbon finishing treatment have very similar mechanical properties, as indicated in Table 6, and in a range that is intrinsic to the raw material.

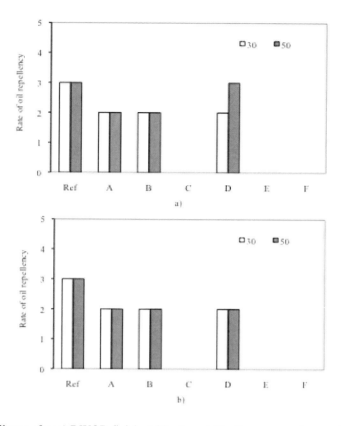

Figure 5. Oil repellency after a) DWOR finish at 30 g/L and 50 g/L concentrations and b) FWOR finish at 30 g/L and 50 g/L concentrations after: application alone (ref); 150 cycles of abrasion (A); 300 cycles of abrasion (B); laundering (5 cycles) (C); laundering (5 cycles) and ironing (D); laundering (15 cycles) (E); laundering (15 cycles) and ironing (F).

Table 6. Mechanical properties of treated fabrics

Cotton		Untreated	DWR		DWOR		FWOR	
Concentration (g/l)		0	100	125	30	50	30	50
GSM (g/m²)		66.80	70.84	70.79	69.29	71.39	69.00	70.05
Weight change (%)		0.0	6.0	6.0	3.7	6.9	4.0	6.7
TS (N)	Warp	223.1	212.2	204.5	235.8	219.3	237.7	220.4
	Weft	144.6	85.5	112.7	152.7	192.3	154.1	189.3
ELB (%)	Warp	7.6	4.5	3.6	4.6	5.5	4.7	5.3
	Weft	14.1	15.8	18.0	23.2	26.0	23.1	25.0

3.3. Color Difference Measurements

Color-difference evaluation in the visible waveband (400-700 nm) computed on the basis of (4) is depicted in Fig. 6 for the three tested finishes on the cotton fabric samples. It was observed that the application of three finishing products have an effect on color change in any color, whether in hue, chroma, or lightness (AATCC).

Table 7. ΔL,* Δa*, Δb* values of reactive dyestuffs used on cotton fabrics

Colour	$\Delta L*$			$\Delta a*$			$\Delta b*$		
	DWR	DWOR	FWOR	DWR	DWOR	FWOR	DWR	DWOR	FWOR
Yellow	0.40	0.84	0.80	0.78	0.30	0.25	0.38	-0.30	-0.29
Red	-0.77	0.37	0.25	-1.86	-1.29	-1.15	-1.01	-1.03	-0.59
Blue	0.34	0.28	-0.39	-0.46	-0.36	0.88	-0.22	-0.15	-0.46

Color difference trends are comparable for all colors, and always show the highest ΔE value for DWR, while the lowest for FWOR finishing. Besides, distinctive behaviours are observed for red color, which presents the highest ΔE for all finishings, and for blue one, whose attitude is decidedly opposing. Color matching as depicted in Fig. 6, and ΔL,* Δa*, Δb* values of dyestuffs selected for dyeing in Table 7 show a remarkable influence of finishing onto color difference of textile.

3.4. Reflectance Difference Measurements

One of the major effects on color difference after finishing can be related to surface roughness change, seen as reflectance difference ΔR between the percentage reflectance of treated fabrics from the one of control fabrics, as indicated in Fig. 7. As Fig. 7 shows, the DWR, DWOR and FWOR finishes have similar curves of reflectance, which stabilize after 530 nm around an almost constant level.

It is shown that the finishing treatment changes the percentage reflectance values of the fabric, depending on the chemical type: samples treated with chemicals that are smaller in particle size (DWOR, FWOR) generally give higher reflectance (%) values for λ>530nm; DWOR and FWOR finishes increase their refractive index, in comparison with the fabric itself, and this behaviour is opposite to the one of DWR. Samples treated with DWR finishing, having greater particle size, exhibit in fact a significant decrease in reflectance (%), especially for short wavelengths, if compared with control fabrics. This pronounced decrease may be related to the height of the surface roughness, which can scatter wavelength of the visible spectrum. The contribution of the particle size distribution and the assignment of the height profile of a treated fabric surface may be correlated to the profile of the reflectance difference plots of the samples [13].

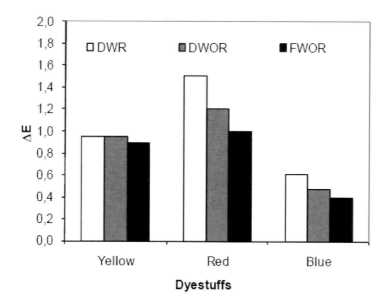

Figure 6. Color differences on cotton fabrics.

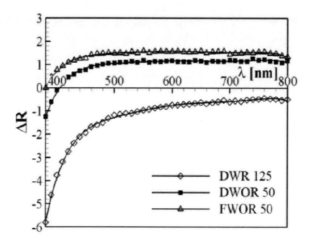

Figure 7. Reflectance differences of three finishing agents at the best concentration selected according to water and oil repellency performance results.

3.5. Absorbance Measurements

Another effect of finishing agents on color difference could be explained by the absorbance plots in the visible range, as shown in Fig. 8. The concentrations selected for the measurement are related with the application concentrations of the products over the textile surface. As shown in Fig. 8, there are two distinct zones for each absorbance curve. The first one is approximately 350-500 nm, in which stronger adsorption of products and a higher difference between each finishing agent occur. The rate of absorbance gradually reduces and so the difference becomes smaller. This is valid under the same concentrations conditions (B, C, D: 0.50 g/L). The absorbance increases for the same product with concentration increasing (e.g., DWR A: 1.25 g/L, D: 0.50 g/L), as expected. The adsorption differences for the three products are depicted in Fig. 9, 10 and 11 for different colors of the same dyestuff group. Equal ratios of the dyestuff and water/product solutions were carried out for each measurement.

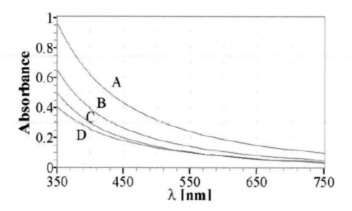

Figure 8. Absorbance spectrum of DWR (A: 1.25 g/L; D:0,5 g/L), DWOR (B: 0.50 g/L), FWOR (C: 0.50 g/L) solutions in the visible range.

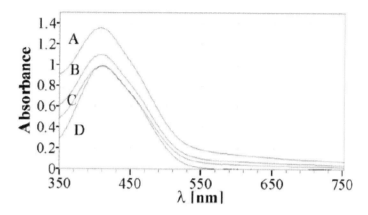

Figure 9. Absorbance spectrum of Yellow Remazol RNL in water solution (D) and in solution with DWR (A: 1.25 g/L), DWOR (B: 0.50 g/L), FWOR (C: 0.50 g/L) products in the visible range.

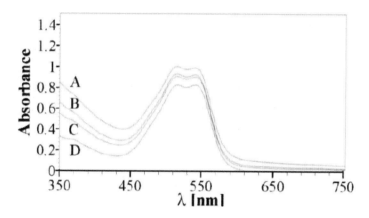

Figure 10. Absorbance spectrum of Red Remazol F3B in water solution (D) and in solution with DWR (A: 1.25 g/L), DWOR (C: 0.50 g/L), FWOR (B: 0.50 g/L) products in the visible range.

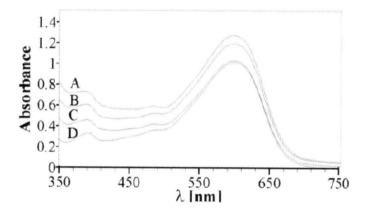

Figure 11. Absorbance spectrum of Black Remazol B in water solution (D:0,1 g/L) and in solution with DWR (A: 1.25 g/L), DWOR (C: 0.50 g/L), FWOR (B: 0.50 g/L) products in the visible range.

For all colors, there is no change in the absorbance spectrum profile of dyestuff solution with finishing agents, but an overlay of the absorbance values in the short wavelength. This means that the dyestuffs do not interact with finishing agents but show a small variation of the λ_{max}, which results with ΔE. Absorbance values at λ_{max} of dyestuff with water/product solutions on measurement concentrations are reported in Table 8.

Table 8. Absorbance values of dyestuff
with water/product solutions at λ_{max}

Absorbance	Dyestuff	DWR	DWOR	FWOR
Yellow Remazol RNL	0.98	1.34	0.98	1.09
Red Remazol F3B	0.82	0.98	0.90	0.92
Black Remazol B	1.00	1.26	1.02	1.18

It was evaluated that red dyestuff has a lower absorbance at λ_{max} than the other two colors. When combined with the finishing products, red dyestuff has lower influence on the increase of λ_{max}, too. The absorbance effect of DWR is the best for all the colors, while DWOR and FWOR show similar absorbance effect except in the case of Black Remazol B. This can be related with the lower adsorption of dendrimer in comparison with fluorocarbon polymers at the same concentration. When fluorocarbon polymers dendrimers are applied, the percentage of fluorocarbon in the product decreases because the same effect of oil and water repellency is achieved with the help of the only dendrimers. That is why absorbance value is always the lowest for DWOR. Looking at absorbance values of FWOR (having only fluorocarbons), fluorocarbons polymers increase the λ_{max} of dyestuff alone more than with dendrimers, and dendrimers alone have the lowest effect on λ_{max}.

3.6. Color Difference Evaluation

Color difference evaluation for the three tested finished, as depicted in Fig. 6, could be also estimated on the base of absorbance and reflectance values considerations. For yellow (λ = 413 nm) dyed and finished cotton fabrics, the effect of absorbance for DWR is the highest when compared with all other colors. This effect is compensated for DWR lower reflectance at its wavelength, and this results in a ΔE similar to the DWOR one. FWOR finishing presents an absorbance value similar to DWOR and the highest reflectance one, so ΔE is lower. For red (λ = 541 nm) dyed and finished cotton fabrics, the absorbance effect is the lowest for all finishes when compared with other colors. Reflectance of DWR is still quite low, but absorbance is lower than the yellow case, so ΔE is the highest and decidedly higher than in yellow dyeing. For all finishes, absorbance values are quite similar, but DWOR and FWOR have an increment of ΔE, due to the reflectance contribution enhancement. For blue (λ = 600 nm) dyed and finished cotton fabrics, absorbance values are intermediate between yellow and red dyeing, while reflectance values are the highest for all finishes. This provokes a fall in DWR, DWOR and FWOR ΔE values. In synthesis, for yellow and blue colors, having high values of absorbance, ΔE increases if reflectance is low, while ΔE decreases with high reflectance values. For red color, having for all finishes the lowest absorbance value and a quite high reflectance value for all finishes, a ΔE enhancement is observed.

Conclusion

Textile fabrics are subjected to various treatments for water/oil repellency, for crease recovery and for a large number of different functionalities, but one of the most innovative effects could be ascribed to the color change induced by finishing operations. Hitherto, the use of dendrimers as textile dyeing auxiliaries is quite unexplored, and their application is still to be optimized; in this context, dendrimer influence on color assessment is worthy of exploration.

This paper has given new results associated with the evaluation of colour assessment of cotton fabric due to nanoparticle sized finishing.

Color matching of dyed samples before and after repellency treatments shows an influence of finishing onto color difference. This can be due to three possible reasons: a change on the surface roughness, an influence of finishing agent's concentration and a dyeing agent/dendrimer interaction or reaction. DWR, DWOR and FWOR finishes particle sizes designate the distribution and orientation of surface layer roughness, and this results in color difference. The reflectance change of the treated cotton fabric seems to be related to the particle size and to the distribution of the chemical applied. In addition, it should be also taken into consideration the absorbance of finishing agents, which in the visible range influences the dyestuff λ_{max}.

Effect of surface roughness of DWR is higher than DWOR and FWOR on color difference because of particle size distribution on the fabric surface. So it was expected to observe the highest ΔE for DWR, if taking into consideration only this major factor. As the absorbance level of DWR is not same for three colors, it was clearly seen the positive effect of absorbance increasing on color difference. Increasing of λ_{max} after finishing chemicals application is always positive (or there is no effect) because of an additional absorbance after applying additives over the textile surface. Although the amount of additives over the cotton surface influences additional absorbance, no interaction with dyes results in the increase on λ_{max}. With a concentration and particle size increasing, the distribution of chemicals over the fabric causes higher surface roughness, so reflectance directly effects on concentration. Because the ability of the human eye to appreciate differences in color differs from individual to individual (as it is a combination of eye characteristics and skill of the operator), for the three different finishes used $\Delta E<1$ values indicated that the color changes were appreciable only by instrumental equipment. Dendrimers finishing on cotton textiles are confirmed as good device for water and oil repellency treatments, and its performances can be compared to the commonly used fluorocarbon products. Finally, mechanical characteristics of the fabric are not considerably altered by finishing treatments; no significant differences were observed in the mechanical properties between treated and control samples.

References

[1]   Nakamura A. (2000) *Fiber science and technology*. Science, USA
[2]   Cerne L. & Simoncic, B. (2004) Influence of Repellent Finishing on the Free Surface Energy of Cellulosic Textile Substrates. *Textile Research Journal, 74*, 426

[3]   Schindler W. D. & Hauser P. J (2004) Chemical Finishing of Textiles. *Woodhead Publishing in Textiles, Cambridge England,* 6 , 79-81

[4]   Tully D. C.& Fréchet, J. M. J. (2001) Dendrimers at Surfaces and Interfaces: Chemistry and Applications., *Chem. Commun.,*1229-1239

[5]   Menger F. M., Peresypkin A. V. & Wu, S. (2001) Do Dendritic Amphiphiles Self-Assemble in Water? A Fourier Transform Pulse-Gradient Spin-Echo NMR Study, *J. Phys. Org. Chem., 14* 392-399

[6]   Mani G., Fan Q., Ugbole S., &Eiff, I.M. (2003) Effect of Nanoparticle Size and Its Distribution on the Dyeability of Polypropylene, *AATCC Rev,* 3, 22

[7]   Qian, L. (2004) Nanotechnology in Textiles: Recent Development and Future Prospect, *AATCC Rev,* 4, 14

[8]   Jang J., Jeong Y. (2006) Nano roughening of PET and PTT fabrics via continuous UV/O_3 irradiation. *Dyes Pigments,* 69 137

[9]   Rosace G, Massafra MR, Iskender A & Sancaktaroglu E. (2008) "Influence of dendrimer finishing on color assessment of CO/PES blended fabrics" Proceedings of the 21st IFATCC International Congress, 6-9/05/2008 Barcellona

[10]   Dykes G.M. (2001) Dendrimers: A Review of Their Appeal and Applications. *Chem. Technol. Biotechnol.,* 76 , 903-918

[11]   Duschek G. (2001) Low-emission and APEO-free Fluorocarbon Finishing, *Melliand English,* 7-8

[12]   Duschek Gunther (2004) "Bionic Finish" *Revista de Quimica Textil* 170, 20-35

[13]   Cem G., Dilek K. & Mehmet O., (2007) Effect of the Particle Size of Finishing Chemicals on the Color Assessment of Treated Cotton Fabrics, *J. of Applied Polymer Science 104*, 2587-2594

In: Handbook of Research on Nanomaterials, Nanochemistry ...   ISBN: 978-1-61942-525-5
Editors: A. K. Haghi and G. E. Zaikov   © 2013 Nova Science Publishers, Inc.

Chapter XVIII

Novel Class of Eco-Flame Retardants Based on the Renewable Raw Materials

A. M. Sakharov, P. A. Sakharov,
S. M. Lomakin and G. E. Zaikov
N.M. Emanuel's Institute of Biochemical Physics
Russian Academy of Sciences, Moscow, Russia

Introduction

Today, as well as it was a lot of centuries back, prevention and extinguishing of fires is one of the global problems standing in front of humanity. In the world annually, thousands of the new reagents appear directed on the decision of this problem. But as we can see, the problems such as, for example, extinguishing of forest fires or creation of low inflammable polymeric materials, are far from the solution.

The combustion of natural and polymer materials, like the combustion process of any other fuel material, is a combination of complex physical and chemical processes, which includes the transformation of initial products. This whole conversion process may be divided into stages, with specific physical and chemical processes occurring in each of these stages. In contrast to the combustion of gases, the combustion process of condensed substances has a multi-phase character. Each stage of the initial transformation of a substance correlates with a corresponding value (combustion wave) with specific physical and chemical properties (state of aggregation, temperature range, concentration of the reacting substances, kinetic parameters of the reaction, etc.).

Char-forming materials often swell and intumesce during their degradation (combustion), and the flame-retardant approach is to promote the formation of such intumescing char. The study of new polymer flame retardants has been directed at finding ways to increase the fire resistance.

There is a strong correlation between char yield and fire resistance. This follows because char is formed at the expense of combustible gases and because the presence of a char inhibits further flame spread by acting as a thermal barrier around the unburned material. The tendency of a polymer to char can be increased with chemical additives and by altering its molecular structure. Polymeric additives (polyvinyl alcohol systems) that promote the formation of char [3-5] have been studied.

These polymeric additives usually produce a highly conjugated system—aromatic structures that char during thermal degradation and/or transform into cross-linking agents at high temperatures.

Char, which is formed in the process of thermal degradation, can play several roles in fire retardancy. The formation of char in and of itself has a significant effect on the degradation because char formation must occur at the expense of other reactions that may form volatiles; thus, char formation may limit the amount of fuel available. An example of this occurs in cellulose, which may degrade either by a series of dehydration reactions that yield water, carbon dioxide, and char, or by a process in which levoglucosan is produced, which eventually leads to the formation of volatiles.

It is believed that the temperature at the surface of a burning polymer is close to the temperature at which extensive thermal degradation occurs (usually 300-600°C). The bottom layer of char, near the polymer surface, is at the same temperature, whereas the upper surface, exposed to the flame, can be as hot as 1,500°C [6].

Therefore, fire-retardancy chemistry is concerned with chars, which may be produced at temperatures between 300° and 1,500°C [7].

Char-forming materials often swell and intumesce during their degradation (combustion), and the flame-retardant approach is to promote the formation of such intumescent char. Intumescent systems, on heating, give a swollen multicellular char capable of protecting the underlying material from the action of the flame. Fire-protective coatings with intumescent properties have been in use for 50 years, whereas incorporation of intumescent additives in polymeric materials is a relatively recent approach [8-10].

On burning, these additives develop the foamed char on the surface of degraded material. The suggested mechanism of fire retardancy assumes that the char acts as a physical barrier against heat transmission and diffusion of oxygen toward the polymer and of combustible degradation products of the polymer toward the flame. Thus, the rate of pyrolysis of the natural and polymeric materials is expected to decrease below flame-feeding requirements, which leads to flame extinguishment. The intumescent behavior, resulting from a combination of charring and foaming of the surface of the burning polymers, is being widely developed for fire retardancy because it is characterized by a low environmental impact.

Apparently, intumescent coatings are the most effective methods of protecting wood and natural materials from fire. When heated, they form a thick, porous carbonaceous layer. This provides an ideal insulation of the protected surface against an excessive increase in temperature and oxygen availability, thus preventing thermal decomposition, which plays a decisive role in retarding the combustion process. Under the heating, intumescent compounds expand up to 200 times their volume and, in some cases, to even 200 times [11]. [Vandersall H L, Intumescent coating systems, their development and chemistry *J Fire Flamm*, 1971, 2 April: 97–140]. Intumescent systems such as paints, lacquers, mastics and linings must contain ingredients that when heated to high temperatures will form large amounts of non-flammable residues. These residues, under the influence of emitted gases, produce foam with good insulating properties. [11,12,13].

Heating of cellulosic materials and wood at 105 -110°C results in removal of moisture. Reactions occurring at this stage proceed slowly and are mostly endothermic. After exceeding the above temperature, a slow thermal decomposition of the components of natural polymers begins, and at 150–200 °C, gas products of the decomposition start to be released.

At 160 °C, decomposition of lignin begins. Lignin under thermal degradation yields phenols from cleavage of ether and carbon–carbon linkages, resulting in more char than in the case of cellulose. Most of the fixed carbon in charcoal originates from lignin. At 180°C, hydrolysis of low molecular weight polysaccharides (hemicelluloses) begins [13]. Thermal stability of hemicelluloses is lower than that of cellulose, and they release more incombustible gases and fewer tarry substances [12]. The most abundant gaseous products contain about 70% of incombustible CO_2 and about 30% of combustible CO. Depending on availability of oxygen, subsequent reactions can be exothermic or endothermic. In the temperature range of 220–260 °C, exothermic reactions begin. They are characterized by evolution of gaseous products of decomposition, release of tarry substances and appearance of ignition areas of hydrocarbons with low boiling points. Cellulose decomposes in the temperature range between 260 and 350°C, and it is primarily responsible for the formation of flammable volatile compounds. [12].

At temperatures about 275–280°C, accelerated release of considerable quantities of heat begins and increased amounts of liquid and gaseous products (280–300°C), including methanol, acetic acid are formed. The amount of evolving carbon monoxide and dioxide decreases, mechanical slackening of wood structure proceeds and ignition occurs. Mass loss of wood reaches about 39% [12]. Tar begins to appear at 290°C. The release of gases still increases and rapid formation of charcoal takes place. This reaction is highly exothermic and proceeds at 280–320°C. Secondary reactions of pyrolysis become predominant result an increased amount of gaseous products. Combustion proceeds in the gas phase at a small distance from the surface rather than on the wood surface itself. The ignition of wood occurs at 300–400°C and depends on its origin and lignin content [13].

At the final stage of combustion of wood (above 500°C), the formation of combustible compounds is small, and the formation of charcoal increases. Charcoal makes an insulating layer that hinders heat transfer, thus preventing the temperature from increasing to the pyrolysis of wood occurs.

Mechanisms of flame retardancy depend on the character of the action of flame retardants' chemical compounds present in fire retardants. In comprehensive review on wood flame retardant, R. Kozlowski and M. Wladyka-Przybylak proposed two general groups of fire retardants for wood: additive and reactive [12]. Additive compounds are those whose interaction with a substrate is only physical in its nature, whereas reactive compounds interact chemically with cellulose, hemicellulose or lignin [12, 14]. The applied additive compounds include mono- and diammonium phosphate, halogenated phosphate esters, phosphonates, inorganic compounds such as antimony oxide and halogens, ammonium salts (ammonium bromide, ammonium fluoroborate, ammonium polyphosphate and ammonium chloride); amino resins (compounds used for their manufacture are dicyandiamide, phosphoric acid, formaldehyde, melamine and urea); hydrated alumina; stannic oxide hydrate; zinc chloride and boron compounds (boric acid, borax, zinc borate, triammonium borate, ethyl and methyl borates) [12].

One of the most effective methods of protecting wood from fire is the use of fire retardant intumescent coatings. When heated, they form a thick, porous carbonaceous layer. This provides an ideal insulation of the protected surface against an excessive increase in temperature and atmospheric oxygen, thus preventing thermal decomposition, which plays a decisive role in retarding the combustion process.

Studies of new flame retardant systems for the protection of natural polymers and wood can be directed to chemical modification of natural polymers and wood and the use of more efficient intumescent systems and fire protectors. Here, we represent the results of research of use of non-polluting reagents from renewed vegetative raw materials as highly effective flame retardants' intumescent/char-forming systems for cellulosic materials and wood, which can find wide applications due to their cheapness and simple technology of production.

Earlier it has been found that starch, cellulose and lignin can be easy oxidized by oxygen in the presence of copper salts and alkali. The basic products of oxidation of polysaccharides and lignin are salts of polyoxyacids [15]. Such salts can be used as components of washing powders, components of boring solutions, gums for use in building, etc. But unexpectedly, it was found that the salts of polyoxyacids are also diminishing the burning of different materials [16]. In the present work, the data on properties of the oxidized polysaccharides and a lignin, used as flame retardants are presented.

Oxidized Polysaccharides and Lignin

The methods of oxidative modifying of polysaccharides with receiving of polyacids as main products are widely used in practice due to availability of initial raw material and high consumer properties of oxidized polysaccharides. Salts of polyacids are widely used as water-soluble glues in production of paper, cardboard, in processes of materials dressing, as components of drilling agents, etc. However, as well as for a lot of other processes of organic compounds, oxidation as oxidizing agent of polysaccharides the hypochlorites and periodates [17,18], hydrogen peroxide and not gaseous oxygen are used until recently that is connected with low activity of oxygen in processes of polysaccharides oxidation.

In the presence of copper complexes and bases, not only simple in their structure alcohols and ketones but also polysaccharides (starches, dextranes, cellulose) may be oxidized by oxygen with high rates. High rates of polysaccharides oxidation by oxygen exceeding oxidation rates by hypochlorites and other oxidizing agents are reached at temperatures 40-90° C [19].

Oxidation of polysaccharides with the greatest efficiency proceeds in the presence of copper and alkali salts. In Fig. 1, kinetic curve of oxygen absorption (curve 1), NaOH consumption (curve 2) and changes of viscosity of initial gel (curve 3) during at of 10% of potato starch gel oxidation are presented. As it is seen from Fig.1, at the first 10-20 minutes of starch, oxidation viscosity of initial gel decreases in hundreds times and final (after three to four hours) viscosity of a solution comes is near to viscosity of water. The molecular weight of an initial polysaccharide at first minutes decreases slightly [20].

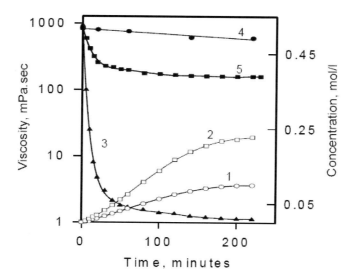

Figure 1. The kinetic curves of O_2 (1) and NaOH (2) consumption and change of viscosity of initial gel in the course of starch oxidation in the presents of cupper ions (3) and change of viscosity of gel in the absence of catalyst (4). Curve 5 – change of the viscosity of initial gel of starch (5 g starch and 0.5 g NaOH in 50 ml H_2O) under inert atmosphere after addition of 1 g dry powder of deep oxidized starch. $[CuCl_2]_o = 5.10^{-3}$ M , 5 g potato starch, 0.5 g NaOH, 50 ml H_2O, 338 K.

Without copper salt, the oxidation of starch does not proceed with measureable rate and viscosity of gel practically does not change (curve 4). But an addition of oxidized starch to the native gel of starch without oxygen drastically (in ten time) diminishes the viscosity of initial gel (curve 5).

Below, the most probable formula of the oxidized polysaccharide (starch) is presented. The chemical formula of the oxidized polysaccharides includes unmodified α-D-glucopyranosyl cycles and the oxidized links of polysaccharides containing carboxyl groups [20].

Copper salts are the most effective catalysts of oxidation of polysaccharides in the presence of alkali. For lack of the catalyst, starch and cellulose practically are not oxidized at temperatures below 373 K. However, lignin, which contains many of high reactive phenolic groups (under high rates), is oxidized by oxygen and without the catalyst with measurable rates.

As well as under oxidation of polyols with low molecular weights [21], the anion forms of polysaccharides reacts with Cu^{2+} ions with $[Cu^{2+}...A^-]$ adducts formation (A^- –

deprotonated polysaccharide). The function of catalyst (Cu^{2+} ions) is to activate deprotonated forms of substrate to oxygen. It is possible to suppose that the role of bivalent copper ions is the oxidation of anion form of substrate and followed interaction of intermediate radicals or anion radicals with O_2. However, high-molecular polysaccharides with a low content of end aldehyde groups in anaerobic medium are redox inactive and the rate of reduction of Cu^2 to Cu^+- ions (due to electron transfer from substrate anion form to Cu^{2+}) proceeds with rates in hundred times lower than the rate of oxygen adsorption during the process of polysaccharides oxidation in alkaline mediums. Apparently, just direct interaction of oxygen with anions A^- in coordination sphere of copper ion leads to formation of hydroxycarboxylates as the main primary products of oxidation. Absorption of oxygen is completely stagnated after neutralization of introduced alkali by polyoxyacids those formed during the oxidation process. Thus, by varying of amount of introduced alkali, we may change the degree of polysaccharides oxidation into polyacids. This fact allows us to change the final products viscosity, bonding ability, solubility in water, etc.

As initial raw materials for receiving of salts of polyoxyacids, not only starch can be used but any of other starch-containing raw materials: corns of maize, oats, rice, etc., including ill-conditioned raw materials (grain-crops affected by various fungus diseases, waste of rice slashing, waste of mill houses, etc.). This fact enables decreasing the prime cost of the final product significantly. Moreover, it was observed the similar to the polysaccharides behavior of lignin under oxidation condition in alkali media with/without the catalyst. Lignin is amongst the most abundant biopolymers on earth. It is estimated that the planet currently contains 3×10^{11} metric tons of lignin with an annual biosynthetic rate of approximately 2×10^{10} tons [22]. Lignin constitutes approximately 30% of the dry weight of softwoods and about 20% of the weight of hardwoods [23]. Lignin is absent from primitive plants such as algae, and fungi, which lack a vascular system and mechanical reinforcement. The presence of lignin within the cellulosic fibre wall, mixed with hemicelluloses, creates a naturally occurring composite material, which imparts strength and rigidity to trees and plants.

Lignin is a random copolymer consisting of phenylpropane units having characteristic side chains. Lignin slightly cross-links and takes an amorphous structure in the solid state. The molecular motion is observed as glass transition by thermal, viscoelastic and spectroscopic measurements. The hydroxyl group of lignin plays a crucial role in interaction with water. Lignin is usually considered as a polyphenolic material having an amorphous structure, which arises from an enzyme-initiated dehydrogenative polymerization of p-coumaryl, coniferyl and sinapyl alcohols. The basic lignin structure is classified into only two components; one is the aromatic part and the other is the C3 chain. The only usable reaction site in lignin is the OH group, which is the case for both phenolic and alcoholic hydroxyl groups.

Lignin is one of the most important bio-resources for the raw material of the synthesis of environmentally compatible polymers. Lignins are derived from renewable resources such as trees, grasses, and agricultural crops. About 30% of wood constituents are lignin. Lignins are nontoxic and extremely versatile in performance. Production of lignin as a by-product of pulping process in the world is over 30 million tons per year [23].

According to the widely accepted concept, lignin may be defined as an amorphous, three-dimensional polyphenolic polymer arising from an enzyme-mediated dehydrogenative polymerization of three phenylpropanoid monomers, coniferyl, sinapyl, and p-coumaryl alcohols. Basically, three major structures of lignin, 4-hydroxyphenyl (1), guaiacyl (2), and syringyl (3) structures are conjugated to produce a lignin polymer in the process of radical-based lignin biosynthesis.

C—C—C   C—C—C   C—C—C

OMe

MeO   OMe

OH   OH   OH

(1)   (2)   (3)

Results of estimation of efficiency of fire protecting action of coverings based on oxidized starch-containing raw materials with different amounts of applied reagent were carried out by standard method (ASTM E136-09 Standard Test Method for Behavior of Materials in a Vertical Tube Furnace) (Fig. 2). It is obvious that all pine samples treated with "one-layer covering" of oxidized starch reagent (OSR) - 100 g/m^2 provide good fire-protection activity (II group) to reach a rank of hardly-inflammable wood. In the case of "multi-layer" covering, 3÷4 covering, (OSR consumption 300÷400 g/m^2) mass losses at fire testing are significantly decreased allow us to obtain the I group of fire-protection − hardly inflammable wood.

Mode of Action of
Oxidized Polysaccharides
(Intumescence Behavior)

Study of thermal decomposition of oxidized starch at dynamic heating in the range of temperatures 25 – 950 °C has shown that formation of the made foam coke occurs at the very first stage at temperature 150 – 280 °C as a result of synchronous processes of reduction of viscosity of polymer at its transition from glass-forming form to viscous-fluid and chemical reactions of decarboxylation and dehydration. Making foam agents are gaseous products of decomposition – water and a carbon dioxide. Reactions of intermolecular dehydration promote the formation of the sewed spatially-mesh structure and stabilization and hardening of the made foam coke. Fig.2 shows the formation of intumescent coke from oxidized starch under the heating at 400°C in air.

The combustibility study of pinewood samples treated with aqueous solution of oxidized starch have shown high efficiency of flame retardant action of this intumescent additive during combustion process.

Fire tests results of plain pinewood samples and the pine samples treated with aqueous solution of oxidized starch according to ASTM E136-09 (Standard Test Method for Behavior of Materials in a Vertical Tube Furnace) (Figs. 3, 4).

Figure 2. Formations of coke after heat treatment of a metal plate with covering oxidized starch.

Figure 3. Photograph of initial pinewood samples before fire test.

Figure 4. Photograph of pinewood samples treated with aqueous solution oxidized starch flame retardant after fire tests (ASTM E136-09).

The effect of fireproof action of high-molecular oxidized starch flame retardant is connected with formation on a surface of wood sufficient foam coke layer. This foam coke shows excellent heat-shielding barrier properties, complicates heat access to the wood surface and hinders the evolution of combustible gaseous products of decomposition of wood.

Fig. 5 shows the actual mass loss pinewood samples treated with aqueous solution of different oxidized polysaccharides after fire tests (ASTM E136-09).

The combustibility of pristine pinewood sample and the treated by oxidized lignin and starch ones was also evaluated by the Mass Loss Calorimeter (ISO 13927) at an incident heat flux of 35 kW/m^2 condition [24]. Mass loss rates (MLR) of pinewood samples acquired at 35 kW/m^2 are presented in Fig. 6.

It is clearly seen that under conditions of initial surface combustion, the mass loss rate of the pinewood samples treated with of oxidized lignin and starch (15% by wt.) and the pinewood sample treated with oxidized starch and its rate are noticeably lower than an adequate value for the neat pinewood samples. The peak of MLR of the pinewood sample treated with of oxidized lignin is 41% lower than that of pure pinewood sample, while that of the pinewood sample treated with oxidized starch is 33% lower. Due to predominant solid-phase mode of action of oxidized lignin and starch flame retardants, the MLR curves are supposed to be similar to the RHR (rate of heat release) curves, so the reduction of the MLR is evidently the primary factor responsible for the lower RHR of the pinewood samples. An improvement in flame resistance of the pinewood samples treated with of oxidized lignin over the pinewood sample happens as a result of the char formation providing a transient protective barrier of oxidized lignin (as polyphenolic carbonizing structure), whereas the pinewood sample treated with oxidized starch indicates the strong intumescent action. An additional observation obtained from the MLR plots is the substantial increase in induction period time of self-ignition (flashpoint) of the pinewood sample treated with oxidized starch as compared with plain pinewood sample and the pinewood sample treated with oxidized lignin due to intumescent behavior of oxidized starch (Fig.6).

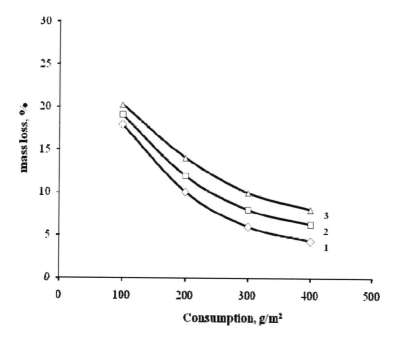

Figure 5. Dependence of mass losses of wood samples on amount of applied fire-protecting coverings based on modified polysaccharides: 1 − oxidized starch (high degree of oxidation), 2 − oxidized rice (high degree of oxidation), 3 − oxidized starch (average degree of oxidation).

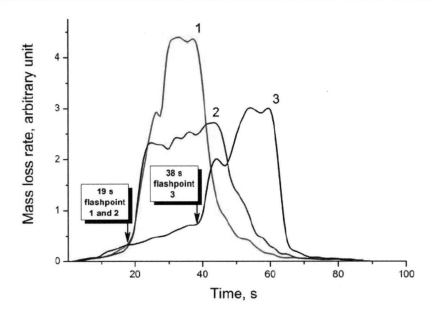

Figure 6. Mass loss rate vs. time for plain pinewood sample - (1), pinewood sample treated with of oxidized lignin (15% by wt.) - (2) and pinewood sample treated with oxidized starch (15% by wt.) – (3).

References

[1] Lomakin S.M., Brown J.E., Breese R.S. and Nyden M.R., *Polymer Degrad. and Stab.* 41, 229-243 (1993).

[2] Factor A., Char Formation in Aromatic Engineering Polymers, in Fire and Polymers, Ed. Nelson G.L. (Ed.), pp. 274-287, *ACS Symposium Series* 425, ASC, Washington DC (1990).

[3] Lomakin S.M., Zaikov G.E. and Artsis M.I., *Intern. J. Polym. Mater.* 32 (1-4) 173-202 (1996).

[4] Lomakin S.M., Zaikov G.E., Artsis M.I., Ruban L.V. and Aseeva R.M., *Oxidation Comm.* 18 (2), 105-112 (1995).

[5] Lomakin S.M., Zaikov G.E. and Artsis M.I., *Intern. J. Polym. Mater.* 26 (3-4), 187-194 (1994).

[6] R.M. Aseeva, G.E. Zaikov. *Combustion of Polymeric Materials.* Moscow: *Nauka,* 1981, pp. 84-135 (in Russian).

[7] S. Levchik and Ch. A. Wilkie , Char Formation, Chapter 6 in *Fire Retardancy of Polymeric Materials,* ed. By A.F. Grand and Ch. A. Wilkie , Marcel Dekker, Inc. New York (2000), 171-216

[8]   G. Camino. Flame retardants: Intumescent systems. In: G Pritchard, ed. *Plastics Additives, London: Chapman Hall,* 1998, pp. 297-306.

[9] G. Camino and R. Delobel, Intumescence, Chapter 7 in *Fire Retardancy of Polymeric Materials,* ed. By A.F. Grand and Ch. A. Wilkie, Marcel Dekker, Inc. New York (2000), 217-244

[10] G. Camino, S. Lomakin. Intumescent materials In: A R Horrocks and D Price, eds., Fire Retardant Materials, CRC Press, New York, Washington DC (2001), 318-335.

[11] Vandersall H L, Intumescent coating systems, their development and chemistry. *J. Fire Flamm*, 1971, 2 April: 97–140

[12] R. Kozlowski and M. Wladyka-Przybylak, Natural polymers, wood and lignocellulosic materials In: A R Horrocks and D Price, eds., *Fire Retardant Materials*, CRC Press, New York, Washington DC (2001), 293-317.

[13] Hurst N W, Jones T A, A review of products evolved from heated coal, wood and PVC, *Fire Mater*, 1985, 9(1), 1–9.

[14] Lewin M, Atlas S M, Pearce E M, *'Flame-Retardant Polymeric Materials'*, vol 1, New York and London, Plenum Press, 1975.

[15]   Sakharov A.M. In: *Chemical and Biochemical Physics: New Frontiers* (2006), Editor: Zaikov G.E., Nova Science Publishers, 113.

[16] Sivenkov A.B., Serkov B.B., Aseeva R.M., Sakharov A.M., Sakharov P.A., Skibida I.P. (2002). *Fire & Explosion Safety*. 1, 39.(in Russian).

[17] Floor M., Kieboom A.P.G., H. van Bekhum. (1989). *Starch,* 41, 303.

[18] Santacesria E., Trully F., Brussani G.F. Gelosa D., Serio M. Di. (1994).

[19] C. Skibida I.P., Sakharov An. M., Sakharov Al. M. (1993). EP 91122164.6, *Carbohydrate Polymers,* 23, 35.

[20] Sakharov A.M., Skibida I.P. (2001). *Chem. Phys.,* 20, 101.

[21] Sakharov A.M., Silakhtaryan N.T., Skibida I.P. (1996). *Kinetics and Catalysis,* 37, 393.

[22] Cereal Straw as a Resource for Sustainable Biomaterials and Biofuels: Chemistry, Extractives, Lignins, Hemicelluloses and Cellulose, Ed. By Run-Cang Sun Elsevier, 2010, 292 p.

[23] H. Hatakeyama and T. Hatakeyama, Lignin Structure, Properties, and Applications. In Biopolymers - Lignin, Proteins, Bioactive Nanocomposites, ed. By A. Abe, K. Dušek and Sh. Kobayashi, Springer, Heidelberg, Dordrecht, London, New York, 2010, p.2-64.

[24] Standard Test Method for Screening Test for Mass Loss and Ignitability of Materials (ASTM E2102). American Society for Testing and Materials, Philadelphia, PA.

In: Handbook of Research on Nanomaterials, Nanochemistry ...   ISBN: 978-1-61942-525-5
Editors: A. K. Haghi and G. E. Zaikov   © 2013 Nova Science Publishers, Inc.

Chapter XIX

Combustion and Thermal Degradation of Polypropylene in the Presence of Multiwalled Carbon Nanotube Composites

**G. E. Zaikov[1], S. M. Lomakin[1],
N. G. Shilkina[2] and R. Kozlowski[3]**

[1] NM Emanuel Institute of Biochemical Physics of
Russian Academy of Sciences, Moscow, Russia
[2] NN Semenov Institute of Chemical Physics of
Russian Academy of Sciences, Moscow, Russia
[3] Institut Inzynierii Materialow Polimerowych I Barwnikow, Torun, Poland

Introduction

At present time, great attention is given to the study of properties of polymeric nanocomposites produced on the basis of well-known thermoplastics (PP, PE, PS, PMMA, polycarbonates, polyamides) and carbon nanotubes (CN). CNs are considered to have the wide set of important properties like thermal stability, reduced combustibility, electroconductivity, etc. [1-7]. Thermoplastic polymer nanocomposites are generally produced with the use of melting technique [1-12].

Development of synthetic methods and the thermal characteristics study of PP/multiwalled carbon nanotube (MWCNT) nanocomposites were taken as an objective in this paper.

A number of papers pointed at synthesis, and research of thermal properties of nanocomposites (atactic polypropylene (aPP)/MWCNT) were reported [10-12]. It is remarkable that PP/MWCNT composites with minor level of nanocarbon content (1-5% by weight) were determined to obtain an increase in thermal and thermal-oxidative stability in the majority of these publications.

Thermal stability of aPP and aPP/MWCNT nanocomposites with the various concentrations of MWCNT was studied in the paper [10]. It was shown that thermal degradation processes are similar for aPP and aPP/MWCNT nanocomposites, and initial degradation temperatures are the same. However, the maximum mass loss rate temperature of PP/MWCNT nanocomposites with 1 and 5% wt of MWCNT raised by 40° - 70°C as compared with pristine aPP.

Kashiwagi et al. published the results of a study of thermal and combustion properties of PP/MWCNT nanocomposites [11, 12]. A significant decrease of maximum heat release rate was detected during combustion research with use of cone calorimeter. A formation of char network structure during the combustion process was considered to be the main reason for combustibility decrease. The carbonization influence upon combustibility of polymeric nanocomposites was widely presented in literature [10-12, 13]. Notably, Kashiwagi et al. [11, 12] were the first to hypothesize that abnormal dependence of maximum heat release rate upon MWCNT concentration is closely related with thermal conductivity growth of PP/MWCNT nanocomposites during high-temperature pyrolysis and combustion.

Experimental

Materials

Isotactic polypropylene (melting flow index = 0.7 g/10 min) was used as a polymer matrix in this paper. Multiwalled carbon nanotubes (MWCNT) (purchased from Shenzhen Nanotechnologies Co. Ltd.) were used as a carbon-containing nanofillers. This product contains low amount of amorphous carbon (less than 0.3 wt%) and could be produced with different size characteristics—different length and different diameter and therefore different diameter-to-length ratio. Size characteristics for three MWCNT used in this paper are given in Table 1. Sizes and structure of initial MWCNT were additionally estimated by SEM (Fig.1).

Table 1. Properties of MWCNT

Designation	D, nm	L, μm	Density, g/cm^3	Specific surface area, m^2/g
MWCNT (K1)	<10	5-15	2	40 - 300
MWCNT (K2)	40-60	1-2	2	40 - 300
MWCNT (K3)	40-60	5-15	2	40 - 300

Nanocomposite Processing

Compositions were prepared by blending carbon nanotubes with melted polymer in a laboratory mixer Brabender at 190°C. TOPANOL® (1,1,3-tris (2-methyl-4-hydroxy-5-t-butylphenyl) butane) and DLTP (dilaurylthiodipropionate) were added in the amount of 0.3 and 0.5 wt% as antioxidants to prevent thermal-oxidative degradation during polymer processing.

A number of different covalent and non-covalent nanotube modifications (organofillization) were reported to be used to achieve greater structure similarity and therefore greater nanotube distribution in a polymer matrix [14-23]. In order to functionalize MWCNT, we used preliminary ozone treatment of MWCNT followed by ammonolysis of epoxy groups on the MWCNT surface. The selective ozonization of MWCNT was carried out with ozone-oxygen mixture (ozone concentration was 2.3×10^{-4} mol/L) in a bubble reactor. Then, the ammonolysis of oxidized MWCNT has been carried out by *tert*-butylamine in the ultrasonic bath (35 kHz) at 50°C for 120 min with following evaporation of *tert*-butylamine excess. IR transmission spectra of tablet specimens of MWCNTs in KBr matrix was analyzed by using Perkin-Elmer 1725X FTIR spectrometer, and the presence of the alkylamine groups at the MWCNT surface was confirmed by the appearance of the characteristic band ~1210 cm^{-1} corresponding to the valency vibration of the bond –C–N←.

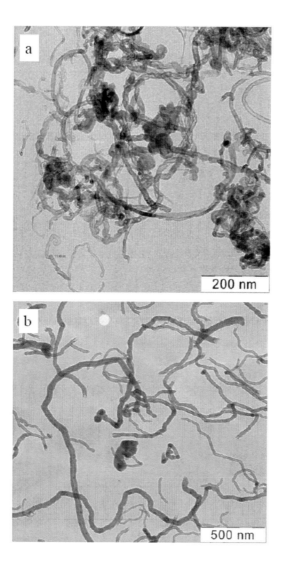

Figure 1. (Continued).

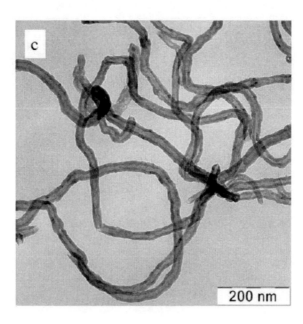

Figure 1. SEM images of original MWCNTs: (a) MWCNT(1); (b) MWCNT(2); (c) MWCNT(3).

Investigation Techniques

Scanning electron microscopy (SEM). The degree of MWCNT distribution in polymer matrix was analyzed with scanning electron microscope JSM-35. Low-temperature chips derived from film-type samples were used for this analysis.

Transmission electron microscopy (TEM). The degree of nanotube dispersion in polymer matrix was studied with transmission electron microscopy (LEO912 AB OMEGA, Germany). Microscopic sections with 70 – 100 nm width prepared with ultramicrotome "Reichert – Jung Ultracut" with diamond cutter at -80°C. Microscopic analysis was made with accelerating potential of about 100 kV without chemical sample staining.

Thermogravimetric analysis (TG). A NETZSCH TG 209 F1 Iris thermomicrobalance has been employed for TGA measurements in oxidizing (oxygen) atmosphere. The measurements were carried out at a heating rate of 20 K/min.

Combustibility characteristics (cone calorimeter) were performed according to the standard procedures ASME E1354/ISO 5660 using a DUAL CONE 2000 cone-calorimeter (FTT). An external radiant heat flux of 35 kW/m^2 was applied. All of the samples having a standard surface area of 70×70 mm and identical masses of 13.0±0.2g were measured in the horizontal position and wrapped with thin aluminum foil except for the irradiated sample surface.

Heat capacity and heat conductivity were determined with the use of NETZSCH 457 MicroFlash.

The electron paramagnetic resonance spectroscopy (EPR) measurements were performed in air with the PP/MWCNT (10 wt%) samples using a Mini-EPR SPIn Co. Ltd spectrometer with 100 kHz field modulation. The g factor and EPR intensity (X-band) were measured with respect to a standard calibrating sample of Mn^{2+} and ultramarine.

Results and Discussion

Nanocomposite Structure

Dispersion analysis of MWNT in nanocomposites. PP/MWNT nanocomposites with original and modified MWNT were produced. Filler concentration varied from 1 to 7 wt% weight percent (0.5-3.5 volume percent, correspondingly). Distribution pattern for composites with modified and non-modified nanotubes was studied with TEM methods (Fig 2). According to TEM images, the addition of 1% by weight leads to sufficiently uniform distribution. However, agglomeration of nanotubes was detected for more concentrated nanocomposites, especially for PP/MWNT with MWNT average diameter less than 10 nm (K1).

TEM images for PP nanocomposites with 5 wt% modified and non-modified MWNT (2.5% by volume) are shown on Fig.2. According to Fig.2, it could be stated that modified nanotubes (K2 and K3) used during melding process are present as individual particles in nanocomposite in most cases. The number and size of agglomerates is reduced due to increased organophility and improved thermodynamic compatibility with nonpolar polymer.

However, preliminary modification doesn't lead to uniform filler distribution for nanocomposites containing thin nanotubes (K1). This could be explained by the fact that interaction energy of CNT is more dependent on nanotube diameter than on its length. Molecular dynamic computation given in paper [24] showed that blending polymers with nanotubes becomes more thermodynamically favorable with increase of nanotubes diameter, owing to the fact that cohesion energy is decreased between the nanotubes and remains almost the same between nanotubes and polymer.

Thus, mixing the thinnest nanotubes (K1) with PP leads to inevitable nanotube agglomeration in nanocomposite sample volume. Nanotube surface modification used in this paper didn't result in complete overcome of nanotube agglomeration tendency for K1 nanotubes [25]. Therefore, the PP/MWNT(K3) nanocomposites presented the main subjects of inquiry in the present study.

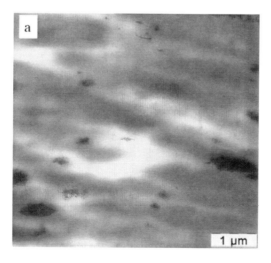

Figure 2. (Continued).

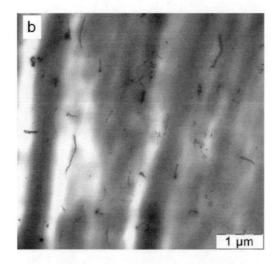

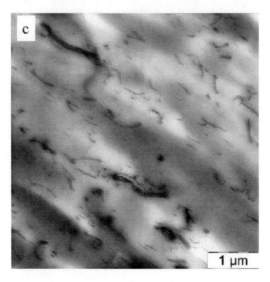

Figure 2. TEM images of PP/MWCNT nanocomposites showing dispersion of MWCNT in a polymer matrix: (a) PP/MWCNT(1); (b) PP/MWCNT(2); (c) PP/MWCNT(3).

Thermal-Oxidative Degradation of PP/MWCNT Nanocomposites

The diverse behavior of PP and PP/MWCNT nanocomposites with 1, 3 and 5 wt% of MWCNT(3) (Fig. 3) shows that the influence of MWCNTs on the thermal-oxidation process resulted in higher thermal-oxidative stability of PP/MWCNT nanocomposites. It is possible to see a regular increase in the temperature values of the maximum mass loss rates (up to 60°C) for the PP/MWCNT as compared to pristine PP (Fig. 3).

Detailed analysis of TGA graphs (Fig. 3) allows claiming that thermal stability increase is achieved even by addition of 1 wt% of MWCNT to PP, while further addition doesn't lead to such fundamental growth. In addition, Fig. 4 shows the comparative results for onset degradation temperatures ($T_{on.}$) and the maximum mass loss temperatures (T_{max}) of PP/MWCNT nanocomposites with the different types and concentrations of MWCNT. One can see nonlinear relation of $(T_{on.})$ and (T_{max}) vs. MWCNT concentration in the PP compositions (Fig. 4).

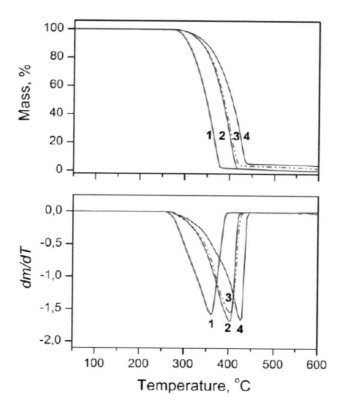

Figure 3. TG and DTG curves for PP (1) and PP/MWCNT(3) composites with 1 (2), 3 (3) and 5 wt% (4) filler loadings.

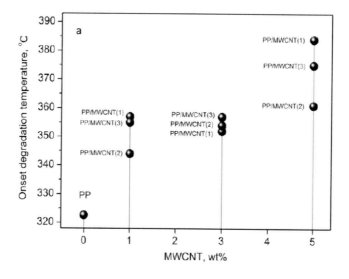

Figure 4. (Continued).

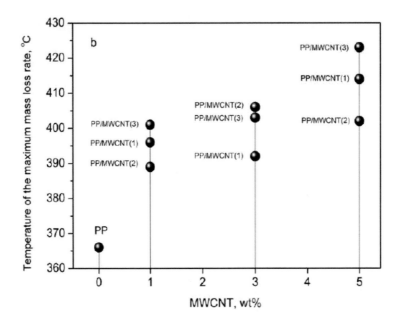

Figure 4. Comparative diagrams showing the onset degradation temperatures (a) and the maximum mass loss temperatures (b) for PP and PP/MWCNT nanocomposites with the different types and concentrations of MWCNT.

At the present time, the nature of thermal stability effect caused by MWCNT addition to polymers is an object of comprehensive study. Most likely, MWCNTs could be considered as high-temperature stabilizers (antioxidants) in reactions of thermal-oxidative degradation by analogy with fullerenes [26]. Stabilizing effect caused by addition of MWCNT was previously detected for PP/MWCNT nanocomposites [8]: the temperature of the maximum mass loss rates of PP/MWCNT (9 wt%) was increased by 50°C as compared with pristine PP.

Results achieved in this study confirm the previous findings of the inhibiting effect of MWNCT upon the PP/MWCNT nanocomposite thermal-oxidative degradation. Obviously, the complex nature of this effect is closely related to radical-acceptor properties of MWCNT, resulting in chain termination reactions of alkyl/alkoxyl radical, which lead to cross-linking and carbonization process in PP/MWCNT nanocomposites. Carbonization phenomenon was reported previously in papers aimed at PP/MWCNT heat resistance and flame retardancy study [11, 12].

Kinetic Analysis of Thermal Degradation of PP/MWNT

Kinetic studies of material degradation have long history, and there exists a long list of data analysis techniques employed for the purpose. Often, TGA is the method of choice for acquiring experimental data for subsequent kinetic calculations, and namely this technique was employed here. It is commonly accepted that the degradation of materials follows the base equation (1) [27]

$$dc/dt = -F(t, T, c, p)$$ (1)

where: t - time, T - temperature, c_o - initial concentration of the reactant, and p - concentration of the final product. The right-hand part of the equation $F(t, T, c, p)$ can be represented by the two separable functions, $k(T)$ and $f(c, p)$:

$$F(t, T, c, p) = k [T(t) \cdot f(c, p)] \tag{2}$$

Arrhenius equation (4) will be assumed to be valid for the following:

$$k(T) = A \cdot exp(-E/RT) \tag{3}$$

Therefore,

$$dc/dt = - A \cdot exp(-E/RT) \cdot f(c,p) \tag{4}$$

All feasible reactions can be subdivided onto classic homogeneous reactions and typical solid-state reactions, which are listed in Table 2. The analytical output must provide good fit to measurements with different temperature profiles by means of a common kinetic model.

Thermogravimetric analysis of PP and PP nanocomposite degradation was carried out in dynamic conditions at the rates of 2.5, 5 and 10 K/min on air. Model-independent estimation of activation energy using Friedman approach [28] was taken to get preliminary model analysis for thermal degradation and selection of initial conditions. According to this evaluation, a two-step process ($A \rightarrow X_1 \rightarrow B \rightarrow X_2 \rightarrow C$) was chosen for PP degradation. Taking into account the carbonization stage, the more complex three-step process ($A \rightarrow X_1 \rightarrow B \rightarrow X_2 \rightarrow C \rightarrow X_3 \rightarrow D$) was selected for PP/MWNT degradation [27, 29].

Table 2. Considered reaction models $dc/dt = -A \cdot exp(-E/RT)f(c, p)$

Reaction models	$f(c, p)$
First order (F_1)	c
Second order (F_2)	c^2
n-order (F_n)	c^n
Two-dimensional phase boundary (R_2)	$2 \cdot c^{1/2}$
Three-dimensional phase boundary (R_3)	$3 \cdot c^{2/3}$
One-dimensional diffusion (D_1)	$0.5/(1 - c)$
Two-dimensional diffusion (D_2)	$-1/ln(c)$
Three-dimensional diffusion, Jander's type (D_3)	$1.5c^{1/3}(c^{-1/3} - 1)$
Three-dimensional diffusion, Ginstling-Brounstein (D_4)	$1.5/(c^{-1/3} - 1)$
One-dimensional diffusion (Fick law) (D_{1F})	-
Three-dimensional diffusion (Fick law) (D_{3F})	-
Prout-Tompkins equation (B_1)	$c \cdot p$
Expanded Prout-Tompkins equation (B_{na})	$c^n \cdot p^a$
First order reaction with autocatalysis by X (C_{1-X})	$c \cdot (1 + K_{cat} X)$
n-order reaction with autocatalysis by X (C_{n-X})	$c^n \cdot (1 + K_{cat} X)$
Two-dimensional nucleation, Avrami-Erofeev equation (A_2)	$2 \cdot c \cdot (-ln(c))^{1/2}$
Three-dimensional nucleation, Avrami-Erofeev equation (A_3)	$3 \cdot c \cdot (-ln(c))^{2/3}$
n- dimensional nucleation, Avrami-Erofeev equation (A_n)	$n \cdot c \cdot (-ln(c))^{(n-1)/n}$

According to the results of nonlinear regression and taking the set of reaction models into consideration, we computed the values of active kinetic parameters, which represent the best approximation of experimental TGA graphs (Fig. 5, Table 3).

Table 3. Kinetic parameters for thermal degradation of (a) PP (Fn →Fn) and (b) PP/MWNT nanocomposite (Fn →D1→ Fn). TGA analysis was performed in airflow with the use of multiple non-linear regression analysis for model processes

(a)

Reaction model	Kinetic parameters	Values	Correlation coefficient
Fn →Fn	$\log A_1$, s^{-1}	9.53	0.9996
	E_1, kJ/mol	110.25	
	n_1	1.89	
	$\log A_2$, s^{-1}	15.25	
	E_2, kJ/mol	150.65	
	n_2	1.50	

(b)

Reaction model	Kinetic parameters	Values	Correlation coefficient
Fn →D1→ Fn	$\log A_1$, s^{-1}	6.3	0.9996
	E_1, kJ/mol	105.1	
	n_1	0.91	
	$\log A_2$, s^{-1}	7.4	
	E_2, kJ/mol	120.4	
	$\log A_3$, s^{-1}	16.7	
	E_3, kJ/mol	229.5	
	n_3	0.5	

Two-step PP thermal-oxidative degradation in dynamic heating conditions was confirmed by obtained data [30]. At the first stage, the values of activation energy and pre-exponential factor are 110.25 kJ/mol and $10^{9.5}$ s^{-1}, correspondingly, while the reaction order is close to 2 (1.89). The values of activation energy and pre-exponential factor are larger on the second stage (E_2 = 150.65 kJ/mol, A_2 = $10^{15.3}$ s^{-1}) with effective reaction order of n_2 = 1.50.

The preferred model for PP/MWNT thermal-oxidative degradation and with respect to statistical analysis of kinetic parameters is composed of three consecutive reactions $F_n \rightarrow D_1 \rightarrow F_n$, where D_1 – one-dimensional diffusion, and F_n – n-order reaction (Fig 4b, Table 3b). In this case, the first step activation energy is equal to 105.1 kJ/mol, reaction order is close to 1 (n_1 = 0.91). On the second step, which is described as one-dimensional diffusion, the value of activation energy is equal 120.4 kJ/mol, while the value is almost twice large for the third step (E_3 =229.5 kJ/mol), with effective reaction order of n_3 = 0.5 (Table 3b).

Comparison of thermal oxidative degradation parameters for PP/MWNT with layered silicate PP//MMT showed that the values of activation energy of the second and the third stages are higher for PP/MWNT:

E_2=120.4 kJ/mol and E_3=229.5 kJ/mol for PP/MWNT;
E_2=100.0 kJ/mol and E_3=199.8 kJ/mol for PP/MPP/MMT, correspondingly [30].

This data may testify to more intensive carbonization in case of PP/MWNT than in case of PP/MMT, which finally leads to decrease in RHR value.

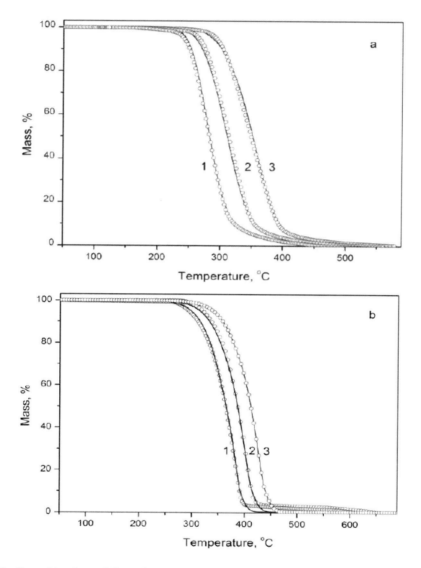

Figure 5. Nonlinear kinetic modeling of a – PP and b – PP/MWCNT(3) thermal-oxidative degradation in air. Comparison between experimental TG data (dots) and the model results (firm lines) at several heating rates: (1) – 2.5, (2) – 5, (3) – 10 K/min.

Combustibility of PP/MWCNT Nanocomposites

Figure 6 depicts the plots of the heat release rate (RHR), as basic flammability characteristic vs. time for PP, as well as for the PP/MWCNT nanocomposites.

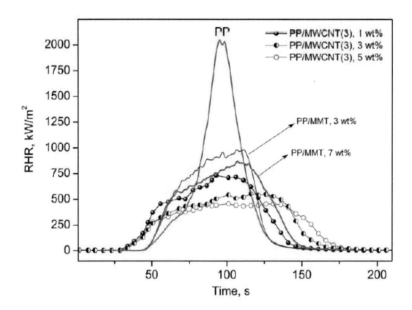

Figure 6. Rate of heat release vs. time for PP, PP/MWCNT and PP/MMT (Cloisite 20A) nanocomposites obtained by cone calorimeter at the incident heat flux of 35 kW m^{-2}.

From Fig. 6, it could be seen that the maximum heat release rate for pristine PP is 2,076 kW/m^2, whereas that for the PP/MWCNT % nanocomposites (1 wt%), PP/MCWNT (3 wt%) and the PP/MCWNT (5 wt.%) RHR values are 729 kW/m^2, 552.8 kW/m^2, and 455.8 kW/m^2, respectively; thus, the peak heat release rate decreases by 65%, 73% and 78%.

The observed flame retardancy effect is associated with solid-phase carbonization reactions, by analogy with layered silicates [31, 32]. In early paper [30], we have found that additions of 3 and 7 wt % of layered silicate (Cloisite 20A) to the PP compositions PP lead to RHR decrease by 51 and 57% as compared with pristine PP (Fig.5).

We believe that a higher carbonization effectiveness of MWCNTs depends on their heat conductivity. It is well known that PP has a low thermal conductivity at standard conditions, and characterized by a minor increase with temperature up to melting point (~0.2 W/m K). On the other hand, the heat conductivity of individual MWCNT is extremely high and equals 3,000 W/m K [33, 34]. During the high temperature pyrolysis of PP/MWCNT composition at temperatures above 300-400°C, the heat conductivity can rise up to 20W/m K [34] due to actual increase of MWCNT concentration in composition caused by volatilization of polypropylene degradation products (Fig.7). The induced heating of PP/MWCNT intensifies a steady carbonization and charring of the samples and leads to decrease of RHR peak value (Fig. 6).

Figures 8 and 9 show graphs for the specific extinction area and effective heat of combustion, correspondingly, for PP and PP/MWCNT(3) nanocomposites. Calculated values of effective heat of combustion for PP and PP/MWNT demonstrate invariant shift of this parameter for these nanocomposites.

In the present study, EPR research were performed to follow formation of stable radicals, responsible for carbonization process, upon isothermal heating of PP/MCWNT (10 wt%) in air at 350°C.

Fig. 10a shows EPR spectrum of the stable paramagnetic centers formed in the samples of PP/MCWNT (10% wt) heating in air at 350°C. When a heated in air PP/MCWNT specimen was placed into an EPR sample tube, a narrow singlet signal with a line width of $\Delta H_{1/2}$ = 0,69 mT and a *g* value of 2.003 was detected due to the stable radicals generation, analogous to those previously registered during polymers carbonization process [35]. No EPR signal similar to that of PP/MCWNT samples were observed in the samples of pristine PP and MCWNT samples heated at 350°C in air. It should be noted that although iron impurity from MWNCT has been mentioned in other studies on pyrolysis of polymer nanocomposites as the radical traps [36, 11], the EPR analyses in the current study showed the presence of paramagnetic centers relating to carbonaceous stable radicals only.

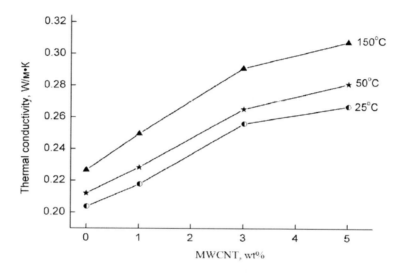

Figure 7. Temperature dependence of the thermal conductivity of PP/MWCNT (3) nanocomposite with different loadings of MWCNT.

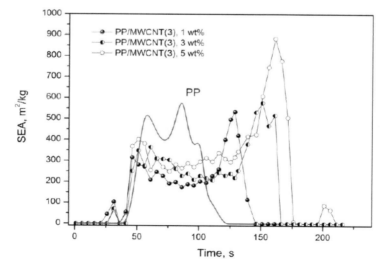

Figure 8. Specific extinction area vs. time for PP and PP/MWCNT(3) nanocomposites obtained by cone calorimeter at the incident heat flux of 35 kW m^{-2}.

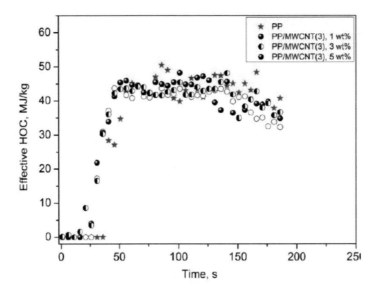

Figure 9. Effective heat of combustion vs. time for PP and PP/MWCNT(3) nanocomposites obtained by cone calorimeter at the incident heat flux of 35 kW m^{-2}.

As it is seen from Fig. 10b, the formation of stabilized radicals occurs with pronounced induction period, which is related to antioxidant properties of MWNCT. Such a type of kinetic dependence is coincided with an oxygen uptake kinetics observed during inhibited polyolefines thermal oxidation. Moreover, no EPR signals were observed in the samples of the PP/MCWNT samples heated at 350°C in inert Ar.

Thus, multiwalled carbon nanotubes are considered to be more effective filling agents than layered silicates in the terms of improvement of thermal properties and flame retardancy of PP matrix. This could be explained by the specific antioxidant properties and high thermal conductivity of MWCNT, which determine the carbonization reactions during thermal-oxidative degradation process.

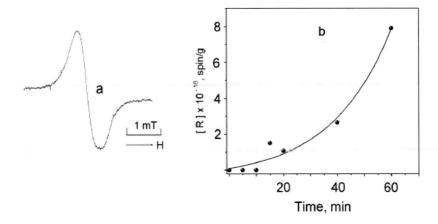

Figure 10. EPR spectrum of the stable paramagnetic centers formed in the samples of PP/MCWNT (10% wt) heating in air at 350°C – (a), kinetic dependence of stable radicals generation from PP/MCWNT (10% wt) under isothermal heating at 350°C in air – (b).

References

[1]   Shaffer MSP, Windle AH. *Adv. Mater.* 1999, 11, 937.

[2]   Qian D, Dickey EC, Andrews R, Rantell T. *Appl. Phys. Lett.* 2000, 76, 2868.

[3]   Jin Z, Pramoda KP, Xu G, Goh SH. *Chem. Phys. Lett.* 2001, 337, 43.

[4]   Thostenson ET, Chou TW. *J. Phys. D: Appl. Phys.* 2002, 35, L77.

[5]   Bin Y, Kitanaka M, Zhu D, Matsuo M. Macromolecules 2003, 36, 6213.

[6]   Potschke P, Dudkin SM, Alig I. *Polymer.* 2003, 44, 5023.

[7]   Safadi B, Andrews R, Grulke EA. *J. Appl. Polym. Sci.* 2002, 84, 2660.

[8]   P.C.P. Watts, P.K. Fearon, W.K. Hsu, N.C. Billingham, H.W. Kroto and D.R.M. Walton, *J. Mater. Chem.* 2003, 13, 491.

[9]   Watts PCP, Hsu WK, Randall DP, Kroto HW and Walton DRM., *Phys. Chem. Chem. Phys.,* 2002, 4, 5655.

[10]   Jun Yang, Yuhan Lin, Jinfeng Wang. *J. Appl. Polym. Sci.* 2005, 98,3,1087.

[11]   Kashiwagi T, Grulke E, Hilding J, Groth K,; Harris RH, Butler K., Shields JR, Kharchenko S, Douglas J. *Polymer.* 2004, 45, 4227–4239.

[12]   Kashiwagi T, Grulke E, Hilding J, Harris Jr RH, Awad WH, Dougls J. *Macromol. Rapid Commun.* 2002, 23, 761, 5.

[13]   Lomakin SM, Novokshonova LA, Brevnov PN, Shchegolikhin AN. *Journal of Materials Science.* 2008, 43, 4, 1340.

[14]   Chen J, Hamon M A, Hu H, Chen Y, Rao A M, Eklund P C and Haddon R C. *Science.* 1998, 282, 95.

[15]   Stevens J L, Huang A Y, Peng H, Chiang I W, Khabashesku V N and Margrave J L. *Nano Lett.,* 2003, 3, 331.

[16]   Eitan A, Jiang K, Dukes D, Andrews R and Schadler L S. *Chem. Mater.,* 2003, 15 3198.

[17]   Hu H, Ni Y, Montana V, Haddon R C and Parpura V. *Nano Lett.,* 2004, 4, 507.

[18]   Kong H, Gao C and Yan D J. *Am. Chem. Soc.,* 2004,126, 412.

[19]   Holzinger M, Vostrowsky O, Hirsch A, Hennrich F, Kappes M, Weiss R and Jellen F Angew. *Chem. Int.* Edn 2001,40, 4002.

[20]   Holzinger M, Abraham J, Whelan P, Graupner R, Ley L, Hennrich F, Kappes M and Hirsch A J. *Am. Chem. Soc.,* 2003, 125, 8566.

[21]   Yao Z, Braidy N, Botton G A and Adronov A. *J. Am.Chem. Soc.* 2003, 125, 16015.

[22]   Ying Y, Saini R K, Liang F, Sadana A K and Billups W E. *Org. Lett.* 2003, 5, 1471.

[23]   Alvaro M, Atienzar P, de la Cruz P, Delgado J L, Garcia H and Langa F, *J. Phys. Chem.* B 2004, 108 12691,

[24]   M R. Nyden, S I. Stoliarov, *Polymer.* 2008,49, 635-641,

[25]   Rakhimkulov AD, Lomakin SM, Alexeyeva OV, Dubnikova IL, Schegolikhin AN, Zaikov GE in *Proceedings of IBCP International Conference* (2006) 56.

[26]   Krusic J, Wasserman E, Keizer PN, Morton JR, Preston KF. *Science.* 1991, 254, 1183,

[27]   Opfermann J. *J. Thermal Anal Cal.* 2000, 60, 3, 641.

[28]   Friedman H.L., *J. Polym. Sci.,* 1965, C6, 175.

[29]   Opfermann J., Kaisersberger E. *Thermochim. Acta.* 1992, 11, 1, 167.

[30]   Lomakin SM, Dubnikova IL, Berezina SM, Zaikov GE, *Polymer Science.* 2006, 48, Ser. A, 1, 72.

[31] Gilman GW, Jackson CL, Morgan AB, Harris RH, Manias E, Giannelis EP, Wuthenow M, Hilton D, Phillips S, *Chem. Mater.* 2000, 12, 1866.

[32] Kashiwagi T, Harris RH Jr., Zhang X, Briber RM, Cipriano BH, Raghavan SR, Awad WH, Shields J R. *Polymer.* 2004, 45, 881.

[33] Kim P, Shi L, Majumdar A, McEuen PL. *Phys. Rev. Lett.* 2001, 87, 215502.

[34] Yi, W.; Lu, L; Zhang, D.L.; Pan, Z.W.; Xie, S.S. *Phys. Rev. B* 1999, 59, R9015–8.

[35] Echevskii GV, Kalinina NG, Anufrienko VF, Poluboyarov VA. *React. Kinet. Catal. Lett.* 1987, 33, 8, 305.

[36] Zhu. J, Uhl F, Morgan AB, Wilkie CA, *Chem. Mater.* 2001, 13,4649–54.

In: Handbook of Research on Nanomaterials, Nanochemistry ...   ISBN: 978-1-61942-525-5
Editors: A. K. Haghi and G. E. Zaikov   © 2013 Nova Science Publishers, Inc.

Chapter XX

Theoretical Structural Model of Nanocomposites Polymer/Organoclay Reinforcement

Georgiy V. Kozlov[1], Boris Zh. Dzhangurazov[1], Stefan Kubica[2], Gennady E. Zaikov[3] and Abdulakh K. Mikitaev[1]
[1] Kabardino-Balkarian State University, Russian Federation
[2] Institut Inzynierii Materialow Polimerowych I Barwnikow, Torun, Poland
[3] N.M. Emanuel Institute of Biochemical Physics of Russian Academy of Sciences, Moscow, Russian Federation

Introduction

At present, nanocomposites polymer/organoclay studies attained very big, widespread interest [1, 2]. However, the majority of works fulfilled on this theme has mainly an applied character and theoretical aspects of polymers reinforcement by organoclays are studied much less. So, the authors [3] developed for this purpose multiscale micromechanical model, in the basis of which representation about organoclay "effective particle" was assumed. The indicated "effective particle" includes in itself both nanofiller platelets and adjoining to them, (or located between them) polymeric matrix layers. Despite the complexity of model [3], it has an essential lacks number. First, the above-indicated complexity results in necessity of parameters large number usage, a part of which is difficult enough and sometimes impossible to determine. Secondly, this model is based on micromechanical models application, which in essence exhausted their resources [4]. Thirdly, at the entire of its complexity, the model [3] does not take into account such basis for polymer nanocomposites factors as a real level of interfacial adhesion nanofiller-polymeric matrix and polymer chain flexibility for nanocomposite matrix.

The other model [2], used for the same purpose, is based on the following percolation relationship application [5]:

$$\frac{E_n}{E_m} = 1 + 11\varphi_n^{1.7}, \tag{1}$$

where E_n and E_m are elasticity modulus of nanocomposite and matrix polymer, accordingly, φ_n is nanofiller volume contents. Let us note that the model [2] is based on the approach, principally different from micromechanical models: it is assumed that polymer composites properties are defined by their matrix structural state only and filler role consists in modification and fixation of matrix polymer structure [6].

Proceeding from the said above, the model [2] correctness checking is the present paper's purpose with using of the experimental data for nanocomposites epoxy polymer/Na$^+$-montmorillonite (EP/MMT) [7] and also the comparison of models [2] and [3] theoretical estimations for the mentioned nanocomposites.

Experimental

As matrix polymer epoxy polymer on the basis of 3,4-epoxycyclohexylmethyl-3,4-epoxycyclohexane carboxylate (epoxy monomer), cured by hexahydro-4-methylphalic anhydride at molar ratio 0.87-1.0, was used. To this mixture, nanofiller Na$^+$-montmorillonite under product name Cloisite 30B was added. Beforehand, nanofiller was mixed with denatured ethanol for removal of any excess of surfactants from silicate platelets surface. This mixture was then centrifuged at 5,000 rpm for ten min prior to decanting of the ethanol. After Cloisite 30B addition, the mixture was stirred in a mixer speedmixer DAC 150 FV at a setting of 2,500 rpm for 45 s and then at 3,000 rpm during the same time [7].

The nanocomposites EP/MMT samples were prepared in three stages. The samples were first cured isothermally for up to eight h at temperatures 353, 373, 393 and 413 K, followed by eight h at 453 K and finally 12 h at 493 K under vacuum. The samples were cooled down at a rate of 1-2 K/min and $20 \times 5 \times 1$ mm^3 pieces for dynamic spectroscopy measurements were machined. Na$^+$-montmorillonite contents in the studied nanocomposites made up 2, 5, 10 and 15 mass. % [7]. The dynamical mechanical tests were carried out by using Rheometric Scientific ARES rheometer in torsional mode at frequency 1 s^{-1} in nitrogen atmosphere.

The nanocomposites structure studies using wide-angle X-ray diffractometry were carried out on Bruker-AXS General Area Diffraction Detection System using Cu irradiation with wave length $\lambda=1.54$ Å. The distance between Na$^+$-montmorillonite platelets (d_{001} spacing) was calculated according to these measurements [7].

Results and Discussion

The model [3] essence consists of the following. Composite continuous simulation found out that properties improvement depends strongly on filler particles individual features: their volume fraction f_p, particle aspect ratio (anisotropy) L/t and ratio of particle and matrix mechanical properties. These important aspects of nanocomposites polymer/organoclay require coordinated and precise definition. The authors [3] used multiscale simulation strategy for the calculation of nanocomposite hierarchical morphology at scale of order of thousand microns, the large particle aspect ration within the matrix limits represents the structure; at scale of several microns, clay particle structure presents itself either fully divided organoclay platelets with thickness3 of order of nanometer, or organoclay parallel platelets packing, separated by interlayer galleries with thickness of nanometer level, and polymeric matrix. In this case, the quantitative structural parameters, obtained by X-ray diffraction and electron microscopy methods (silicate platelets number N in organoclay bundle, spacing d_{001} between silicate platelets), were used by the authors [3] for definition of clay "particles" geometrical features including L/t and ratio f_p to nanofiller mass content W_n. These geometrical features together with silicate platelet stiffness estimations, obtained from molecular dynamics simulations, give basis for prediction of organoclay particle effective mechanical properties. It is easy to see that it is just such approach that defines the model [3] complexity.

Unlike many mineral fillers, used at plastics production (talcum powder, mica, etc.), organoclays, in particular montmorillonite, are capable of stratifying and dispersing into separate platelets with thickness of about 1 nm [3]. The montmorillonite platelets bundles, inseparable after the introduction into polymer, are often called tactoids. The term "intercalation" describes the case when small polymer amounts penetrate into galleries between silicate platelets, which causes these platelets separation on the value ~ 2-3 nm. Exfoliation or stratification occurs at the distance between platelets (in X-raying this distance is accepted to be called the spacing d_{001}) of order of 8-10 nm. Well-stratified and dispersed nanocomposite includes separate organoclay platelets, uniformly distributed in polymeric matrix [1]. The equation (1) usage for the prediction of nanocomposites reinforcement degree E_n/E_m has shown the necessity of its following modification [1, 2]. First of all, it has been found out that not only in the full sense nanofiller (organoclay platelets), but also interfacial regions, formed on its platelets surface, with relative fraction φ_{if} are nanocomposites structure reinforcing or strengthening element. Such situation is due to interfacial layers higher stiffness in comparison with bulk polymeric matrix in virtue of strong interactions polymer – organoclay and molecular mobility suppression in the mentioned layers [2]. Thus, Na^+-montmorillonite platelets and formed on their surface interfacial regions also present themselves organoclay "effective particle" with volume concept $(\varphi_n+\varphi_{if})$, where φ_n is nanofiller volume contents. Then the equation (1) can be rewritten as follows [1]:

$$\frac{E_n}{E_m} = 1 + 11\left(\varphi_n + \varphi_{if}\right)^{1.7}.$$

(2)

Besides, within the frameworks of fractal model of interfacial layers formation in polymer nanocomposites, it has been shown that between φ_{if} and φ_n, the following relationship exists [8]:

$$\varphi_{if} = 0.955\varphi_n b$$

(3)

for intercalated organoclay and

$$\varphi_{if} = 1.910\varphi_n b \tag{4}$$

for exfoliated one, where b is parameter, characterizing the nanofiller – polymeric matrix interfacial adhesion level [9]. Let us note that the equations (3) and (4) were obtained by accounting for organoclay platelets strong anisotropy. For example, for particulate (approximately spherical) nanofiller particles, the similar equation has the following form [2]:

$$\varphi_{if} = 0.102\varphi_n b . \tag{5}$$

The equations (2)-(4) combination allows obtaining the final variant of the formulae for nanocomposites polymer/organoclay reinforcement degree determination [2]:

$$\frac{E_n}{E_m} = 1 + 11(1.955\varphi_n b)^{1.7} \tag{6}$$

for intercalated organoclay and

$$\frac{E_n}{E_m} = 1 + 11(2.910\varphi_n b)^{1.7} \tag{7}$$

for exfoliated one.

Therefore, the equations (2)-(7) comparison demonstrates that the model [2] accounts for nanofiller volume contents, its particles anisotropy degree and interfacial adhesion level but does not account for nanofiller characteristics (for example, its elasticity modulus), which differs principally this model from the treatment [3].

The nanofiller volume fraction φ_n can be estimated as follows. At it is known [5], the following interconnection exists between nanofiller volume φ_n and mass W_n contents:

$$\varphi_n = \frac{W_n}{\rho_n}, \tag{8}$$

where ρ_n is the nanofiller density, determined according to the equation [5]:

$$\rho_n = \frac{6}{S_u D_p}, \tag{9}$$

where S_u is the filler specific surface, which is equal to $\sim 74 \times 10^3$ m^2/kg for Na$^+$-montmorillonite [11], D_p is its particles size. Since Na$^+$-montmorillonite particle length is anisotropic and has length is ~ 100 nm, width is ~ 35 nm and thickness is ~ 1 nm [10], then as D_p, these sizes arithmetical mean was chosen. Then the value ρ_n=1790 kg/m^3.

In paper [7], it has been found out experimentally that for nanocomposites EP/MMT with W_n=2 and 5 mass. %, the interlayer spacing value $d_{001} \geq 10$ nm and for the same nanocomposites with W_n=10 and 15 mass. % d_{001}=6.3-9.8 nm. Therefore, according to the above-adduced classification, nanocomposites with W_n=2 and 5 mass. % are to be defined as having exfoliated organoclay and with W_n=10 and 15 mass. % – as intercalated organoclay. In Fig. 1, the curves $E_n(W_n)$, calculated for the above-indicated cases according to the equations (7) and (6), accordingly, at the condition b=1 (perfect adhesion by Kerner) and also the experimental data for nanocomposites EP/MMT (points) are adduced. As one can see, the good correspondence of the data for EP/MMT with W_n=10 and 15 mass. % and calculation according to the equation (6) and with W_n=2 and 5 mass. % – to the equation (7), which is provided for the model [2]. Therefore, the mentioned model reflects well both nanofiller contents and its platelets stratification degree and in addition, it is much simpler and physically much clearer than multiscale model [3].

One more aspect of the model [2] concerns of the parameter b determination, which characterizes interfacial adhesion nanofiller – polymeric matrix level. The condition b=1 was accepted for the value E_n estimation (Fig. 1), but this parameter can also be estimated more precisely according to the equations (6) and (7), using experimental values E_n, E_m and φ_n independent estimation of interfacial adhesion level for the studied nanocomposites can be obtained with the aid of the parameter A, determined according to the equation [12]:

$$A = \frac{1}{1-\varphi_n} \frac{\mathrm{tg}\,\delta_n}{\mathrm{tg}\,\delta_m} - 1,$$

(10)

where tg δ_n and tg δ_m are mechanical losses angle tangents for nanocomposite and matrix polymer, accordingly, the values of which are adduced in paper [7].

As it is known [12], A decrease means interfacial adhesion level increase. In Fig. 2, the comparison of parameters A and b, characterizing interfacial adhesion level, is adduced for nanocomposites EP/MMT. As it has been expected, A decrease according to the above-indicated reasons corresponds to b growth, i.e., determined according to the equations (6) and (7) parameter b characterizes the real level of interfacial adhesion in nanocomposites polymer/organoclay. As it was assumed above, the greater part of the obtained by the indicated mode values b is close to a unit.

Further, let us compare organoclay "effective particle" characteristics in models [2] and [3]. The authors [3] calculated the dependences of silicate relative fraction χ in "effective particle" and plotted its dependences on ratio d_{001}/d_{pl} for different N, where d_{pl} is silicate platelet thickness (d_{pl}=1 nm), N is these platelets number in "effective particle." In Fig. 3, theoretical calculations according to the model [3] are shown by solid lines for N=1, 2 and 4, and experimental dependences of $\varphi_n/(\varphi_n+\varphi_{if})$ in the EP/MMT – by points. As one can see, the similar trends of χ and $\varphi_n/(\varphi_n+\varphi_{if})$ change as a function of d_{001}/d_{pl} are obtained, which indicates the identity of "effective particle" definition in the models [2] and [3]. The comparison of theoretical curves and experimental points in Fig. 3 shows that for nanocomposites EP/MMT, the value N is somewhat smaller than 2, i.e., organoclay in these nanocomposites is stratified enough.

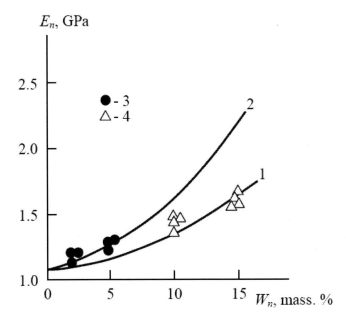

Figure 1. The comparison of calculated according to the equations (6) (1) and (7) (2) and experimental of exfoliated (3) and intercalated (4) organoclay dependences of elasticity modulus E_n on organoclay mass contents W_n for nanocomposites EP/MMT.

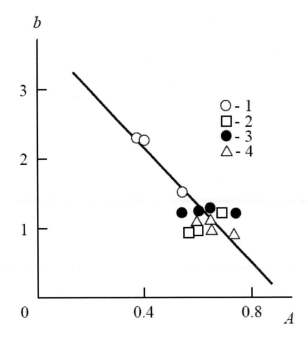

Figure 2. The relation of parameters b and A, characterizing interfacial adhesion level, for nanocomposites EP/MMT with organoclay mass contents W_n=2 (1), 5 (2), 10 (3) and 15 (4) mass. %.

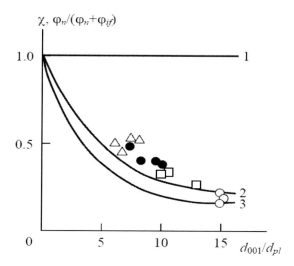

Figure 3. The theoretical dependences of organoclay fraction in "effective particle" χ for silicate platelets number in it $N=1$ (1), 2 (2) and 4 (3) and parameter $\varphi_n/(\varphi_n+\varphi_{if})$ for nanocomposites EP/MMT (points) on the ratio d_{001}/d_{pl} value. Conventional signs are the same that in Fig. 2.

In Fig. 4, theoretical dependences of "effective porticles" volume fraction and nanofiller mass contents ratio f_p/W_n on ratio d_{001}/d_{pl} value, shown by straight lines for $N=1$, 2, 3 and 4, are adduced. In the same Figure, the dependences of ratio of sum $(\varphi_n+\varphi_{if})$, which in the model [2] is f_p analogue, and W_n, obtained according to the equations (6) and (7), for nanocomposites EP/MMT (points) are given. The similarity of trends of the indicated ratios as a function of d_{001}/d_{pl} change is observed again, which confirms again "effective particle" definition in the models [2] and [3] identity. And as earlier, for nanocomposites EP/MMT, the value N can be estimated as somewhat smaller than $N=2$.

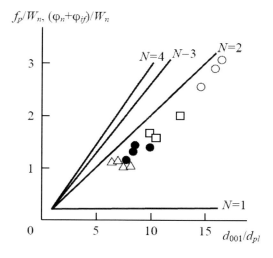

Figure 4. The dependences of organoclay fraction in "effective particles" volume and mass fractions ratio f_p/W_n in the model [3] for silicate platelets number in it $N=1$ (1), 2 (2), 3 (3) and 4 (4) and $(\varphi_n+\varphi_{if})/W_n$ in the model [2] for nanocomposites EP/MMT (points) on the ratio d_{001}/d_{pl} value. Conventional signs are the same that in Fig. 2.

Conclusion

As the present paper's results have shown, the authors [2] offered the model of organoclay "effective particle," which describes nanocomposites polymer/organoclay structure and properties so well, as the multiscale micromechanical model, elaborated in paper [3]. However, unlike latter, the model [2] is much simpler, has clear physical significance, does not require micromechanical models application and, hence, organoclay and interfacial regions characteristics usage but accounts for interfacial adhesion polymeric matrix – nanofiller real level. The indicated factors make the model [2] more suitable for applied calculations and clearer from the point of view of physical treatment in comparison with the model [3].

References

[1]   Malamatov A.Kh., Kozlov G.V., Mikitaev M.A. *The Reinforcement Mechanisms of Polymer Nanocomposites.* Moscow, Publishers the D.I. Mendeleev RKhTU, 2006, 240 p.

[2]   Mikitaev A.K., Kozlov G.V., Zaikov G.E. *Polymer Nanocomposites: Variety of Structural Forms and Applications.* New York, Nova Science Publishers, Inc., 2008, 319 p.

[3]   Sheng N., Boyce M.C., Parks D.M., Rutledge G.C., Abes J.I., Cohen R.E. Multiscale micromechanical modeling of polymer/clay nanocomposites and the effective clay particle. // *Polymer*, 2004, v. 45, № 2, pp. 487-506.

[4]   Ahmed S., Jones F.R. A review of particulate reinforcement theories of polymer composites. // *J. Mater. Sci.,* 1990, v. 25, № 12, pp. 4933-4942.

[5]   Bobryshev A.N., Kozomazov V.N., Babin L.O., Solomatov V.I. *Synergetics of Composite Materials.* Lipetsk, NPO ORIUS, 1994, 154 p.

[6]   Novikov V.U., Kozlov G.V. The fractal parametrization of filled polymers structure. // *Mekhanika Kompozitnykh Materialov*, 1999, v. 35, № 3, pp. 269-290.

[7]   Chen J.-S., Poliks M.D., Ober C.K., Zhang Y., Wiesner U., Giannelis E. Study of the interlayer expansion mechanism and thermal-mechanical properties of surface-initiated epoxy nanocomposites. // *Polymer,* 2002, v. 43, № 14, pp. 4895-4904.

[8]   Kozlov G.V., Malamatov A.Kh., Antipov E.M., Karnet Yu.N., Yanovskii Yu.G. Structure and mechanical properties of polymer nanocomposites within the frameworks of fractal concepts. // *Mekhanika Kompozitsionnykh Materialov i Konstruktsii,* 2006, v. 12, № 1, pp. 99-140.

[9]   Kozlov G.V., Aphashagova Z.Kh., Burya A.I., Lipatov Yu.S. Nanoadhesion and reinforcement mechanism of particulate-filled polymer nanocomposites. // *Inzhenernaya Fizika,* 2008, № 1, pp. 47-50.

[10] Dennis H.R., Hunter D.L., Chang D., Kim S., White J.L., Cho J.W., Paul D.R. Effect of melt processing conditions on extent of exfoliation in organoclay-based nanocomposites. // *Polymer,* 2001, v. 42, № 24, pp. 9513-9522.

[11] Pernyeszi T., Dekany I. Surface fractal and structural properties of layered clay minerals monitored by small-angle X-ray scattering and low-temperature nitrogen adsorption experiments. // *Colloid Polymer Sci.*, 2003, v. 281, № 1, pp. 73-78.

[12] Kubat J., Rigdahl M., Welander M. Characterization of interfacial interactions in high density polyethylene filled with glass spheres using dynamic-mechanical analysis. // *J. Appl. Polymer Sci.*, 1990, v. 39, № 9, pp. 1527-1539.

In: Handbook of Research on Nanomaterials, Nanochemistry ...   ISBN: 978-1-61942-525-5
Editors: A. K. Haghi and G. E. Zaikov   © 2013 Nova Science Publishers, Inc.

Chapter XXI

Theoretical Estimation Acid Force of Fluorine-Containing Pyrimidines

V. A. Babkin[1], D. S. Andreev[1], E. S. Titova[2],
I. Yu. Kameneva[2], A. I. Rakhimov[2],
S. Kubica[3] and G. E. Zaikov[4]

[1] Volgograd State Architect-Build University Sebrykov Department,
Michailovka, Volgograd region, Russia
[2] Volgograd State Pedagogical University, Volgograd, Russia
[3] Institut Inzynierii Materialow Polimerowych I Barwnikow, Torun, Poland
[4] N.M. Emanuel Institute of Biochemical Physics Russian Academy of Sciences;
Moscow, Russia

Introduction

Quantum chemical calculation of a molecules of 2-methylsulfanil-4-difluoro methoxypyrimidine, 2-ethylsulfanil-4-difluoromethoxypyrimidine, 2-isopropyl sulfanil-4-difluoromethoxypyrimidine, 2-isobutylsulfanil-4-difluoromethoxypyrimidine, 2-methyl sulfanil-4-difluoromethoxy-6-methylpyrimidine, 2-ethylsulfanil-4-difluoromethoxy-6-methyl pyrimidine, 2-isobutylsulfanil-4-difluoromethoxy-6-methylpyrimidine is executed by methods MNDO and AB INITIO in base 6-311G** with optimization of geometry on all parameters for the first time. The Optimized geometrical and electronic structure of these compounds is received.

Acid force of these compounds is theoretically appreciated (MNDO: $26 \leq pKa \leq 27$, AB INITIO: $28 \leq pKa \leq 32$). It is established, than 2-methylsulfanil-4-difluoromethoxypyrimidine, 2-ethylsulfanil-4-difluoromethoxypyrimidine, 2-isopropyl sulfanil-4-difluoro methoxy pyrimidine, 2-isobutylsulfanil-4-difluoromethoxypyrimidine, 2-methylsulfanil-4-difluoro methoxy-6-methylpyrimidine, 2-ethylsulfanil-4-difluoromethoxy-6-methylpyrimidine, 2-isobutylsulfanil-4-difluoromethoxy-6-methylpyrimidine to relate to a class of weak H-acids ($pKa > 14$).

Aims and Backgrounds

The aim of this work is a quantum chemical calculation of molecules 2-methylsulfanil-4-difluoromethoxypyrimidine, 2-ethylsulfanil-4-difluoro-methoxypyrimidine, 2-isopropyl sulfanil-4-difluoromethoxypyrimidine, 2-isobutylsulfanil-4-difluoromethoxypyrimidine, 2-methyl sulfanil-4-difluoro-methoxy-6-methylpyrimidine, 2-ethylsulfanil-4-difluoromethoxy-6-methyl-pyrimidine, 2-isopropylsulfanil-4-difluoromethoxy-6-methylpyrimidine, 2-isobutyl sulfanil-4-difluoromethoxy-6-methylpyrimidine[1] and theoretical estimation their acid power by methods MNDO and AB INITIO in base 6-311G**.

The calculation was done with optimization of all parameters by standard gradient method built-in in PC GAMESS [2]. The calculation was executed in approach the insulated molecule in gas phase. Program MacMolPlt was used for visual presentation of models of molecules [3].

Results of Calculation

Optimized geometric and electronic structures, general and electronic energies of molecules 2-methylsulfanil-4-difluoromethoxypyrimidine, 2-ethylsulfanil-4-difluoro methoxy pyrimidine, 2-isopropylsulfanil-4-difluoro-methoxypyrimidine, 2-isobutylsulfanil-4-difluoro methoxypyrimidine, 2-methylsulfanil-4-difluoromethoxy-6-methylpyrimidine, 2-ethyl sulfanil-4-difluoromethoxy-6-methyl pyrimidine, 2-isopropylsulfanil-4-difluoromethoxy-6-methyl pyrimidine, 2-isobutylsulfanil-4-difluoromethoxy-6-methyl pyrimidine were received by methods MNDO and AB INITIO in base 6-311G** and are shown on Fig. 1-16 and in Table 1-17. The universal factor of acidity was calculated for method MNDO by formula [4]: pKa=42.11-147.18q_{max}^{H+} (where, +0.10≤q_{max}^{H+} ≤ +0.11 − a maximum positive charge on atom of the hydrogen) for method AB INITIO: pKa=49.04-134.61 q_{max}^{H+}, which used with success, for example in [5-31]. (where, +0.13≤q_{max}^{H+} ≤ +0.16). 26≤pKa≤27 (MNDO), 28≤pKa≤32 (AB INITIO).

Quantum-chemical calculation of molecules 2-methylsulfanil-4-difluoro metho xypyrimidine, 2-ethylsulfanil-4-difluoromethoxypyrimidine, 2-isopropylsulfanil-4-difluoro methoxypyrimidine, 2-isobutylsulfanil-4-difluoromethoxypyrimidine, 2-methylsulfanil-4-difluoromethoxy-6-methylpyrimidine, 2-ethylsulfanil-4-difluoromethoxy-6-methyl pyrimi dine, 2-isobutylsulfanil-4-difluoromethoxy-6-methylpyrimidine by methods MNDO and AB INITIO in base 6-311G** was executed for the first time. Optimized geometric and electronic structures of these compounds were received.

Acid power of studied molecules was theoretically evaluated (26≤pKa≤27 (MNDO), 28≤pKa≤32 (AB INITIO)). These compounds pertain to class of very weak H-acids (pKa>14).

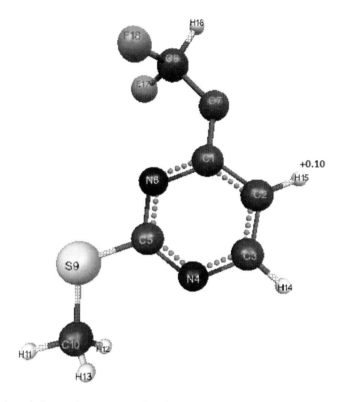

Figure 1. Geometric and electronic structure of molecule of 2-methylsulfanil-4-difluoromethoxypyrimidine (method MNDO); (E_0= -267467 kDg/mol, E_{el}= -1146348 kDg/mol, pKa = 27).

Table 1. Optimized bond lengths, valence corners and charges on atoms of the molecule of 2-methylsulfanil-4-difluoromethoxypyrimidine (method MNDO)

Bond lengths	R,A	Valence corners	Grad	Atom	Charge (by Milliken)
C(2)-C(1)	1.42	C(1)-C(2)-C(3)	116	C(1)	+0.29
C(3)-C(2)	1.41	C(2)-C(3)-N(4)	122	C(2)	-0.18
N(4)-C(3)	1.35	C(3)-N(4)-C(5)	118	C(3)	+0.13
C(5)-N(4)	1.36	C(1)-N(6)-C(5)	117	N(4)	-0.30
C(5)-N(6)	1.36	C(2)-C(1)-N(6)	123	C(5)	+0.05
N(6)-C(1)	1.36	C(2)-C(1)-O(7)	119	N(6)	-0.30
O(7)-C(1)	1.35	C(1)-O(7)-C(8)	129	O(7)	-0.32
C(8)-O(7)	1.41	N(4)-C(5)-S(9)	121	C(8)	+0.59
S(9)-C(5)	1.68	C(5)-S(9)-C(10)	110	S(9)	+0.21
C(10)-S(9)	1.73	S(9)-C(10)-H(11)	106	C(10)	-0.06
H(11)-C(10)	1.11	S(9)-C(10)-H(12)	113	H(11)	+0.03
H(12)-C(10)	1.11	S(9)-C(10)-H(13)	113	H(12)	+0.02
H(13)-C(10)	1.11	C(2)-C(3)-H(14)	122	H(13)	+0.02
H(14)-C(3)	1.10	C(1)-C(2)-H(15)	123	H(14)	+0.09
H(15)-C(2)	1.09	O(7)-C(8)-H(16)	105	**H(15)**	**+0.10**
H(16)-C(8)	1.14	O(7)-C(8)-F(17)	110	H(16)	+0.09
F(17)-C(8)	1.35	O(7)-C(8)-F(18)	110	F(17)	-0.23
F(18)-C(8)	1.35			F(18)	-0.23

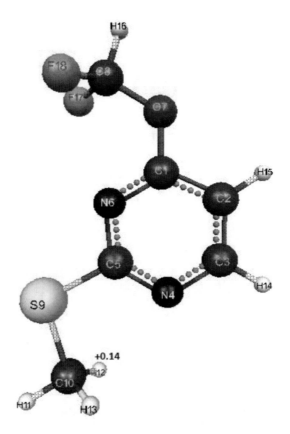

Figure 2. Geometric and electronic structure of molecule of 2-methylsulfanil-4-difluoromethoxypyrimidine (method AB INITIO); (E_0= -2649786 kDg/mol, E_{el}= -4795277 kDg/mol, pKa = 30).

Table 2. Optimized bond lengths, valence corners and charges on atoms of the molecule of 2-methylsulfanil-4-difluoromethoxypyrimidine (method AB INITIO)

Bond lengths	R,A	Valence corners	Grad	Atom	Charge
C(2)-C(1)	1.39	C(1)-C(2)-C(3)	115	C(1)	+0.52
C(3)-C(2)	1.37	C(2)-C(3)-N(4)	123	C(2)	-0.27
N(4)-C(3)	1.33	C(3)-N(4)-C(5)	116	C(3)	+0.18
C(5)-N(4)	1.31	C(1)-N(6)-C(5)	116	N(4)	-0.42
C(5)-N(6)	1.33	C(2)-C(1)-N(6)	123	C(5)	+0.25
N(6)-C(1)	1.29	C(2)-C(1)-O(7)	116	N(6)	-0.44
O(7)-C(1)	1.35	C(1)-O(7)-C(8)	123	O(7)	-0.46
C(8)-O(7)	1.36	N(4)-C(5)-S(9)	120	C(8)	+0.71
S(9)-C(5)	1.76	C(5)-S(9)-C(10)	102	S(9)	+0.12
C(10)-S(9)	1.81	S(9)-C(10)-H(11)	105	C(10)	-0.38
H(11)-C(10)	1.08	S(9)-C(10)-H(12)	111	H(11)	+0.12
H(12)-C(10)	1.08	S(9)-C(10)-H(13)	111	**H(12)**	**+0.14**
H(13)-C(10)	1.08	C(2)-C(3)-H(14)	121	H(13)	+0.14
H(14)-C(3)	1.08	C(1)-C(2)-H(15)	122	H(14)	+0.12
H(15)-C(2)	1.07	O(7)-C(8)-H(16)	106	H(15)	+0.12
H(16)-C(8)	1.08	O(7)-C(8)-F(17)	112	H(16)	+0.10
F(17)-C(8)	1.32	O(7)-C(8)-F(18)	112	F(17)	-0.28
F(18)-C(8)	1.32			F(18)	-0.28

Table 3. Optimized bond lengths, valence corners and charges on atoms of the molecule of 2-ethylsulfanil-4-difluoromethoxypyrimidine (method MNDO)

Bond lengths	R,A	Valence corners	Grad	Atom	Charge
C(2)-C(1)	1.42	C(1)-C(2)-C(3)	116	C(1)	+0.29
C(3)-C(2)	1.40	C(2)-C(3)-N(4)	122	C(2)	-0.19
N(4)-C(3)	1.35	C(3)-N(4)-C(5)	118	C(3)	+0.13
C(5)-N(4)	1.36	C(1)-N(6)-C(5)	117	N(4)	-0.31
C(5)-N(6)	1.36	C(2)-C(1)-N(6)	122	C(5)	+0.06
N(6)-C(1)	1.36	C(2)-C(1)-O(7)	120	N(6)	-0.33
O(7)-C(1)	1.35	C(1)-O(7)-C(8)	125	O(7)	-0.30
C(8)-O(7)	1.42	N(4)-C(5)-S(9)	121	C(8)	+0.58
S(9)-C(5)	1.68	C(5)-S(9)-C(10)	110	S(9)	+0.21
C(10)-S(9)	1.74	S(9)-C(10)-C(11)	110	C(10)	-0.10
C(11)-C(10)	1.53	C(10)-C(11)-H(12)	110	C(11)	+0.03
H(12)-C(11)	1.11	S(9)-C(10)-H(13)	110	H(12)	0.00
H(13)-C(10)	1.11	S(9)-C(10)-H(14)	110	H(13)	+0.03
H(14)-C(10)	1.11	C(2)-C(3)-H(15)	122	H(14)	+0.03
H(15)-C(3)	1.10	C(1)-C(2)-H(16)	122	H(15)	+0.10
H(16)-C(2)	1.09	O(7)-C(8)-H(17)	115	**H(16)**	**+0.11**
H(17)-C(8)	1.13	C(10)-C(11)-H(18)	112	H(17)	+0.10
H(18)-C(11)	1.11	C(10)-C(11)-H(19)	112	H(18)	+0.01
H(19)-C(11)	1.11	O(7)-C(8)-F(20)	107	H(19)	+0.01
F(20)-C(8)	1.35	O(7)-C(8)-F(21)	104	F(20)	-0.23
F(21)-C(8)	1.35			F(21)	-0.22

Table 4. Optimized bond lengths, valence corners and charges on atoms of the molecule of 2-ethylsulfanil-4-difluoromethoxypyrimidine (method AB INITIO)

Bond lengths	R,A	Valence corners	Grad	Atom	Charge
C(2)-C(1)	1.39	C(1)-C(2)-C(3)	115	C(1)	+0.52
C(3)-C(2)	1.37	C(2)-C(3)-N(4)	123	C(2)	-0.27
N(4)-C(3)	1.33	C(3)-N(4)-C(5)	116	C(3)	+0.18
C(5)-N(4)	1.31	C(1)-N(6)-C(5)	117	N(4)	-0.42
C(5)-N(6)	1.33	C(2) C(1) N(6)	123	C(5)	+0.25
N(6)-C(1)	1.30	C(2)-C(1)-O(7)	117	N(6)	-0.48
O(7)-C(1)	1.34	C(1)-O(7)-C(8)	120	O(7)	-0.42
C(8)-O(7)	1.37	N(4)-C(5)-S(9)	120	C(8)	+0.68
S(9)-C(5)	1.76	C(5)-S(9)-C(10)	103	S(9)	+0.13
C(10)-S(9)	1.82	S(9)-C(10)-C(11)	109	C(10)	-0.30
C(11)-C(10)	1.53	C(10)-C(11)-H(12)	109	C(11)	-0.26
H(12)-C(11)	1.09	S(9)-C(10)-H(13)	108	H(12)	+0.10
H(13)-C(10)	1.08	S(9)-C(10)-H(14)	109	**H(13)**	**+0.14**
H(14)-C(10)	1.08	C(2)-C(3)-H(15)	121	H(14)	+0.14
H(15)-C(3)	1.08	C(1)-C(2)-H(16)	122	H(15)	+0.13
H(16)-C(2)	1.07	O(7)-C(8)-H(17)	113	H(16)	+0.13
H(17)-C(8)	1.07	C(10)-C(11)-H(18)	111	H(17)	+0.10
H(18)-C(11)	1.09	C(10)-C(11)-H(19)	111	H(18)	+0.11
H(19)-C(11)	1.09	O(7)-C(8)-F(20)	110	H(19)	+0.10
F(20)-C(8)	1.32	O(7)-C(8)-F(21)	106	F(20)	-0.29
F(21)-C(8)	1.32			F(21)	-0.27

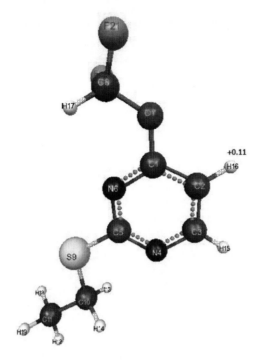

Figure 3. Geometric and electronic structure of molecule of 2-ethylsulfanil-4-difluoromethoxypyrimidine (method MNDO). (E_0= -282544 kDg/mol, E_{el}= -1262764 kDg/mol) pKa = 26.

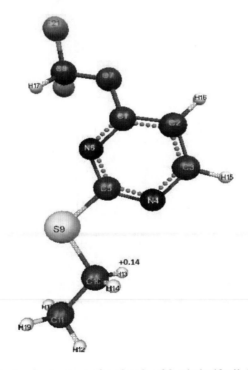

Figure 4. Geometric and electronic structure of molecule of 2-ethylsulfanil-4-difluoromethoxypyrimidine (method AB INITIO). (E_0= -2752131 kDg/mol, E_{el}= -5120463 kDg/mol) pKa = 30.

Table 5. Optimized bond lengths, valence corners and charges on atoms of the molecule of 2-isopropylsulfanil-4-difluoromethoxypyrimidine (method MNDO)

Bond lengths	R,A	Valence corners	Grad	Atom	Charge
C(2)-C(1)	1.42	C(1)-C(2)-C(3)	116	C(1)	+0.28
C(3)-C(2)	1.40	C(2)-C(3)-N(4)	122	C(2)	-0.19
N(4)-C(3)	1.35	C(3)-N(4)-C(5)	118	C(3)	+0.12
C(5)-N(4)	1.36	C(1)-N(6)-C(5)	117	N(4)	-0.31
C(5)-N(6)	1.36	C(2)-C(1)-N(6)	122	C(5)	+0.05
N(6)-C(1)	1.36	C(2)-C(1)-O(7)	120	N(6)	-0.33
O(7)-C(1)	1.35	C(1)-O(7)-C(8)	125	O(7)	-0.30
C(8)-O(7)	1.42	N(4)-C(5)-S(9)	122	C(8)	+0.58
S(9)-C(5)	1.68	C(5)-S(9)-C(10)	112	S(9)	+0.21
C(10)-S(9)	1.76	S(9)-C(10)-C(11)	107	C(10)	-0.14
C(11)-C(10)	1.54	C(10)-C(11)-H(12)	112	C(11)	+0.04
H(12)-C(11)	1.11	C(10)-C(11)-H(13)	110	H(12)	+0.01
H(13)-C(11)	1.11	C(10)-C(11)-H(14)	111	H(13)	0.00
H(14)-C(11)	1.11	S(9)-C(10)-H(15)	110	H(14)	+0.01
H(15)-C(10)	1.11	C(2)-C(3)-H(16)	122	H(15)	+0.05
H(16)-C(3)	1.10	C(1)-C(2)-H(17)	122	H(16)	+0.10
H(17)-C(2)	1.09	O(7)-C(8)-H(18)	115	**H(17)**	**+0.11**
H(18)-C(8)	1.13	S(9)-C(10)-C(19)	112	H(18)	+0.10
C(19)-C(10)	1.54	C(10)-C(19)-H(20)	111	C(19)	+0.04
H(20)-C(19)	1.11	C(10)-C(19)-H(21)	112	H(20)	0.00
H(21)-C(19)	1.11	C(10)-C(19)-H(22)	111	H(21)	+0.01
H(22)-C(19)	1.11	O(7)-C(8)-F(23)	107	H(22)	+0.01
F(23)-C(8)	1.35	O(7)-C(8)-F(24)	104	F(23)	-0.23
F(24)-C(8)	1.35			F(24)	-0.22

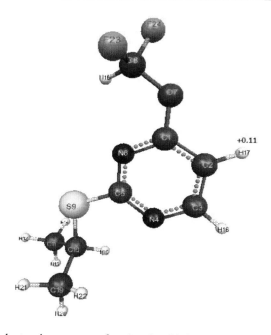

Figure 5. Geometric and electronic structure of molecule of 2-isopropylsulfanil-4-difluoromethoxypyrimidine (method MNDO). (E_0= -297601 kDg/mol, E_{el}= -1415574 kDg/mol) pKa = 26.

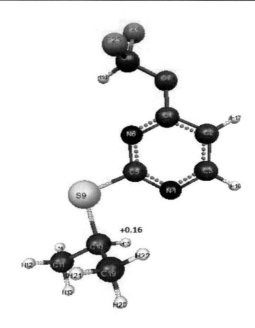

Figure 6. Geometric and electronic structure of molecule of 2-isopropylsulfanil-4-difluoromethoxypyrimidine (method AB INITIO). (E_0= -2854460 kDg/mol, E_{el}= -5531650 kDg/mol) pKa = 28.

Table 6. Optimized bond lengths, valence corners and charges on atoms of the molecule of 2-isopropylsulfanil-4-difluoromethoxypyrimidine (method AB INITIO)

Bond lengths	R,A	Valence corners	Grad	Atom	Charge
C(2)-C(1)	1.39	C(1)-C(2)-C(3)	115	C(1)	+0.53
C(3)-C(2)	1.37	C(2)-C(3)-N(4)	123	C(2)	-0.27
N(4)-C(3)	1.33	C(3)-N(4)-C(5)	116	C(3)	+0.18
C(5)-N(4)	1.31	C(1)-N(6)-C(5)	117	N(4)	-0.42
C(5)-N(6)	1.33	C(2)-C(1)-N(6)	123	C(5)	+0.26
N(6)-C(1)	1.30	C(2)-C(1)-O(7)	117	N(6)	-0.48
O(7)-C(1)	1.34	C(1)-O(7)-C(8)	120	O(7)	-0.42
C(8)-O(7)	1.37	N(4)-C(5)-S(9)	121	C(8)	+0.68
S(9)-C(5)	1.76	C(5)-S(9)-C(10)	104	S(9)	+0.13
C(10)-S(9)	1.84	S(9)-C(10)-C(11)	107	C(10)	-0.32
C(11)-C(10)	1.53	C(10)-C(11)-H(12)	111	C(11)	-0.24
H(12)-C(11)	1.09	C(10)-C(11)-H(13)	109	H(12)	+0.10
H(13)-C(11)	1.09	C(10)-C(11)-H(14)	112	H(13)	+0.10
H(14)-C(11)	1.08	S(9)-C(10)-H(15)	106	H(14)	+0.11
H(15)-C(10)	1.08	C(2)-C(3)-H(16)	121	**H(15)**	**+0.16**
H(16)-C(3)	1.08	C(1)-C(2)-H(17)	122	H(16)	+0.12
H(17)-C(2)	1.07	O(7)-C(8)-H(18)	113	H(17)	+0.13
H(18)-C(8)	1.07	S(9)-C(10)-C(19)	112	H(18)	+0.10
C(19)-C(10)	1.53	C(10)-C(19)-H(20)	110	C(19)	-0.19
H(20)-C(19)	1.09	C(10)-C(19)-H(21)	111	H(20)	+0.09
H(21)-C(19)	1.09	C(10)-C(19)-H(22)	111	H(21)	+0.10
H(22)-C(19)	1.08	O(7)-C(8)-F(23)	110	H(22)	+0.11
F(23)-C(8)	1.32	O(7)-C(8)-F(24)	106	F(23)	-0.29
F(24)-C(8)	1.32			F(24)	-0.27

Table 7. Optimized bond lengths, valence corners and charges on atoms of the molecule of 2-isobutylsulfanil-4-difluoromethoxypyrimidine (method MNDO)

Bond lengths	R,A	Valence corners	Grad	Atom	Charge
C(2)-C(1)	1.42	C(1)-C(2)-C(3)	116	C(1)	+0.29
C(3)-C(2)	1.40	C(2)-C(3)-N(4)	122	C(2)	-0.18
N(4)-C(3)	1.36	C(3)-N(4)-C(5)	118	C(3)	+0.13
C(5)-N(4)	1.36	C(1)-N(6)-C(5)	117	N(4)	-0.31
C(5)-N(6)	1.36	C(2)-C(1)-N(6)	122	C(5)	+0.06
N(6)-C(1)	1.36	C(2)-C(1)-O(7)	120	N(6)	-0.34
O(7)-C(1)	1.35	C(1)-O(7)-C(8)	125	O(7)	-0.30
C(8)-O(7)	1.42	N(4)-C(5)-S(9)	123	C(8)	+0.58
S(9)-C(5)	1.68	C(5)-S(9)-C(10)	116	S(9)	+0.21
C(10)-S(9)	1.77	S(9)-C(10)-C(11)	102	C(10)	-0.17
C(11)-C(10)	1.56	C(10)-C(11)-H(12)	112	C(11)	+0.05
H(12)-C(11)	1.11	C(10)-C(11)-H(13)	111	H(12)	+0.01
H(13)-C(11)	1.11	C(10)-C(11)-H(14)	112	H(13)	0.00
H(14)-C(11)	1.11	C(2)-C(3)-H(15)	122	H(14)	+0.01
H(15)-C(3)	1.10	C(1)-C(2)-H(16)	122	**H(15)**	**+0.10**
H(16)-C(2)	1.09	O(7)-C(8)-H(17)	115	H(16)	+0.10
H(17)-C(8)	1.13	S(9)-C(10)-C(18)	112	H(17)	+0.10
C(18)-C(10)	1.55	C(10)-C(18)-H(19)	111	C(18)	+0.04
H(19)-C(18)	1.11	C(10)-C(18)-H(20)	111	H(19)	0.00
H(20)-C(18)	1.11	C(10)-C(18)-H(21)	113	H(20)	0.00
H(21)-C(18)	1.11	S(9)-C(10)-C(22)	112	H(21)	+0.02
C(22)-C(10)	1.55	C(10)-C(22)-H(23)	111	C(22)	+0.04
H(23)-C(22)	1.11	C(10)-C(22)-H(24)	111	H(23)	0.00
H(24)-C(22)	1.11	C(10)-C(22)-H(25)	113	H(24)	0.00
H(25)-C(22)	1.11	O(7)-C(8)-F(26)	107	H(25)	+0.01
F(26)-C(8)	1.35	O(7)-C(8)-F(27)	104	F(26)	-0.23
F(27)-C(8)	1.35			F(27)	-0.22

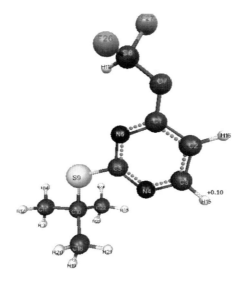

Figure 7. Geometric and electronic structure of molecule of 2-isobutylsulfanil-4-difluoromethoxypyrimidine (method MNDO). (E_0= -312632 kDg/mol, E_{el}= -1587688 kDg/mol) pKa = 27.

Figure 8. Geometric and electronic structure of molecule of 2-isobutylsulfanil-4-difluoromethoxypyrimidine (method AB INITIO). (E_0= -2956778 kDg/mol, E_{el}= -5971860 kDg/mol) pKa = 32.

Table 8. Optimized bond lengths, valence corners and charges on atoms of the molecule of 2-isobutylsulfanil-4-difluoromethoxypyrimidine (method AB INITIO)

Bond lengths	R,A	Valence corners	Grad	Atom	Charge
C(2)-C(1)	1.39	C(1)-C(2)-C(3)	114	C(1)	+0.52
C(3)-C(2)	1.37	C(2)-C(3)-N(4)	123	C(2)	-0.27
N(4)-C(3)	1.33	C(3)-N(4)-C(5)	116	C(3)	+0.18
C(5)-N(4)	1.31	C(1)-N(6)-C(5)	117	N(4)	-0.42
C(5)-N(6)	1.33	C(2)-C(1)-N(6)	123	C(5)	+0.25
N(6)-C(1)	1.30	C(2)-C(1)-O(7)	117	N(6)	-0.48
O(7)-C(1)	1.34	C(1)-O(7)-C(8)	120	O(7)	-0.42
C(8)-O(7)	1.37	N(4)-C(5)-S(9)	123	C(8)	+0.68
S(9)-C(5)	1.76	C(5)-S(9)-C(10)	109	S(9)	+0.15
C(10)-S(9)	1.86	S(9)-C(10)-C(11)	102	C(10)	-0.35
C(11)-C(10)	1.54	C(10)-C(11)-H(12)	112	C(11)	-0.23
H(12)-C(11)	1.09	C(10)-C(11)-H(13)	109	H(12)	+0.11
H(13)-C(11)	1.09	C(10)-C(11)-H(14)	112	H(13)	+0.10
H(14)-C(11)	1.09	C(2)-C(3)-H(15)	121	H(14)	+0.11
H(15)-C(3)	1.08	C(1)-C(2)-H(16)	122	H(15)	+0.12
H(16)-C(2)	1.07	O(7)-C(8)-H(17)	113	**H(16)**	**+0.13**
H(17)-C(8)	1.07	S(9)-C(10)-C(18)	112	H(17)	+0.10
C(18)-C(10)	1.53	C(10)-C(18)-H(19)	109	C(18)	-0.18
H(19)-C(18)	1.09	C(10)-C(18)-H(20)	111	H(19)	+0.09
H(20)-C(18)	1.08	C(10)-C(18)-H(21)	111	H(20)	+0.10
H(21)-C(18)	1.08	S(9)-C(10)-C(22)	112	H(21)	+0.13
C(22)-C(10)	1.53	C(10)-C(22)-H(23)	109	C(22)	-0.18
H(23)-C(22)	1.09	C(10)-C(22)-H(24)	111	H(23)	+0.09
H(24)-C(22)	1.08	C(10)-C(22)-H(25)	111	H(24)	+0.10
H(25)-C(22)	1.08	O(7)-C(8)-F(26)	110	H(25)	+0.13
F(26)-C(8)	1.32	O(7)-C(8)-F(27)	106	F(26)	-0.29
F(27)-C(8)	1.32			F(27)	-0.27

Table 9. Optimized bond lengths, valence corners and charges on atoms of the molecule of 2-methylsulfanil-4-difluoromethoxy-6-methylpyrimidine (method MNDO)

Bond lengths	R,A	Valence corners	Grad	Atom	Charge
C(2)-C(1)	1.42	C(1)-C(2)-C(3)	117	C(1)	+0.29
C(3)-C(2)	1.41	C(2)-C(3)-N(4)	121	C(2)	-0.18
N(4)-C(3)	1.36	C(3)-N(4)-C(5)	119	C(3)	+0.10
C(5)-N(4)	1.36	C(1)-N(6)-C(5)	117	N(4)	-0.30
C(5)-N(6)	1.36	C(2)-C(1)-N(6)	123	C(5)	+0.06
N(6)-C(1)	1.36	C(2)-C(1)-O(7)	119	N(6)	-0.33
O(7)-C(1)	1.35	C(1)-O(7)-C(8)	125	O(7)	-0.31
C(8)-O(7)	1.42	N(4)-C(5)-S(9)	121	C(8)	+0.58
S(9)-C(5)	1.68	C(5)-S(9)-C(10)	111	S(9)	+0.20
C(10)-S(9)	1.73	S(9)-C(10)-H(11)	106	C(10)	-0.05
H(11)-C(10)	1.11	C(1)-C(2)-H(12)	122	H(11)	+0.03
H(12)-C(2)	1.09	O(7)-C(8)-H(13)	115	**H(12)**	**+0.10**
H(13)-C(8)	1.13	S(9)-C(10)-H(14)	113	H(13)	+0.10
H(14)-C(10)	1.11	S(9)-C(10)-H(15)	113	H(14)	+0.02
H(15)-C(10)	1.11	C(2)-C(3)-C(16)	123	H(15)	+0.02
C(16)-C(3)	1.51	C(3)-C(16)-H(17)	113	C(16)	+0.07
H(17)-C(16)	1.11	C(3)-C(16)-H(18)	110	H(17)	+0.03
H(18)-C(16)	1.11	C(3)-C(16)-H(19)	110	H(18)	+0.01
H(19)-C(16)	1.11	O(7)-C(8)-F(20)	104	H(19)	+0.01
F(20)-C(8)	1.35	O(7)-C(8)-F(21)	107	F(20)	-0.22
F(21)-C(8)	1.35			F(21)	-0.23

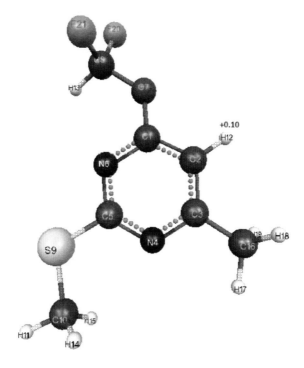

Figure 9. Geometric and electronic structure of molecule of 2-methylsulfanil-4-difluoromethoxy-6-methylpyrimidine (method MNDO). (E_0- -282558 kDg/mol, $E_{el}=$ 1279166 kDg/mol) pKa = 27.

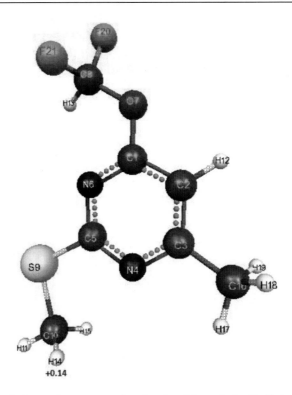

Figure 10. Geometric and electronic structure of molecule of 2-methylsulfanil-4-difluoromethoxy-6-methylpyrimidine (method AB INITIO). (E_0= -2752149 kDg/mol, E_{el}= -5147744 kDg/mol) pKa = 30.

Table 10. Optimized bond lengths, valence corners and charges on atoms of the molecule of 2-methylsulfanil-4-difluoromethoxy-6-methylpyrimidine (method AB INITIO)

Bond lengths	R,A	Valence corners	Grad	Atom	Charge
C(2)-C(1)	1.39	C(1)-C(2)-C(3)	116	C(1)	+0.53
C(3)-C(2)	1.38	C(2)-C(3)-N(4)	122	C(2)	-0.24
N(4)-C(3)	1.33	C(3)-N(4)-C(5)	117	C(3)	+0.17
C(5)-N(4)	1.31	C(1)-N(6)-C(5)	116	N(4)	-0.45
C(5)-N(6)	1.33	C(2)-C(1)-N(6)	124	C(5)	+0.27
N(6)-C(1)	1.30	C(2)-C(1)-O(7)	117	N(6)	-0.49
O(7)-C(1)	1.34	C(1)-O(7)-C(8)	120	O(7)	-0.43
C(8)-O(7)	1.37	N(4)-C(5)-S(9)	120	C(8)	+0.68
S(9)-C(5)	1.76	C(5)-S(9)-C(10)	102	S(9)	+0.11
C(10)-S(9)	1.81	S(9)-C(10)-H(11)	105	C(10)	-0.38
H(11)-C(10)	1.08	C(1)-C(2)-H(12)	121	H(11)	+0.13
H(12)-C(2)	1.07	O(7)-C(8)-H(13)	113	H(12)	+0.12
H(13)-C(8)	1.07	S(9)-C(10)-H(14)	111	H(13)	+0.10
H(14)-C(10)	1.08	S(9)-C(10)-H(15)	111	**H(14)**	**+0.14**
H(15)-C(10)	1.08	C(2)-C(3)-C(16)	121	H(15)	+0.14
C(16)-C(3)	1.50	C(3)-C(16)-H(17)	110	C(16)	-0.21
H(17)-C(16)	1.08	C(3)-C(16)-H(18)	110	H(17)	+0.13
H(18)-C(16)	1.09	C(3)-C(16)-H(19)	110	H(18)	+0.12
H(19)-C(16)	1.09	O(7)-C(8)-F(20)	106	H(19)	+0.12
F(20)-C(8)	1.32	O(7)-C(8)-F(21)	110	F(20)	-0.27
F(21)-C(8)	1.32			F(21)	-0.29

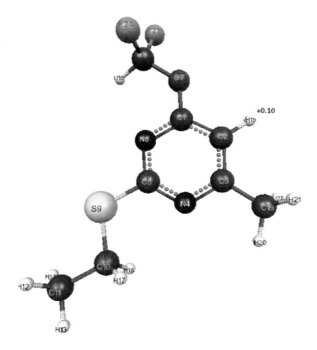

Figure 11. Geometric and electronic structure of molecule of 2-ethylsulfanil-4-difluoromethoxy-6-methylpyrimidine (method MNDO). (E_0= -297630 kDg/mol, E_{el}= -1416522 kDg/mol) pKa = 27.

Table 11. Optimized bond lengths, valence corners and charges on atoms of the molecule of 2-ethylsulfanil-4-difluoromethoxy-6-methylpyrimidine (method MNDO)

Bond lengths	R,A	Valence corners	Grad	Atom	Charge
C(2)-C(1)	1.42	C(1)-C(2)-C(3)	117	C(1)	+0.29
C(3)-C(2)	1.41	C(2)-C(3)-N(4)	121	C(2)	-0.18
N(4)-C(3)	1.36	C(3)-N(4)-C(5)	119	C(3)	+0.10
C(5)-N(4)	1.36	C(1)-N(6)-C(5)	117	N(4)	-0.30
C(5)-N(6)	1.36	C(2)-C(1)-N(6)	123	C(5)	+0.06
N(6)-C(1)	1.36	C(2)-C(1)-O(7)	119	N(6)	-0.33
O(7)-C(1)	1.35	C(1)-O(7)-C(8)	125	O(7)	-0.31
C(8)-O(7)	1.42	N(4)-C(5)-S(9)	121	C(8)	+0.58
S(9)-C(5)	1.68	C(5)-S(9)-C(10)	110	S(9)	+0.21
C(10)-S(9)	1.74	S(9)-C(10)-C(11)	110	C(10)	-0.10
C(11)-C(10)	1.53	C(10)-C(11)-H(12)	112	C(11)	+0.03
H(12)-C(11)	1.11	C(10)-C(11)-H(13)	110	H(12)	+0.01
H(13)-C(11)	1.11	C(10)-C(11)-H(14)	112	H(13)	0.00
H(14)-C(11)	1.11	C(1)-C(2)-H(15)	122	H(14)	+0.01
H(15)-C(2)	1.09	O(7)-C(8)-H(16)	115	**H(15)**	**+0.10**
H(16)-C(8)	1.13	S(9)-C(10)-H(17)	110	H(16)	+0.10
H(17)-C(10)	1.11	S(9)-C(10)-H(18)	110	H(17)	+0.03
H(18)-C(10)	1.11	C(2)-C(3)-C(19)	123	H(18)	+0.03
C(19)-C(3)	1.51	C(3)-C(19)-H(20)	113	C(19)	+0.07
H(20)-C(19)	1.11	C(3)-C(19)-H(21)	110	H(20)	+0.03
H(21)-C(19)	1.11	C(3)-C(19)-H(22)	110	H(21)	+0.01
H(22)-C(19)	1.11	O(7)-C(8)-F(23)	104	H(22)	+0.01
F(23)-C(8)	1.35	O(7)-C(8)-F(24)	107	F(23)	-0.22
F(24)-C(8)	1.35			F(24)	-0.23

Table 12. Optimized bond lengths, valence corners and charges on atoms of the molecule of 2-ethylsulfanil-4-difluoromethoxy-6-methylpyrimidine (method AB INITIO)

Bond lengths	R,A	Valence corners	Grad	Atom	Charge
C(2)-C(1)	1.39	C(1)-C(2)-C(3)	115	C(1)	+0.54
C(3)-C(2)	1.37	C(2)-C(3)-N(4)	122	C(2)	-0.24
N(4)-C(3)	1.34	C(3)-N(4)-C(5)	117	C(3)	+0.15
C(5)-N(4)	1.30	C(1)-N(6)-C(5)	116	N(4)	-0.46
C(5)-N(6)	1.33	C(2)-C(1)-N(6)	124	C(5)	+0.26
N(6)-C(1)	1.30	C(2)-C(1)-O(7)	117	N(6)	-0.49
O(7)-C(1)	1.34	C(1)-O(7)-C(8)	120	O(7)	-0.43
C(8)-O(7)	1.37	N(4)-C(5)-S(9)	120	C(8)	+0.68
S(9)-C(5)	1.76	C(5)-S(9)-C(10)	103	S(9)	+0.13
C(10)-S(9)	1.82	S(9)-C(10)-C(11)	109	C(10)	-0.30
C(11)-C(10)	1.53	C(10)-C(11)-H(12)	111	C(11)	-0.26
H(12)-C(11)	1.09	C(10)-C(11)-H(13)	109	H(12)	+0.10
H(13)-C(11)	1.09	C(10)-C(11)-H(14)	111	H(13)	+0.10
H(14)-C(11)	1.09	C(1)-C(2)-H(15)	121	H(14)	+0.10
H(15)-C(2)	1.07	O(7)-C(8)-H(16)	113	H(15)	+0.12
H(16)-C(8)	1.07	S(9)-C(10)-H(17)	109	H(16)	+0.10
H(17)-C(10)	1.08	S(9)-C(10)-H(18)	109	**H(17)**	**+0.14**
H(18)-C(10)	1.08	C(2)-C(3)-C(19)	122	H(18)	+0.14
C(19)-C(3)	1.50	C(3)-C(19)-H(20)	109	C(19)	-0.16
H(20)-C(19)	1.08	C(3)-C(19)-H(21)	109	H(20)	+0.12
H(21)-C(19)	1.08	C(3)-C(19)-H(22)	112	H(21)	+0.12
H(22)-C(19)	1.08	O(7)-C(8)-F(23)	106	H(22)	+0.10
F(23)-C(8)	1.32	O(7)-C(8)-F(24)	110	F(23)	-0.27
F(24)-C(8)	1.32			F(24)	-0.29

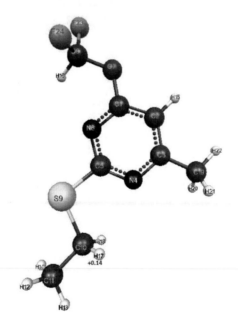

Figure 12. Geometric and electronic structure of molecule of 2-ethylsulfanil-4-difluoromethoxy-6-methylpyrimidine (method AB INITIO). (E_0= -2854483 kDg/mol, E_{el}= -5519953 kDg/mol) pKa = 30.

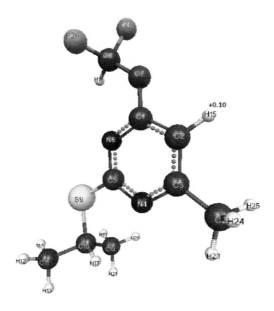

Figure 13. Geometric and electronic structure of molecule of 2-isopropylsulfanil-4-difluoromethoxy-6-methylpyrimidine (method MNDO). (E_0= -312687 kDg/mol, E_{el}= -1578119 kDg/mol) pKa = 27.

Table 13. Optimized bond lengths, valence corners and charges on atoms of the molecule of 2-isopropylsulfanil-4-difluoromethoxy-6-methylpyrimidine (method MNDO)

Bond lengths	R,A	Valence corners	Grad	Atom	Charge
C(2)-C(1)	1.42	C(1)-C(2)-C(3)	116	C(1)	+0.28
C(3)-C(2)	1.41	C(2)-C(3)-N(4)	121	C(2)	-0.18
N(4)-C(3)	1.36	C(3)-N(4)-C(5)	119	C(3)	+0.10
C(5)-N(4)	1.35	C(1)-N(6)-C(5)	117	N(4)	-0.30
C(5)-N(6)	1.36	C(2)-C(1)-N(6)	123	C(5)	+0.06
N(6)-C(1)	1.36	C(2)-C(1)-O(7)	119	N(6)	-0.33
O(7)-C(1)	1.35	C(1)-O(7)-C(8)	125	O(7)	-0.31
C(8)-O(7)	1.42	N(4)-C(5)-S(9)	122	C(8)	+0.58
S(9)-C(5)	1.68	C(5) S(9) C(10)	112	S(9)	+0.21
C(10)-S(9)	1.76	S(9)-C(10)-C(11)	107	C(10)	-0.14
C(11)-C(10)	1.54	C(10)-C(11)-H(12)	111	C(11)	+0.04
H(12)-C(11)	1.11	C(10)-C(11)-H(13)	110	H(12)	+0.01
H(13)-C(11)	1.11	C(10)-C(11)-H(14)	112	H(13)	0.00
II(14)-C(11)	1.11	C(1)-C(2)-H(15)	122	H(14)	+0.01
H(15)-C(2)	1.09	O(7)-C(8)-H(16)	115	**H(15)**	**+0.10**
H(16)-C(8)	1.13	S(9)-C(10)-H(17)	110	H(16)	+0.10
H(17)-C(10)	1.11	S(9)-C(10)-C(18)	112	H(17)	+0.05
C(18)-C(10)	1.54	C(10)-C(18)-H(19)	112	C(18)	+0.04
H(19)-C(18)	1.11	C(10)-C(18)-H(20)	111	H(19)	+0.01
H(20)-C(18)	1.11	C(10)-C(18)-H(21)	111	H(20)	0.00
H(21)-C(18)	1.11	C(2)-C(3)-C(22)	123	H(21)	0.00
C(22)-C(3)	1.51	C(3)-C(22)-H(23)	113	C(22)	+0.07
H(23)-C(22)	1.11	C(3)-C(22)-H(24)	110	H(23)	+0.03
H(24)-C(22)	1.11	C(3)-C(22)-H(25)	110	H(24)	+0.01
H(25)-C(22)	1.11	O(7)-C(8)-F(26)	107	H(25)	+0.01
F(26)-C(8)	1.35	O(7)-C(8)-F(27)	104	F(26)	-0.23
F(27)-C(8)	1.35			F(27)	-0.22

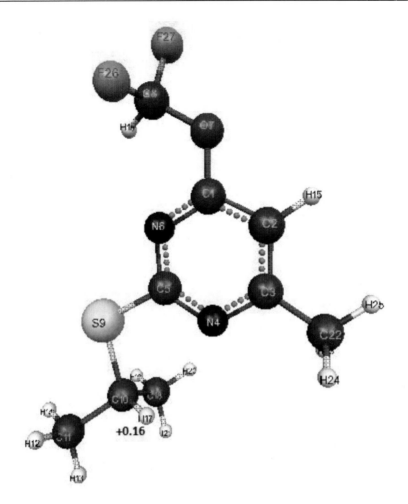

Figure 14. Geometric and electronic structure of molecule of 2-isopropylsulfanil-4-difluoromethoxy-6-methylpyrimidine (method AB INITIO). (E_0= -2956812 kDg/mol, E_{el}= -5946904 kDg/mol) pKa = 28.

Table 14. Optimized bond lengths, valence corners and charges on atoms of the molecule of 2-isopropylsulfanil-4-difluoromethoxy-6-methylpyrimidine (method AB INITIO)

Bond lengths	R,A	Valence corners	Grad	Atom	Charge
C(2)-C(1)	1.39	C(1)-C(2)-C(3)	115	C(1)	+0.54
C(3)-C(2)	1.37	C(2)-C(3)-N(4)	122	C(2)	-0.24
N(4)-C(3)	1.34	C(3)-N(4)-C(5)	117	C(3)	+0.15
C(5)-N(4)	1.30	C(1)-N(6)-C(5)	116	N(4)	-0.46
C(5)-N(6)	1.33	C(2)-C(1)-N(6)	124	C(5)	+0.26
N(6)-C(1)	1.30	C(2)-C(1)-O(7)	117	N(6)	-0.49
O(7)-C(1)	1.34	C(1)-O(7)-C(8)	120	O(7)	-0.43
C(8)-O(7)	1.37	N(4)-C(5)-S(9)	121	C(8)	+0.68
S(9)-C(5)	1.76	C(5)-S(9)-C(10)	104	S(9)	+0.13
C(10)-S(9)	1.83	S(9)-C(10)-C(11)	107	C(10)	-0.32
C(11)-C(10)	1.53	C(10)-C(11)-H(12)	111	C(11)	-0.24

Bond lengths	R,A	Valence corners	Grad	Atom	Charge
H(12)-C(11)	1.08	C(10)-C(11)-H(13)	109	H(12)	+0.11
H(13)-C(11)	1.09	C(10)-C(11)-H(14)	111	H(13)	+0.10
H(14)-C(11)	1.09	C(1)-C(2)-H(15)	121	H(14)	+0.10
H(15)-C(2)	1.07	O(7)-C(8)-H(16)	113	H(15)	+0.12
H(16)-C(8)	1.07	S(9)-C(10)-H(17)	106	H(16)	+0.10
H(17)-C(10)	1.08	S(9)-C(10)-C(18)	112	**H(17)**	**+0.16**
C(18)-C(10)	1.53	C(10)-C(18)-H(19)	111	C(18)	-0.19
H(19)-C(18)	1.09	C(10)-C(18)-H(20)	111	H(19)	+0.09
H(20)-C(18)	1.08	C(10)-C(18)-H(21)	110	H(20)	+0.11
H(21)-C(18)	1.09	C(2)-C(3)-C(22)	122	H(21)	+0.09
C(22)-C(3)	1.50	C(3)-C(22)-H(23)	109	C(22)	-0.16
H(23)-C(22)	1.08	C(3)-C(22)-H(24)	109	H(23)	+0.12
H(24)-C(22)	1.08	C(3)-C(22)-H(25)	112	H(24)	+0.12
H(25)-C(22)	1.08	O(7)-C(8)-F(26)	110	H(25)	+0.10
F(26)-C(8)	1.32	O(7)-C(8)-F(27)	106	F(26)	-0.29
F(27)-C(8)	1.32			F(27)	-0.27

Table 15. Optimized bond lengths, valence corners and charges on atoms of the molecule of 2-isobutylsulfanil-4-difluoromethoxy-6-methylpyrimidine (method MNDO)

Bond lengths	R,A	Valence corners	Grad	Atom	Charge
C(2)-C(1)	1.42	C(1)-C(2)-C(3)	116	C(1)	+0.29
C(3)-C(2)	1.41	C(2)-C(3)-N(4)	121	C(2)	-0.18
N(4)-C(3)	1.36	C(3)-N(4)-C(5)	119	C(3)	+0.10
C(5)-N(4)	1.35	C(1)-N(6)-C(5)	117	N(4)	-0.30
C(5)-N(6)	1.36	C(2)-C(1)-N(6)	123	C(5)	+0.06
N(6)-C(1)	1.36	C(2)-C(1)-O(7)	119	N(6)	-0.33
O(7)-C(1)	1.35	C(1)-O(7)-C(8)	125	O(7)	-0.31
C(8)-O(7)	1.42	N(4)-C(5)-S(9)	123	C(8)	+0.58
S(9)-C(5)	1.68	C(5)-S(9)-C(10)	116	S(9)	+0.21
C(10)-S(9)	1.77	S(9)-C(10)-C(11)	102	C(10)	-0.17
C(11)-C(10)	1.56	C(10)-C(11)-H(12)	112	C(11)	+0.05
H(12)-C(11)	1.11	C(10)-C(11)-H(13)	111	H(12)	+0.01
H(13)-C(11)	1.11	C(10)-C(11)-H(14)	112	H(13)	0.00
H(14)-C(11)	1.11	C(1)-C(2)-H(15)	122	H(14)	+0.01
H(15)-C(2)	1.09	O(7)-C(8)-H(16)	115	**H(15)**	**+0.10**
H(16)-C(8)	1.13	S(9)-C(10)-C(17)	112	H(16)	+0.10
C(17)-C(10)	1.55	C(10)-C(17)-H(18)	111	C(17)	+0.04
H(18)-C(17)	1.11	C(10)-C(17)-H(19)	111	H(18)	0.00
H(19)-C(17)	1.11	C(10)-C(17)-H(20)	113	H(19)	0.00
H(20)-C(17)	1.11	S(9)-C(10)-C(21)	112	H(20)	+0.01
C(21)-C(10)	1.55	C(10)-C(21)-H(22)	111	C(21)	+0.04
H(22)-C(21)	1.11	C(10)-C(21)-H(23)	111	H(22)	0.00
H(23)-C(21)	1.11	C(10)-C(21)-H(24)	112	H(23)	0.00
H(24)-C(21)	1.11	C(2)-C(3)-C(25)	123	H(24)	+0.01
C(25)-C(3)	1.51	C(3)-C(25)-H(26)	113	C(25)	+0.07
H(26)-C(25)	1.11	C(3)-C(25)-H(27)	110	H(26)	+0.03
H(27)-C(25)	1.11	C(3)-C(25)-H(28)	110	H(27)	+0.01
H(28)-C(25)	1.11	O(7)-C(8)-F(29)	104	H(28)	+0.01
F(29)-C(8)	1.35	O(7)-C(8)-F(30)	107	F(29)	-0.22
F(30)-C(8)	1.35			F(30)	-0.23

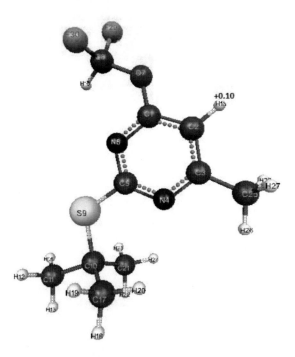

Figure 15. Geometric and electronic structure of molecule of 2-isobutylsulfanil-4-difluoromethoxy-6-methylpyrimidine (method MNDO). (E_0= -327720 kDg/mol, E_{el}= -1758670 kDg/mol) **pKa** = 27.

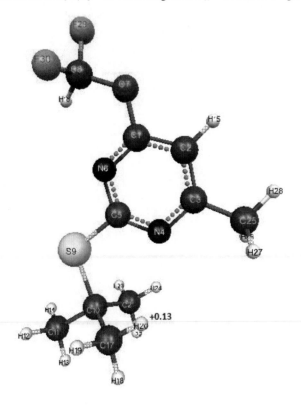

Figure 16. Geometric and electronic structure of molecule of 2-isobutylsulfanil-4-difluoromethoxy-6-methylpyrimidine (method AB INITIO). (E_0= -3059131 kDg/mol, E_{el}= -6403461 kDg/mol) **pKa** = 32.

Table 16. Optimized bond lengths, valence corners and charges on atoms of the molecule of 2-isobutylsulfanil-4-difluoromethoxy-6-methylpyrimidine (method AB INITIO)

Bond lengths	R,A	Valence corners	Grad	Atom	Charge
C(2)-C(1)	1.39	C(1)-C(2)-C(3)	115	C(1)	+0.54
C(3)-C(2)	1.37	C(2)-C(3)-N(4)	122	C(2)	-0.24
N(4)-C(3)	1.34	C(3)-N(4)-C(5)	117	C(3)	+0.15
C(5)-N(4)	1.30	C(1)-N(6)-C(5)	116	N(4)	-0.46
C(5)-N(6)	1.33	C(2)-C(1)-N(6)	124	C(5)	+0.25
N(6)-C(1)	1.30	C(2)-C(1)-O(7)	117	N(6)	-0.49
O(7)-C(1)	1.34	C(1)-O(7)-C(8)	120	O(7)	-0.43
C(8)-O(7)	1.37	N(4)-C(5)-S(9)	122	C(8)	+0.68
S(9)-C(5)	1.76	C(5)-S(9)-C(10)	109	S(9)	+0.14
C(10)-S(9)	1.86	S(9)-C(10)-C(11)	103	C(10)	-0.35
C(11)-C(10)	1.54	C(10)-C(11)-H(12)	112	C(11)	-0.23
H(12)-C(11)	1.09	C(10)-C(11)-H(13)	109	H(12)	+0.11
H(13)-C(11)	1.09	C(10)-C(11)-H(14)	112	H(13)	+0.10
H(14)-C(11)	1.09	C(1)-C(2)-H(15)	121	H(14)	+0.11
H(15)-C(2)	1.07	O(7)-C(8)-H(16)	113	H(15)	+0.12
H(16)-C(8)	1.07	S(9)-C(10)-C(17)	112	H(16)	+0.10
C(17)-C(10)	1.53	C(10)-C(17)-H(18)	109	C(17)	-0.18
H(18)-C(17)	1.09	C(10)-C(17)-H(19)	111	H(18)	+0.09
H(19)-C(17)	1.08	C(10)-C(17)-H(20)	111	H(19)	+0.10
H(20)-C(17)	1.08	S(9)-C(10)-C(21)	112	**H(20)**	**+0.13**
C(21)-C(10)	1.53	C(10)-C(21)-H(22)	109	C(21)	-0.18
H(22)-C(21)	1.09	C(10)-C(21)-H(23)	111	H(22)	+0.09
H(23)-C(21)	1.08	C(10)-C(21)-H(24)	111	H(23)	+0.10
H(24)-C(21)	1.08	C(2)-C(3)-C(25)	122	H(24)	+0.13
C(25)-C(3)	1.50	C(3)-C(25)-H(26)	110	C(25)	-0.16
H(26)-C(25)	1.08	C(3)-C(25)-H(27)	109	H(26)	+0.12
H(27)-C(25)	1.08	C(3)-C(25)-H(28)	112	H(27)	+0.12
H(28)-C(25)	1.08	O(7)-C(8)-F(29)	106	H(28)	+0.10
F(29)-C(8)	1.32	O(7)-C(8)-F(30)	110	F(29)	-0.27
F(30)-C(8)	1.32			F(30)	-0.29

Table 17. General energy (E_0), electronic energy (E_{el}), maximum positive charge on atom of the hydrogen (q_{max}^{H+}), the universal factor of acidity (pK_a) molecules of fluorine-containing pyrimidines

№	Pyrimidine	Method	-E_0 kDg/mol	-E_{el} kDg/mol	q_{max}^{H+}	pKa
1	2-methylsulfanil-4-difluoromethoxypyrimidine	MNDO	267467	1146348	+0.10	27
		AB INITIO	2649786	4795277	+0.14	30
2	2-ethylsulfanil-4-difluoromethoxypyrimidine	MNDO	282544	1262764	+0.11	26
		AB INITIO	2752131	5120463	+0.14	30
3	2-isopropylsulfanil-4-difluoromethoxypyrimidine	MNDO	297601	1415574	+0.11	26
		AB INITIO	2854460	5531650	+0.16	28
4	2-isobutylsulfanil-4-difluoromethoxypyrimidine	MNDO	312632	1587688	+0.10	27
		AB INITIO	2956778	5971860	+0.13	32

Table 17. (Continued)

№	Pyrimidine	Method	$-E_0$ kDg/mol	$-E_{el}$ kDg/mol	q_{max}^{H+}	pKa
5	2-methylsulfanil-4-difluoromethoxy-6-methylpyrimidine	MNDO	282558	1279166	+0.10	27
		AB INITIO	2752149	5147744	+0.14	30
6	2-ethylsulfanil-4-difluoromethoxy-6-methylpyrimidine	MNDO	297630	1416522	+0.10	27
		AB INITIO	2854483	5519953	+0.14	30
7	2-isopropylsulfanil-4-difluoromethoxy-6-methylpyrimidine	MNDO	312687	1578119	+0.10	27
		AB INITIO	2956812	5946904	+0.16	28
8	2-isobutylsulfanil-4-difluoromethoxy-6-methylpyrimidine	MNDO	327720	1758670	+0.10	27
		AB INITIO	3059131	6403461	+0.13	32

References

[1] V. A. Babkin, D. S. Andreev, E. S. Titova, I. Yu. Kameneva, A. I. Rakhimov. Quantum chemical calculation of fluorine-containing pyrimidines. In *material IV All-Russian science-practical conference,* c. Volgorad –c. Mihailovka, 22-23 October, 2009, pp 221-224.

[2] Shmidt M. W., Baldridge K. K., Elbert J. A., Gordon M. S., Ensen J. H., Koseki S., Matsunaga N., Nguen K. A., Su S. J., Winds T. L.. Together with Montgomery J. A. General Atomic and Molecular Electronic Structure Systems. *J. Comput. Chem.* 1993. 14, pp 1347-1363.

[3] Bode B.M. and Gordon M.S., "MacMolPlt: A Graphical User Interface for GAMESS," *J. Molec. Graphics.,* 1998,16,pp 133-138.

[4] Babkin V.A., Fedunov R.G., Minsker K.S. and others. *Oxidation communication,* 2002, №1, 25, 21-47.

[5] Babkin V.A., Andreev D.S. Quantum chemical calculation of molecule allylcyclohexane by method AB INITIO. In book: *Quantum chemical calculation of unique molecular system.* Vol. II. Publisher VolSU, c. Volgograd, 2010, pp. 88-90.

[6] Babkin V.A., Andreev D.S. Quantum chemical calculation of molecule vinylcyclohexane by method AB INITIO. In book: *Quantum chemical calculation of unique molecular system.* Vol. II. Publisher VolSU, c. Volgograd, 2010, pp. 91-93.

[7] Babkin V.A., Andreev D.S. Quantum chemical calculation of molecule 4-methylmethylencyclohexane by method AB INITIO. In book: *Quantum chemical calculation of unique molecular system.* Vol. II. Publisher VolSU, c. Volgograd, 2010, pp. 93-96.

[8] Babkin V.A., Andreev D.S. Quantum chemical calculation of molecule methylencyclopentane by method AB INITIO. In book: *Quantum chemical calculation of unique molecular system.* Vol. II. Publisher VolSU, c. Volgograd, 2010, pp. 96-98.

[9] Babkin V.A., Andreev D.S. Quantum chemical calculation of molecule 1-methylcyclopentene by method AB INITIO. In book: *Quantum chemical calculation of unique molecular system.* Vol. II. Publisher VolSU, c. Volgograd, 2010, pp. 98-101.

[10] Babkin V.A., Andreev D.S. Quantum chemical calculation of molecule 3-methylcyclopentene by method AB INITIO. In book: *Quantum chemical calculation of unique molecular system.* Vol. II. Publisher VolSU, c. Volgograd, 2010, pp. 101-103.

[11] Babkin V.A., Andreev D.S. Quantum chemical calculation of molecule cyclopentene by method AB INITIO. In book: *Quantum chemical calculation of unique molecular system.* Vol. II. Publisher VolSU, c. Volgograd, 2010, pp. 103-105.

[12] Babkin V.A., Andreev D.S. Quantum chemical calculation of molecule metylencyclobutane by method AB INITIO. In book: *Quantum chemical calculation of unique molecular system.* Vol. II. Publisher VolSU, c. Volgograd, 2010, pp. 105-107.

[13] Babkin V.A., Andreev D.S. Quantum chemical calculation of molecule 1,2-dicyclopropylethylene by method AB INITIO. In book*: Quantum chemical calculation of unique molecular system.* Vol. II. Publisher VolSU, c. Volgograd, 2010, pp. 108-110.

[14] Babkin V.A., Andreev D.S. Quantum chemical calculation of molecule isopropenylcyclopropane by method AB INITIO. In book: *Quantum chemical calculation of unique molecular system.* Vol. II. Publisher VolSU, c. Volgograd, 2010, pp. 110-112.

[15] Babkin V.A., Dmitriev V.Yu., Zaikov G.E. Quantum chemical calculation of molecule heterolytic base uracyl. In book: *Quantum chemical calculation of unique molecular system.* Vol. I. Publisher VolSU, c. Volgograd, 2010, pp. 43-45.

[16] Babkin V.A., Dmitriev V.Yu., Zaikov G.E. Quantum chemical calculation of molecule heterolytic base adenin. In book*: Quantum chemical calculation of unique molecular system.* Vol. I. Publisher VolSU, c. Volgograd, 2010, pp. 45-47.

[17] Babkin V.A., Dmitriev V.Yu., Zaikov G.E. Quantum chemical calculation of molecule heterolytic base guanin. In book: *Quantum chemical calculation of unique molecular system.* Vol. I. Publisher VolSU, c. Volgograd, 2010, pp. 47-49.

[18] Babkin V.A., Dmitriev V.Yu., Zaikov G.E. Quantum chemical calculation of molecule heterolytic base timin. In book: *Quantum chemical calculation of unique molecular system.* Vol. I. Publisher VolSU, c. Volgograd, 2010, pp. 49-51.

[19] Babkin V.A., Dmitriev V.Yu., Zaikov G.E. Quantum chemical calculation of molecule heterolytic base cytozin. In book: *Quantum chemical calculation of unique molecular system.* Vol. I. Publisher VolSU, c. Volgograd, 2010, pp. 51-53.

[20] Babkin V.A., Dmitriev V.Yu., Zaikov G.E. Quantum chemical calculation of molecule hexene-1. In book: *Quantum chemical calculation of unique molecular system.* Vol. I. Publisher VolSU, c. Volgograd, 2010, pp. 71-73.

[21] Babkin V.A., Dmitriev V.Yu., Zaikov G.E. Quantum chemical calculation of molecule heptene-1. In book: *Quantum chemical calculation of unique molecular system.* Vol. I. Publisher VolSU, c. Volgograd, 2010, pp. 73-75.

[22] Babkin V.A., Dmitriev V.Yu., Zaikov G.E. Quantum chemical calculation of molecule decene-1. In book: *Quantum chemical calculation of unique molecular system.* Vol. I. Publisher VolSU, c. Volgograd, 2010, pp. 75-77.

[23] Babkin V.A., Dmitriev V.Yu., Zaikov G.E. Quantum chemical calculation of molecule nonene-1. In book: *Quantum chemical calculation of unique molecular system.* Vol. I. Publisher VolSU, c. Volgograd, 2010, pp. 78-80.

[24] Babkin V.A., Dmitriev V.Yu., Zaikov G.E. Quantum chemical calculation of molecule hexene-1 by method MNDO. In book*: Quantum chemical calculation of unique molecular system.* Vol. I. Publisher VolSU, c. Volgograd, 2010, pp. 93-95.

[25] Babkin V.A., Dmitriev V.Yu., Zaikov G.E. Quantum chemical calculation of molecule heptene-1 by method MNDO. In book: *Quantum chemical calculation of unique molecular system.* Vol. I. Publisher VolSU, c. Volgograd, 2010, pp. 95-97.

[26] Babkin V.A., Dmitriev V.Yu., Zaikov G.E. Quantum chemical calculation of molecule decene-1 by method MNDO. In book: *Quantum chemical calculation of unique molecular system.* Vol. I. Publisher VolSU, c. Volgograd, 2010, pp. 97-99.

[27] Babkin V.A., Dmitriev V.Yu., Zaikov G.E. Quantum chemical calculation of molecule nonene-1 by method MNDO. In book: *Quantum chemical calculation of unique molecular system.* Vol. I. Publisher VolSU, c. Volgograd, 2010, pp. 99-102.

[28] Babkin V.A., Andreev D.S. Quantum chemical calculation of molecule isobutylene by method MNDO. In book: *Quantum chemical calculation of unique molecular system.* Vol. I. Publisher VolSU, c. Volgograd, 2010, pp. 176-177.

[29] Babkin V.A., Andreev D.S. Quantum chemical calculation of molecule 2-methylbutene-1 by method MNDO. In book: *Quantum chemical calculation of unique molecular system.* Vol. I. Publisher VolSU, c. Volgograd, 2010, pp. 177-179.

[30] Babkin V.A., Andreev D.S. Quantum chemical calculation of molecule 2-methylbutene-2 by method MNDO. In book: *Quantum chemical calculation of unique molecular system.* Vol. I. Publisher VolSU, c. Volgograd, 2010, pp. 179-180.

[31] Babkin V.A., Andreev D.S. Quantum chemical calculation of molecule 2-methylpentene-1 by method MNDO. In book: *Quantum chemical calculation of unique molecular system.* Vol. I. Publisher VolSU, c. Volgograd, 2010, pp. 181-182.

In: Handbook of Research on Nanomaterials, Nanochemistry ...   ISBN: 978-1-61942-525-5
Editors: A. K. Haghi and G. E. Zaikov   © 2013 Nova Science Publishers, Inc.

Chapter XXII

On Principles of the Addition of Interactions Energy Characteristics

G. A. Korablev[1]*, V. I. Kodolov[1]* and G. E. Zaikov[2]*

[1] Basic Research-Educational Center of Chemical Physics and Mesoscopy
Udmurt Research Center, Ural Division, Russian Academy of Science, Izhevsk, Russia
[2] N.M. Emanuel Institute of Biochemical Physics,
Russian Academy of Science, Moscow, Russia

Introduction

It is demonstrated that for two-particle interactions, the principle of adding reciprocals of energy characteristics of subsystems is fulfilled for the processes flowing along the potential gradient, and the principle of their algebraic addition – for the processes against the potential gradient.

1. Directedness of the Interaction Processes

The analysis of the kinetics of various physical and chemical processes shows that in many cases, the reciprocals of velocities, kinetic or energy characteristics of the corresponding interactions are added.

* Korablev Grigory Andreevich, Dr.Chem.Sc., Professor, Head of Physics Department at Izhevsk State Agricultural Academy. 426052, Izhevsk, 30 let Pobedy St., 98-14, tel.: +7 (3412) 591946, e-mail: biakaa@mail.ru, korablev@udm.net.

* Kodolov Vladimir Ivanovich, Dr.Chem.Sc., Professor, Head of Department of Chemistry and Chemical Engineering at Izhevsk State Technical University. 426069, Izhevsk, Studencheskaya St., 7, tel.: +7 (3412)582438; e-mail: kodol@istu.ru

* Zaikov Gennady Efremovich, Dr.Chem.Sc., Professor, N.M. Emmanuel Institute of Biochemical Physics RAS, Russia, Moscow, 119991, Kosygina St., 4, tel.: +7(495)9397320, e-mail: chembio@sky.chph.ras.ru.

Some examples: ambipolar diffusion, resulting velocity of topochemical reaction, change in the light velocity during the transition from vacuum into the given medium, effective permeability of bio-membranes.

In particular, such supposition is confirmed by the formula of electron transport possibility (W_∞) due to the overlapping of wave functions 1 and 2 (in steady state) during electron-conformation interactions:

$$W_\infty = \frac{1}{2} \frac{W_1 W_2}{W_1 + W_2}.$$ (1)

Equation (1) is used when evaluating the characteristics of diffusion processes followed by non-radiating transport of electrons in proteins [1].

And also: "From classical mechanics, it is known that the relative motion of two particles with the interaction energy U(r) takes place as the motion of material point with the reduced mass μ:

$$\frac{1}{\mu} = \frac{1}{m_1} + \frac{1}{m_2}$$ (2)

in the field of central force U(r), and general translational motion – as a free motion of material point with the mass:

$$m = m_1 + m_2.$$ (3)

Such things take place in quantum mechanics as well" [2].

The task of two-particle interactions taking place along the bond line was solved in the times of Newton and Lagrange:

$$E = \frac{m_1 v_1^2}{2} + \frac{m_2 v_2^2}{2} + U(\bar{r}_2 - \bar{r}_1),$$ (4)

where E – total energy of the system, first and second elements – kinetic energies of the particles, third element – potential energy between particles 1 and 2, vectors $\bar{r}_2$ and $\bar{r}_1$ – characterize the distance between the particles in final and initial states.

For moving thermodynamic systems, the first commencement of thermodynamics is as follows [3]:

$$\delta E = d\left(U + \frac{mv^2}{2}\right) \pm \delta A,$$ (5)

where: δE – amount of energy transferred to the system;

element $d\left(U + \dfrac{mv^2}{2}\right)$

– characterize the changes in internal and kinetic energies of the system;
$+\delta A$ – work performed by the system;
$-\delta A$ – work performed with the system.

As the work value numerically equals the change in the potential energy, then:

$$+\delta A = -\Delta U \text{ и } -\delta A = +\Delta U \qquad\qquad (6,7)$$

It is probable that not only in thermodynamic but in many other processes in the dynamics of moving particles, interaction not only the value of potential energy is critical but its change as well.

Therefore, similar to the equation (4), the following should be fulfilled for two-particle interactions:

$$\delta E = d\left(\frac{m_1 v_1^2}{2} + \frac{m_2 v_2^2}{2}\right) \pm \Delta U \,. \qquad\qquad (8)$$

Here $\Delta U = U_2 - U_1 ,$ \qquad\qquad (9)

where U_2 and U_1 – potential energies of the system in final and initial states.

At the same time, the total energy (E) and kinetic energy $\left(\dfrac{mv^2}{2}\right)$ can be calculated from their zero value; then only the last element is modified in the equation (4).

The character of the change in the potential energy value $\left(\Delta U\right)$ was analyzed by its sign for various potential fields, and the results are given in Table 1.

From the Table, it is seen that the values $-\Delta U$ and accordingly $+\delta A$ (positive work) correspond to the interactions taking place along the potential gradient, and ΔU and $-\delta A$ (negative work) occur during the interactions against the potential gradient.

The solution of two-particle task of the interaction of two material points with masses m_1 and m_2 obtained under the condition of the absence of external forces, corresponds to the interactions flowing along the gradient, the positive work is performed by the system (similar to the attraction process in the gravitation field).

Table 1. Directedness of the interaction processes

No	Systems	Type of potential field	Process	U	$\dfrac{r_2}{r_1}\left(\dfrac{x_2}{x_1}\right)$	$\dfrac{U_2}{U_1}$	Sign ΔU	Sign δA	Process directedness in potential field
1	opposite electrical charges	electrostatic	attraction	$-k\dfrac{q_1 q_2}{r}$	$r_2 < r_1$	$U_2 > U_1$	-	+	along the gradient
			repulsion	$-k\dfrac{q_1 q_2}{r}$	$r_2 > r_1$	$U_2 < U_1$	+	-	against the gradient
2	similar electrical charges	electrostatic	attraction	$k\dfrac{q_1 q_2}{r}$	$r_2 < r_1$	$U_2 > U_1$	+	-	against the gradient
			repulsion	$k\dfrac{q_1 q_2}{r}$	$r_2 > r_1$	$U_2 < U_1$	-	+	along the gradient
3	elementary masses m_1 and m_2	gravitational	attraction	$-\gamma\dfrac{m_1 m_2}{r}$	$r_2 < r_1$	$U_2 > U_1$	-	+	along the gradient
			repulsion	$-\gamma\dfrac{m_1 m_2}{r}$	$r_2 > r_1$	$U_2 < U_1$	+	-	against the gradient
4	spring deformation	field of elastic forces	compression	$k\dfrac{\Delta x^2}{2}$	$x_2 < x_1$	$U_2 > U_1$	+	-	against the gradient
			extension	$k\dfrac{\Delta x^2}{2}$	$x_2 > x_1$	$U_2 > U_1$	+	-	against the gradient
5	photoeffect	electrostatic	repulsion	$k\dfrac{q_1 q_2}{r}$	$r_2 > r_1$	$U_2 < U_1$	-	+	along the gradient
6	transfer systems	field transfer	passive	$I=-\sigma\dfrac{\Delta\varphi}{\Delta x}$	$x_2 > x_1$	$U_2 > U_1$	-	+	along the gradient
			active	and other	$x_2 < x_1$	$U_2 < U_1$	+	-	against the gradient

The solution of this equation via the reduced mass (μ) is [4] the Lagrange equation for the relative motion of the isolated system of two interacting material points with masses m_1 and m_2, which in coordinate x is as follows:

$$\mu \cdot x'' = -\frac{\partial U}{\partial x}; \quad \frac{1}{\mu} = \frac{1}{m_1} + \frac{1}{m_2}.$$

Here, U – mutual potential energy of material points; μ – reduced mass. At the same time, $x'' = a$ (feature of the system acceleration). For elementary portions of the interactions, Δx can be taken as follows:

$$\frac{\partial U}{\partial x} \approx \frac{\Delta U}{\Delta x} \quad \text{That is: } \mu a \Delta x = -\Delta U. \text{ Then:}$$

$$\frac{1}{1/(a\Delta x)} \frac{1}{(1/m_1 + 1/m_2)} \approx -\Delta U \quad ; \quad \frac{1}{1/(m_1 a\Delta x) + 1/(m_2 a\Delta x)} \approx -\Delta U$$

$$\text{Or: } \frac{1}{\Delta U} \approx \frac{1}{\Delta U_1} + \frac{1}{\Delta U_2}, \tag{10}$$

where ΔU_1 and ΔU_2 – potential energies of material points on the elementary portion of interactions, ΔU – resulting (mutual) potential energy of this interactions.

Thus:

1. In the systems in which the interactions proceed along the potential gradient (positive performance), the Lagrangian is performed, and the resulting potential energy is found based on the principle of adding reciprocals of the corresponding energies of subsystems [5]. Similarly, the reduced mass for the relative motion of two-particle system is calculated.

2. In the systems in which the interactions proceed against the potential gradient (negative performance), the algebraic addition of their masses as well as the corresponding energies of subsystems is performed (by the analogy with Hamiltonian).

2. Spatial-Energy Parameter

From the equation (10), it is seen that the resulting energy characteristic of the system of two material points interaction is found based on the principle of adding reciprocals of initial energies of interacting subsystems.

"Electron with the mass m moving near the proton with the mass M is equivalent to the particle with the mass: $\mu = \dfrac{mM}{m+M}$ " [6].

In this system, the energy characteristics of subsystems are: electron orbital energy (W_i) and nucleus effective energy taking screening effects into account (by Clementi).

Therefore, when modifying the equation (10), we can assume that the energy of atom valence orbitals (responsible for interatomic interactions) can be calculated [5] by the principle of adding reciprocals of some initial energy components based on the following equations:

$$\frac{1}{q^2/r_i} + \frac{1}{W_i n_i} = \frac{1}{P_E}$$

$$\text{or } \frac{1}{P_0} = \frac{1}{q^2} + \frac{1}{(Wrn)_i}; \ P_E = P_0/r_i \ . \tag{11),(12),(13}$$

Here: W_i – electron orbital energy [7]; r_i – orbital radius of i–orbital [8]; $q=Z*/n*$ [9,10], n_i – number of electrons of the given orbital, $Z*$ and $n*$ – nucleus effective charge and effective main quantum number, r – bond dimensional characteristics.

P_0 was called a spatial-energy parameter (SEP), and P_E – effective P–parameter (effective SEP). Effective SEP has a physical sense of some averaged energy of valence electrons in the atom and is measured in energy units, e.g., electron-volts (eV).

The values of P_0-parameter are tabulated constants for the electrons of the given atom orbital.

For dimensionality, SEP can be written down as follows:

$$[P_0] = [q^2] = [E] \cdot [r] = [h] \cdot [v] = \frac{kg \cdot m^3}{s^2} = J \cdot m,$$

where [E], [h] and [v] – dimensions of energy, Planck constant and velocity.

The introduction of P-parameter should be considered as further development of quasi-classical notions using quantum-mechanical data on atom structure to obtain the criteria of energy conditions of phase-formation. At the same time, for the systems of similarly charged (e.g., orbitals in the given atom), homogeneous systems the principle of algebraic addition of such parameters is preserved:

$$; \sum P_E = \frac{\sum P_0}{r} \tag{14),(15}$$

$$\text{or: } \sum P_0 = P_0' + P_0'' + P_0''' + \dots; \ r \sum P_э = \sum P_0 \ . \tag{16),(17}$$

Here, P-parameters are summed on all atom valence orbitals.

In accordance with the conclusions obtained, in these systems the interactions take place against the potential gradient.

The principle of adding the reciprocals of energy characteristics of heterogeneous, oppositely charged systems by the equations (11,12,13) is performed for the interactions proceeding along the potential gradient.

To calculate the values of P_E-parameter at the given distance from the nucleus depending on the bond type, either atom radius (R) or ion radius (r_I) can be used instead of r.

The reliability of such approach is briefly discussed in [5, 11]. The calculations demonstrated that the value of P_E-parameters is numerically equal (within 2%) to total energy of valence electrons (U) by the atom statistic model. Using the known correlation between the electron density (β) and interatomic potential by the atom statistic model, we can obtain the direct dependence of P_E-parameter upon the electron density at the distance r_i from the nucleus:

$$\beta_i^{2/3} = \frac{AP_0}{r_i} = AP_E,$$

where A – constant.

The rationality of such approach is confirmed by the calculation of electron density using wave functions of Clementi and its comparison with the value of electron density calculated via the value of P_E-parameter.

The correlations of the modules of maximum values of radial part of Ψ-function with the valued of P_0-parameter are carried out, and the linear dependence between these values is found. Using some properties of wave function with regard to P-parameter, we obtain the wave equation of P-parameter formally analogous to the equation of Ψ-function.

3. Analogous Comparison of Lagrange and Hamilton Functions with Spatial-Energy Parameter

Lagrange function is the difference between kinetic (T) and potential (U) energies of the system:

$$L = T - U.$$

Hamilton function can be considered as the sum of potential and kinetic energies, i.e., as the total mechanic energy of the system:

$$H = T + U.$$

From this equation and in accordance with energy conservation law, we can see that:

$$H + L = 2T, \tag{18}$$

$$H - L = 2U. \tag{19}$$

Let us try to assess the movement of the isolated system of a free atom as a relative movement of its two subsystems: nucleus and orbital.

The atom structure is formed of oppositely charged masses of nucleus and electrons. In this system, the energy characteristics of subsystems are the orbital energy of electrons (W_i) and effective energy of atom nucleus taking screening effects into account.

In a free atom, its electrons move in Coulomb field of nucleus charge. The affective nucleus charge characterizing the potential energy of such subsystem taking screening effects into account equals: q^2/r_i,

where $q = Z^*/n^*$.

Here: Z* and n* – effective nucleus charge and effective main quantum number [9, 10]; r_i – orbital radius [8].

It can be presumed that orbital energy of electrons during their motion in Coulomb field of atom nucleus is mainly defined by the value of kinetic energy of such motion.

Thus, it is assumed that:

$$T \sim W \quad \text{and} \quad U \sim q^2/r_i.$$

In such approach, the total of the values W and q^2/r_i is analogous to Hamilton function (H):

$$W + q^2/r_i \sim H \tag{20}$$

The analogous comparison of P-parameter with Lagrange function can be carried out when investigating Lagrange equation for relative motion of isolated system of two interacting material points with masses m_1 and m_2 in coordinate x, – equations (11-13).

Thus, it is presumed that: $P_E \sim L$.

Then equation (18) is as follows:

$$\left(W + \frac{q^2}{r_i} \right) + P_E \approx 2W. \tag{21}$$

Using the values of electron bond energy as the orbital electron energy [7], we calculated the values of P_E-parameters of free atoms (Table 2) by equations (11,12). When calculating the values of effective P_E-parameter, mainly the atom radius values by Belov-Bokiy or covalent radii (for non-metals) were applied as dimensional characteristics of atom (R).

Mutual overlapping of orbitals of valence electrons in the atom in a first approximation can be considered via the averaging of their total energy dividing it by a number of valence electrons considered (N):

$$\left(\frac{q^2}{r_i} + W \right) \frac{1}{N} + P_E \approx 2W \tag{22}$$

That is, the expression $\left(\dfrac{q^2}{r_i}+W\right)\dfrac{1}{N}$ characterizes the averaged value of total energy of i-orbital in terms of one valence electron and in this approach, it is the analog of Hamilton function – H.

In free atoms of 1a group of periodic system, S-orbital is the only valence orbital whose electron cloud deformations for three elements of long periods (K, Rb, Cs) were considered via the introduction of the coefficient $K = n/n^*$, where n – main quantum number, n^* – effective main quantum number:

$$\left(\frac{q^2}{r_i}+W\right)\frac{1}{KN}+P_E \approx 2W .$$

(23)

This means that K=$\dfrac{4}{3,7};\dfrac{5}{4};\dfrac{6}{4,2}$ – for K, Rb and Cs, respectively. For the others, K=1.

Besides, for the elements of 1a group, the value $2r_i$ (i.e., the orbital radius of i-orbital) was used as a dimensional characteristic in the first component of equation (23). For hydrogen atom, the ion radius was applied.

By the initial equations (22,23), the values $\left[\left(\dfrac{q^2}{r_i}+W\right)\dfrac{1}{N}+P_E\right]$ and $2W$ were calculated

and compared for the majority of elements. Their results are given in Table 2. Besides, the equation (22) can be transformed as:

$$\left(\frac{q^2}{r_i}+W\right)\frac{1}{N}-P_E \approx 2(W-P_E)$$

(24)

and the correlation obtained is analogous to the equation (19). Actually, by the given analogy, $W \sim T$ and $P_E \sim L$.

That is $(W-P_E)\sim(T-L)$, as $T-L=U$, then $(W-P_E)\sim U$ and equation (24) in general is analogous to the equation (19).

The analysis of the data given in Table 2 reveals that the proximity of the values investigated is mostly within 5% and only sometimes exceeds 5% but never exceeds 10%.

All this indicates that there is a certain analogy of equations (18) and (22), (19) and (24), and in a first approximation the value of P_E-parameter can be considered the analog of Lagrange function and value $\left(\dfrac{q^2}{r_i}+W\right)\dfrac{1}{N}$ – analog of Hamilton function [11].

Table 2. Comparison of energy atomic characteristics

Element	Valence electrons	W (eV)	r_i (Å)	q^2 (eVÅ)	P_0 (eVÅ)	R (Å)	$P_E=P_0/R$ (eV)	N	$\left(\dfrac{\frac{2}{q}+W}{r_i}\right)\dfrac{1}{N}+P_E$ (eV)	2W (eV)
1	2	3	4	5	6	7	8	9	10	11
H	1S¹	13.595	$r_M=$ 1.36	14.394	8.0933	1.36·2	2.9755	1	27.155	27.190
Li	2S¹	5.3416	1.5862	5.8892	3.475	1.55	2.2419	1	9.440	10.683
Be	2S²	8.4157	1.040	13.159	7.512	1.13	6.6478	2	17.182	16.831
B	2P¹	8.4315	0.776	21.105	4.9945	0.91	5.4885	3	17.365	16.863
	2S²	13.462	0.769	23.890	11.092	0.91	12.189	3	27.032	26.924
C	2P²	11.792	0.596	35.395	10.061	0.77	13.066	6	24.923	23.584
	2S²	19.201	0.620	37.240	14.524	0.77	18.862	4	38.824	38.402
N	2P¹	15.445	0.4875	52.912	6.5916	0.55	11.985	7	29.562	30.890
	2S²	25.724	0.521	53.283	17.833	0.71	25.117	5	50.716	51.448
O	2P¹	17.195	0.4135	71.383	6.4663	0.59	10.960	8	34.089	34.390
	2S²	33.859	0.450	72.620	21.466	0.59	36.383	6	68.923	67.718
F	2P¹	19.864	0.3595	93.625	6.6350	0.64	10.367	9	41.511	39.728
Na	3S¹	4.9552	1.713·2	10.058	4.6034	1.89	2.4357	1	10.327	9.9104
Mg	3S¹	6.8859	1.279	17.501	5.8588	1.60	3.6618	2	13.946	13.772
Al	3P¹	5.713	1.312	26.443	5.840	1.43	4.084	3	12.707	11.426
	3S²	10.706	1.044	27.119	12.253	1.43	8.5685	3	20.796	21.412
Si	3P²	8.0848	1.068	29.377	10.876	1.34	8.1169	4	17.014	16.170
	3S²	14.690	0.904	38.462	15.711	1.17	13.428	4	17.737	29.380
	3P²+3S²								44.75	45.50
P	3P³	10.659	0.9175	38.199	16.594	1.10	15.085	7	21.551	21.318
	3S¹	18.951	0.803	50.922	11.716	1.10	10.651	3	38.106	37.902
S	3P²	11.901	0.808	48.108	13.740	1.04	13.215	6	25.122	23.802
	3S²	23.933	0.723	64.852	22.565	1.04	21.697	4	50.105	47.866
Cl	3P¹	13.780	0.7235	59.844	8.5461	1.00	8.5461	5	27.845	27.560
K	4S¹	$4.0130\,\dfrac{3,7}{4}$	2.612·2	$10.993\,\dfrac{3,7}{4}$	4.8490	2.36	2.0547	1	7.7137	8.026
Ca	4S¹	5.3212	1.690	17.406	5.929	1.97	3.0096	2	10.820	10.642
Sc	4S²	5.7174	1.570	19.311	9.3035	1.64	5.6729	3	11.679	11.435

Element	Valence electrons	W (eV)	r_i (Å)	$q2$ (eVÅ)	P_0 (eVÅ)	R (Å)	$PE=P0/R$ (eV)	N	$\left(\dfrac{2}{r_i}q+W\right)\dfrac{1}{N}+P_E$	2W (eV)
Ti	4S²	6.0082	1.477	20.879	9.5934	1.46	6.5708	4	11.230	12.016
1	2	3	4	5	6	7	8	9	10	11
V	4S²	6.2755	1.401	22.328	9.8361	1.34	7.3404	4	12.894	12.549
	4S²	6.2755	1.401	22.328	9.8361	1.34	7.3404	5	11.783	12.549
Cr	4S¹	6.5238	1.453	23.712	6.7720	1.27	5.3323	3	12.947	13.048
	4S²	6.5238	1.453	23.712	10.535	1.27	8.2953	5	12.864	13.048
Mn	4S²	6.7451	1.278	25.118	10.223	1.30	7.8638	5	13.144	13.490
Fe	4S¹	7.0256	1.227	26.572	6.5089	1.26	5.1658	3	14.716	14.051
	4S²	7.0256	1.227	26.572	10.456	1.26	8.2984	5	14.035	14.051
Co	4S²	7.2770	1.181	27.973	10.648	1.25	8.5184	5	14.705	14.544
Ni	4S²	7.5176	1.139	29.248	10.815	1.24	8.7218	5	15.361	15.035
Cu	4S²	7.7485	1.191	30.117	11.444	1.28	8.9406	5	15.548	15.497
Zn	4S²	7.9594	1.065	32.021	11.085	1.37	8.0912	5	15.580	15.919
Ga	4P¹	5.6736	1.254	34.833	5.9081	1.39	4.2504	5	10.941	11.347
	4S²	11.544	0.960	44.940	14.852	1.39	10.685	5	22.356	23.086
Ge	4P²	7.8190	1.090	41.372	12.072	1.39	8.6649	6	16.294	15.636
	4S²	15.059	0.886	58.223	18.298	1.39	13.164	5	29.319	30.116
	4P²+4S²								45.613	45.752
As	4P¹	10.054	0.992	49.936	8.275	1.17	7.0726	5	19.152	20.108
	4S²	18.664	0.826	71.987	21.587	1.40	15.419	5	36.582	37.328
Se	4P¹	10.963	0.909	61.803	8.5811	1.14	7.5272	6	20.656	21.924
	4S²	22.787	0.775	85.678	25.010	1.17	21.376	6	43.599	45.474
Br	4P¹	12.438	0.8425	73.346	9.1690	1.96	4.6781	5	24.577	24.876
Rb	5S¹	$3.7511\,\tfrac{4}{5}$	2.2872	$14.309\,\tfrac{4}{5}$	5.3630	2.48	2.1625	1	7.666	7.5022
Sr	5S¹	4.8559	1.836	21.224	6.2790	2.15·2	1.4663	2	9.6713	9.7118
Y	5S²	6.3376	1.693	22.540	10.030	1.81	5.5417	3	12.092	12.675
Zr	5S²	5.6414	1.593	23.926	10.263	1.60	6.4146	4	11.580	11.283
Nb (5S²4d³)	5S²	5.8947	1.589	25.822	10.857	1.45	7.4876	5	11.917	11.789
Mo (5S²4d⁴)	5S²	6.1140	1.520	28.027	11.175	1.39	8.0396	6	12.132	12.223

Table 2. (Continued)

Element	Valence electrons	W (eV)	r_i (Å)	q^2 (eVÅ)	P_0 (eVÅ)	R (Å)	$P_E = P_0/R$ (eV)	N	$\left(\dfrac{q^2}{r_i}+W\right)\dfrac{1}{N}+P_E$	2W (eV)
1	2	3	4	5	6	7	8	9	10	11
Tc	5S²	6.2942	1.391	30.076	11.067	1.36	8.1376	7	12.126	12.588
Ru (5S²4d⁶)	5S²	6.5294	1.410	31.986	11.686	1.34	8.7208	8	12.373	13.059
Rh (5S²4d⁷)	5S²	6.7240	1.364	33.643	11.871	1.34	8.8588	8	12.782	13.448
Pd (5S²4d⁸)	5S²	6.9026	35.377	1.325	12.057	1.37	8.8007	8	13.00	13.805
Ag (5S²4d¹⁰)	5S¹	7.0655	1.286	26.283	6.752	1.44	4.6889	3	13.867	14.131
Cd	5S¹	7.2070	1.184	38.649	6.9898	1.56	4.4806	4	14.443	14.414
Jn	5P¹	5.3684	1.382	41.318	6.2896	1.66	3.7869	5	10.842	10.737
	5S²	10.141	1.093	52.103	15.551	1.66	9.3681	5	20.930	20.282
Sn	5P¹	7.2124	1.240	47.714	7.5313	1.42	5.3037	5	14.442	14.424
	5S²	12.965	1.027	65.062	18.896	1.42	13.307	6	26.027	25.930
Sb	5P¹	9.1072	1.1665	57.530	8.9676	1.39	6.5519	5	18.237	18.214
	5S²	15.833	0.969	77.644	21.993	1.39	15.822	5	31.816	31.666
Te	5P¹	9.7907	1.087	67.285	9.1894	1.37	6.7076	6	18.656	19.581
	5S²	19.064	0.920	90.537	25.283	1.37	18.454	6	38.033	38.128
J (5S²5p⁵)	5P¹	10.971	1.0215	77.651	9.7936	1.35	7.2545	5 / 7	⟨21,753⟩	21.942
Cs	6S¹	3.3647 6/4,2	2.518·2	16.193 6/4,2	5.5628	2.68	2.0757	1	6.6914	6.7294
Ba	6S²	4.2872	2.060	22.950	9.9812	2.21	4.5164	4	8.3734	8.5744
La (6S²5p¹4f¹)	6S¹	4.3528	1.915	34.681	6.7203	1.87	3.5937	4	9.2094	8.706
Hf	6S¹	5.6863	1.476	33.590	6.7151	1.59	4.2233	4	11.334	11.373
Ta	6S¹	5.9192	1.413	36.285	6.7971	1.46	4.6555	5	10.975	11.838
W(V)	6S¹	6.1184	1.360	38.838	6.8528	1.40	4.8949	5	11.830	12.237
Re(V)	6S¹	6.2783	1.310	40.928	6.8483	1.37	4.9988	5	12.503	12.557
Os(VI)	6S¹	6.4995	1.266	42.620	7.7344	1.35	5.7292	6	12.423	12.999

Element	Valence electrons	W (eV)	r_i (Å)	q^2 (eVÅ)	P_0 (eVÅ)	R (Å)	$P_E=P_0/R$ (eV)	N	$\left(\dfrac{2}{q^2\,r_i} + W\right)\dfrac{1}{N} + P_E$	2W (eV)
1	2	3	4	5	6	7	8	9	10	11
Ir(VI)	$6S^1$	6.6788	1.227	44.655	7.7691	1.35	5.7549	6	12.934	13.358
Pt(VI)	$6S^1$	6.8377	1.221	46.231	7.0718	1.38	5.1245	6	12.575	13.675
Au	$6S^1$	6.9820	1.187	47.849	7.0641	1.44	4.9056	5	14.364	13.964
	$6S^2$	6.9820	1.187	47.849	12.311	1.44	8.5491	8	14.461	13.964
Hg	$6S^1$	7.1037	1.126	49.432	6.8849	1.60	4.3031	5	14.504	14.207
	$6S^2$	7.1037	1.126	49.432	12.086	1.60	7.5538	8	13.929	14.207
Tl	$6P^1$	5.2354	1.319	60.054	6.1933	1.71	3.6218	5	19.567	19.654
	$6S^1$	9.8268	1.060	65.728	8.9912	1.71	5.1995	8	9.9675	10.421
Pb	$6P^2$	7.0913	1.215	61.417	13.460	1.75	7.6914	8	14.897	14.183
	$6S^2$	12.416	1.010	79.515	19.066	1.75	12.711	8	24.104	29.832
Bi	$6P^1$	8.7076	1.2125	71.171	9.0406	1.82	4.9674	5	18.448	17.45
	$6S^1$	15.012	0.963	92.892	12.50	1.51	8.2841	5	30.578	30.024
	$6S^2$	15.012	0.963	92.892	22.050	1.51	14.603	8	28.537	30.024
P_0	$6P^1$	9.2887	1.1385	80.881	9.3523	1.50	6.2349	6 8	$\langle 17{,}950\rangle$	18.577
	$6S^1$	17.909	0.923	106.65	14.312	1.50	9.5413	5	36.233	35.818
	$6S^2$	17.909	0.923	106.65	25.237	1.50	16.825	8	33.507	35.818
At	$6P^1$	10.337	1.078	91.958	9.9074	1.39	7.1276	8	19.083	20.674
	$6S^2$	20.828	0.885	119.70	28.185	1.39	20.277	8	39.787	41.756

Conclusion

1. Two main types of adding energy characteristics of subsystems are defined by the directions of their structural interactions – along the potential gradient or against it.

2. Spatial-energy parameter obtained based on the principle of adding reciprocals of energy characteristics of subsystems can be assessed as Lagrangian analog.

References

[1]   Rubin A.B. Biophysics. Book 1. Theoretical biophysics // *Vysshaya shkola,* M., 1987, 319p.

[2]   Blokhintsev D.I. Basics of quantum mechanics // *Vysshaya shkola,* M., 1961, 512p.

[3]   Yavorsky B.M., Detlaf A.A. Reference-book in physics // *Nauka,* M., 1968, 939p.

[4]   Christy P., Pitti A. Substance structure: Introduction to modern physics. Translated from English. // *Nauka,* M., 1969, 596p.

[5]   Korablev G.A. Spatial-Energy Principles of Complex Structures Formation // *Brill Academic Publishers and VSP*, Netherlands, 2005, 426pp. (Monograph).

[6]   Airing G., Walter J., Kimbal J. *Quantum chemistry* // I.L., M., 1948, 528p.

[7]   Fischer C.F. Average-Energy of Configuration Hartree-Fock Results for the Atoms Helium to Radon //*Atomic Data,* 1972, № 4. pp. 301-399.

[8]   Waber J.T., Cromer D.T. Orbital Radii of Atoms and Ions // *J. Chem. Phys.,* 1965, V 42, №12, pp. 4116-4123.

[9]   Clementi E., Raimondi D.L. Atomic Screening constants from S.C.F. Functions, 1 // *J.Chem. Phys.,* 1963, V.38, №11, pp. 2686-2689.

[10]   Clementi E., Raimondi D.L. Atomic Screening constants from S.C.F. Functions, 1 // *J.Chem. Phys.,* 1967, V.47, №14, pp. 1300-1307.

[11]   Korablev G.A., Zaikov G.E. Energy of chemical bond and spatial-energy principles of hybridization of atom orbitalls // *J. of Applied Polymer Science,* V.101, №3, Ang.5, 2006, pp. 2101-2107.

In: Handbook of Research on Nanomaterials, Nanochemistry ...   ISBN: 978-1-61942-525-5
Editors: A. K. Haghi and G. E. Zaikov   © 2013 Nova Science Publishers, Inc.

Chapter XXIII

Exchange Spatial-Energy Interactions

G. A. Korablev[1*] and G. E. Zaikov[2*]

[1] Izhevsk State Agricultural Academy, Basic Research and Educational Center of
Chemical, Physics and Mesoscopy, Russia, Izhevsk
[2] N.M. Emanuel Institute of Biochemical Physics, RAS, Moscow, Russia

Introduction

The notion of spatial-energy parameter (P-parameter) is introduced based on the modified Lagrangian equation for relative motion of two interacting material points and is a complex characteristic of important atomic values responsible for interatomic interactions and having the direct connection with electron density inside an atom. Wave properties of P-parameter are found; its wave equation having a formal analogy with the equation of Ψ-function is given.

With the help of P-parameter technique, numerous calculations of exchange structural interactions are done, the applicability of the model for the evaluation of the intensity of fundamental interactions is demonstrated, and initial theses of quark screw model are given.

Keywords: Lagrangian equations, wave functions, spatial-energy parameter, electron density, elementary particles, quarks

1. Spatial-Energy Parameter

When oppositely charged heterogeneous systems interact, a certain compensation of the volume energy, which results in the decrease in the resultant energy (e.g., during the

---

[*] Korablev Grigory Andreevich, Doctor of Chemical Science, Professor, Head of Department of Physics at Izhevsk State Agricultural Academy. Izhevsk 426052, 30 let Pobedy St., 98-14, tel.. +7(3412) 591946, e-mail. biakaa@mail.ru, korablevga@udm.net.
[*] Zaikov Gennady Efremovich, Doctor of Chemical Science, Professor of N.M. Emanuel Institute of Biochemical Physics, RAS, Russia, Moscow 119991, Kosygina St., 4, tel.: +7(495)9397320, e-mail: chembio@chph.ras.ru.

hybridization of atom orbitals), takes place [1]. But this is not the direct algebraic deduction of corresponding energies. The comparison of numerous regularities of physical and chemical processes lets us assume that in such and similar cases, the principle of adding reverse values of volume energies or kinetic parameters of interacting structures is observed. For instance:

1) During the ambipolar diffusion: when joint motion of oppositely charged particles is observed in the given medium (in plasma or electrolyte), the diffusion coefficient (D) is found as follows:

$$\frac{\eta}{D} = \frac{1}{a_+} + \frac{1}{a_-}$$

where a_+ and a_- - charge mobility of both atoms, η – constant coefficient.

2) Total velocity of topochemical reaction (v) between the solid and gas is found as follows:

$$\frac{1}{v} = \frac{1}{v_1} + \frac{1}{v_2}$$

where v_1 – reagent diffusion velocity, v_2 – velocity of reaction between the gaseous reagent and solid.

3) Change in the light velocity (Δv) when moving from the vacuum into the given medium is calculated by the principle of algebraic deduction of reverse values of the corresponding velocities:

$$\frac{1}{\Delta v} = \frac{1}{v} - \frac{1}{c},$$

where c – light velocity in vacuum.

4) Lagrangian equation for relative motion of the system of two interacting material points with masses m_1 and m_2 in coordinate x is as follows:

$$m_{np} x'' = -\frac{\partial U}{\partial x} \quad \text{where} \quad \frac{1}{m_r} = \frac{1}{m_1} + \frac{1}{m_2} \tag{1),(1a}$$

here, U – mutual potential energy of material points; m_r – reduced mass. At the same time, $x'' = a$ (characteristic of system acceleration).

For elementary interaction areas, Δx: $\dfrac{\partial U}{\partial x} \approx \dfrac{\Delta U}{\Delta x}$

Then: $m_r a \Delta x = -\Delta U$; $\dfrac{1}{1/(a\Delta x)} \cdot \dfrac{1}{(1/m_1 + 1/m_2)} \approx -\Delta U$ or:

$$\frac{1}{1/(m_1 a\Delta x)+1/(m_2 a\Delta x)} \approx -\Delta U$$

Since in its physical sense the product $m_i a\Delta x$ equals the potential energy of each material point $(-\Delta U_i)$, then:

$$\frac{1}{\Delta U} \approx \frac{1}{\Delta U_1} + \frac{1}{\Delta U_2} \tag{2}$$

Thus, the resultant energy characteristic of the interaction system of two material points is found by the principle of adding the reverse values of initial energies of interacting subsystems.

Therefore, assuming that the energy of atom valent orbitals (responsible for interatomic interactions) can be calculated by the principle of adding the reverse values of some initial energy components, the introduction of P-parameter as the averaged energy characteristic of valent orbitals is postulated based on the following equations:

$$\frac{1}{q^2/r_i} + \frac{1}{W_i n_i} = \frac{1}{P_E} \text{ or}$$

$$\frac{1}{P_0} = \frac{1}{q^2} + \frac{1}{(Wrn)_i} \; ; \; P_E = P_0/r_i \tag{3),(4),(5}$$

here: W_i – orbital energy of electrons; [2] r_i – orbital radius of i–orbital [3]; $q=Z^*/n^*$ [4,5], n_i – number of electrons of the given orbital, Z^* and n^* - effective charge of the nucleus and effective main quantum number, r – bond dimensional characteristics.

The value P_O will be called spatial-energy parameter (SEP), and value P_E – effective P–parameter (effective SEP). Effective SEP has a physical sense of some averaged energy of valent electrons in the atom and is measured in the energy units, e.g., in electron-volts (eV).

Values of P_0-parameter are tabulated constant values for the electrons of the atom given orbital.

For the dimensionality, SEP can be written down as follows:

$$[P_0] = \left[q^2\right] = [E]\cdot[r] = [h]\cdot[\upsilon] = \frac{kgm^3}{s^2} = Jm$$

where [E], [h] and [υ] – dimensionalities of energy, Plank's constant and velocity.

The introduction of P-parameter should be considered as further development of quasi-classic notions using quantum-mechanical data on the atom structure to obtain the criteria of phase-formation energy conditions. At the same time, for similarly charged systems (e.g., orbitals in the given atom) and homogeneous systems, the principle of algebraic addition of these parameters will be preserved:

$$\sum P_E = \sum (P_0 / r_i) \, ;$$

$$\sum P_E = \frac{\sum P_0}{r} \qquad\qquad\qquad (6),(7)$$

or: $\sum P_0 = P_0' + P_0'' + P_0''' + \dots \, ;$

$$r \sum P_E = \sum P_0 \qquad\qquad\qquad (8),(9)$$

Here, P-parameters are summed by all atom valent orbitals.

To calculate the values of P_E-parameter at the given distance from the nucleus, either atomic radius (R) or ionic radius (r_i) can be used instead of r depending on the bond type.

Applying the equation (8) to hydrogen atom, we can write down the following:

$$K(\frac{e}{n})_1^2 = K(\frac{e}{n})_2^2 + mc^2 \lambda \qquad\qquad\qquad (10)$$

where: e – elementary charge, n_1 and n_2 – main quantum numbers, m – electron mass, c – electromagnetic wave velocity, λ - wave length, K - constant.

Using the known correlations $\nu = C/\lambda$ and $\lambda = h/mc$ (where h – Plank's constant, ν – wave frequency) from the formula (10), the equation of spectral regularities in hydrogen atom can be obtained, in which $2\pi^2 e^2 / hC = K$.

2. Effective Energy of Valent Electrons in Atom and Its Comparison with Statistic Model

The modified Thomas-Fermi equation converted to a simple form by introducing dimensionless variables [6] is as follows:

$$U = e(V_i - V_0 + \tau_0^2) \qquad\qquad\qquad (11)$$

where: V_o – countdown potential; e – elementary charge; τ_0 - exchange and correlation corrections; V_i – interatomic potential at the distance r_i from the nucleus, U – total energy of valent electrons.

For the 21st element, the comparisons of the given value U with the values of Pe-parameter are partially given in Table 1.

As it is seen form the Table 1, the parameter values of U and Pe are practically the same (in most cases with the deviation not exceeding 1-2%) *without any transition coefficients*. Multiple corrections introduced into the statistic model are compensated with the application of simple rules of adding reverse values of energy parameters, and SEP quite precisely conveys the known solutions of Thomas-Fermi equation for interatomic potential of atoms at the distance r_i from the nucleus. Namely, the following equality takes place:

$$U = P_E = e \, (V_i\text{-}V_0 + \tau^2_0) \qquad (12)$$

Using the known correlation [6] between the electron density (β_i) and interatomic potential (V_i), we have:

$$\beta^{2/3}_i \approx (3e/5)\cdot(V_i - V_0); \; \beta^{2/3}_i \approx Ae\cdot(V_i - V_0 + \tau^2_0) =$$
$$[Ae\cdot r_i\cdot(V_i - V_0 + \tau^2_0)]/r_i \qquad (13)$$

where A - constant. According to the formulas (12 and 13), we have the following correlation:

$$\beta^{2/3}_i = A \, P_0/r_i \qquad (14)$$

setting the connection between P_0-parameter and electron density in the atom at the distance r_i from the nucleus.

Since the value $e(V_i\text{-}V_0 + \tau^2_0)$ in Thomas-Fermi model, there is a function of charge density, P_0-parameter is a direct characteristic of electron charge density in atom.

This is confirmed by an additional check of equality correctness (14) using Clementi function [7]. A good correspondence between the values β_i, calculated via the value P_0 and obtained from atomic functions (Fig. 1) is observed.

3. Wave Equation of P-Parameter

For the characteristic of atom spatial-energy properties, two types of P-parameters with simple correlation between them are introduced:

$$P_E = \frac{P_0}{R}$$

where R — atom dimension characteristic. Taking into account additional quantum characteristics of sublevels in the atom, this equation in coordinate x can be written down as follows:

$$\Delta P_E \approx \frac{\Delta P_0}{\Delta x} \; \text{ or } \; \partial P_E = \frac{\partial P_0}{\partial x}$$

where the value ΔP equals the difference between P_0-parameter of i-orbital and P_{CD}– countdown parameter (parameter of basic state at the given set of quantum numbers).

Table 1. Comparison of total energy of valent electrons in atom calculated in Thomas-Fermi statistic atom model (U) and with the help of approximation

Atom	Valent electrons	r_i (Å)	X	φ(X)	U(eV)	W_i(eV)	n	q^2(eV Å)	P_E (eV)
1	2	3	4	5	6	7	8	9	10
Ar	$3P^4$	0.639	3.548	0.09-0.084	35.36-33.02	12+	4	73.196	33.45
	$3S^2$	0.607	3.258	0.122-0.105	47.81 / 44.81	34.8 (t) / 29.0	2 / 2	96.107 / 96.107	48.44 / 42.45
V	$2P^4$	0.146	0.785	0.47	834.25	246	4	706.3	817.12
	$4S^2$	1.401	8.538 / 8.23	0.0325 / 0.0345	7.680 / 8.151	7.5	2	22.33	7.730
Cr	$4S^2$	1.453	8.95 / 8.70	0.0295 / 0.0313	7.013 / 7.440	7	2	23.712	7.754
Mn	$4S^2$	1.278	7.75	0.0256	10.89	6.6 (t) / 7.5	2 / 2	25.12 / 25.12	7.895 / 10.87
Fe	$4S^2$	1.227	7.552	0.0282	8.598	8.00 / 7.20 (t)	2 / 2	26.57 / 26.57	9.201 / 8.647
Co	$4S^2$	1.181	7.555 / 7.378	0.02813 / 0.03075	9.255 / 10.127	8 / 7.5 (t)	2 / 2	27.98 / 27.98	10.062 / 9.187
Ni	$4S^2$	1.139	7.2102	0.02596	9.183	9 / 7.7 (t)	2 / 2	29.348 / 29.348	10.60 / 9.640
Cu	$4S^2$	1.191	7.633	0.0272	9.530	7.7	2	30.717	9.639
Zr	$5S^2$	1.093	8.424 / 8.309	0.033 / 0.03415	21.30 / 22.03*	11.7	2	238.3	21.8
	$4d^{10}$	0.4805	3.704	0.106	155.6	20	10	258.23	145.8
Te	$5p^4$	1.063	8.654 / 8.256	0.0335 / 0.0346	23.59 / 24.37*	9.8	4	67.28	24.54
	$5S^2$	0.920	7.239 / 7.146	0.0326 / 0.0341	26.54 / 27.72*	19 / 17	2 / 2	90.577 / 90.537	27.41 / 25.24

Note: 1) Bond energies of electrons W_i are obtained: "t" – theoretically (by Hartry-Fock method), "+" – by XPS method, all the rest – by the results of optic measurements; 2) "*" – energy of valent electrons (U) calculated without Fermi-Amaldi amendment.

According to the established rule [8] of adding P-parameters of similarly charged or homogeneous systems for two orbitals in the given atom with different quantum characteristics and in accordance with the law of energy conservation, we have:

$$\Delta P_E^{''} - \Delta P_E^{'} = P_{E,\lambda}$$

where $P_{E,\lambda}$ – spatial-energy parameter of quantum transition.

Taking as the dimension characteristic of the interaction $\Delta\lambda = \Delta x$, we have:

$$\frac{\Delta P_0^{''}}{\Delta\lambda} - \frac{\Delta P_0^{'}}{\Delta\lambda} = \frac{P_0}{\Delta\lambda} \quad \text{or:} \quad \frac{\Delta P_0^{'}}{\Delta\lambda} - \frac{\Delta P_0^{''}}{\Delta\lambda} = -\frac{P_0\lambda}{\Delta\lambda}$$

We divide term wise by $\Delta\lambda$:

$$\left(\frac{\Delta P_0^{'}}{\Delta\lambda} - \frac{\Delta P_0^{''}}{\Delta\lambda}\right)\bigg/\Delta\lambda = -\frac{P_0}{\Delta\lambda^2}, \text{ where: } \left(\frac{\Delta P_0^{'}}{\Delta\lambda} - \frac{\Delta P_0^{''}}{\Delta\lambda}\right)\bigg/\Delta\lambda \sim \frac{d^2 P_0}{d\lambda^2}, \text{ i.e.: } \frac{d^2 P_0}{d\lambda^2} + \frac{P_0}{\Delta\lambda^2} \approx 0$$

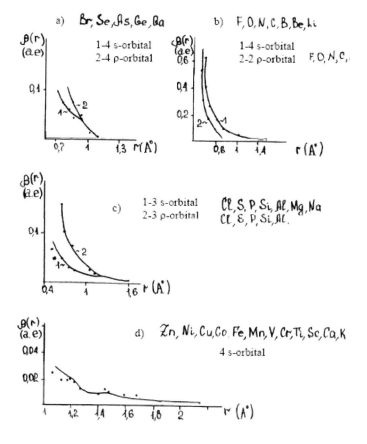

Figure 1. Electron density at the distance r_i, calculated via Clementi functions (solid lines) and with the help of P-parameter (dots).

Taking into account the interactions where $2\pi\Delta x = \Delta\lambda$ (closed oscillator), we have the following equation:

$$\frac{d^2 P_0}{dx^2} + 4\pi^2 \frac{P_0}{\Delta\lambda^2} \approx 0$$

As $\Delta\lambda = \frac{h}{mv}$, then:

$$\frac{d^2 P_0}{dx^2} + 4\pi^2 \frac{P_0}{h^2} m^2 v^2 \approx 0$$

or

$$\frac{d^2 P_0}{dx^2} + \frac{8\pi^2 m}{h^2} P_0 E_k = 0 \qquad (15)$$

where

$$E_k = \frac{mV^2}{2} \text{ - electron kinetic energy.}$$

Schrödinger equation for stationary state in coordinate x:

$$\frac{d^2 \psi}{dx^2} + \frac{8\pi^2 m}{h^2} \psi E_k = 0 \qquad (16)$$

Comparing the equations (15 and 16), we can see that P_0-parameter correlates numerically with the value of Ψ-function:

$$P_0 \approx \Psi$$

and in general case is proportional to it: $P_0 \sim \Psi$. Taking into account wide practical application of P-parameter methodology, we can consider this criterion the materialized analog of Ψ-function. Since P_0-parameters like Ψ-function possess wave properties, the principles of superposition should be executed for them, thus determining the linear character of equations of adding and changing P-parameters.

4. Wave Properties of P-Parameters and Principles of Their Addition

Since P-parameter possesses wave properties (by the analogy with Ψ'-function), then the regularities of the interference of corresponding waves should be mainly executed with structural interactions.

Interference minimum, oscillation attenuation (in anti-phase) takes place if the difference in wave motion (Δ) equals the odd number of semi-waves:

$$\Delta = (2n+1)\frac{\lambda}{2} = \lambda(n + \frac{1}{2}), \quad \text{where } n = 0, 1, 2, 3, \ldots \tag{17}$$

As applied to P-parameters, this rules means that interaction minimum occurs if P-parameters of interacting structures are also "in anti-phase" – there is an interaction either between oppositely charged systems or heterogeneous atoms (for example, during the formation of valent-active radicals CH, CH_2, CH_3, NO_2 ..., etc).

In this case, the summation of P-parameters takes place by the principle of adding the reverse values of P-parameters – equations (3,4).

The difference in wave motion (Δ) for P-parameters can be evaluated via their relative value ($\gamma = \frac{P_2}{P_1}$) or via the relative difference in P-parameters (coefficient α), which with the minimum of interactions produce an odd number:

$$\gamma = \frac{P_2}{P_1} = (n + \frac{1}{2}) = \frac{1}{2}; \frac{3}{2}; \frac{5}{2} \ldots$$

When n=0 (main state) $\dfrac{P_2}{P_1} = \dfrac{1}{2}$ \hfill (18)

Let us mention that for stationary levels of one-dimensional harmonic oscillator, the energy of these levels $\varepsilon = h\nu(n + \frac{1}{2})$, therefore in quantum oscillator, in contrast to a classical one, the minimum possible energy value does not equal zero.

In this model, the interaction minimum does not produce the zero energy, corresponding to the principle of adding the reverse values of P-parameters – equations (3,4). *Interference maximum,* oscillation amplification (in phase) takes place if the difference in wave motion equals the even number of semi-waves: $\Delta = 2n\frac{\lambda}{2} = \lambda n$ or $\Delta = \lambda(n+1)$

As applied to P-parameters, the maximum amplification of interactions in the phase corresponds to the interactions of similarly charged systems or systems homogeneous in their properties and functions (for example, between the fragments and blocks of complex organic structures, such as CH_2 and NNO_2 in octogen).

Then:

$$\gamma = \frac{P_2}{P_1} = (n+1) \tag{19}$$

By the analogy, for "degenerated" systems (with similar values of functions) of two-dimensional harmonic oscillator the energy of stationary states: $\varepsilon = h\nu(n+1)$.

In this model, the interaction maximum corresponds to the principle of algebraic addition of P-parameters – equations (6-8). When n=0 (basic state) we have $P_2 = P_1$, or: interaction maximum of structures takes place when their P-parameters equal. This postulate can be used as [8] the main condition of isomorphic replacements.

5. Structural Exchange Spatial-Energy Interactions

In the process of solution formation and other structural interactions, the single electron density should be set in the points of atom-component contact. This process is accompanied by the redistribution of electron density between the valent areas of both particles and transition of the part of electrons from some external spheres into the neighboring ones. Apparently, frame atom electrons do not take part in such exchange.

Obviously, when electron densities in free atom-components are similar, the transfer processes between boundary atoms of particles are minimal; this will be favorable for the formation of a new structure. Thus, the evaluation of the degree of structural interactions in many cases means the comparative assessment of the electron density of valent electrons in free atoms (on averaged orbitals) participating in the process.

The less the difference ($P'_0/r'_i - P''_0/r''_i$), the more favorable is the formation of a new structure or solid solution from the energy point.

In this regard, the maximum total solubility, evaluated via the coefficient of structural interaction α, is determined by the condition of minimum value α, which represents the relative difference of effective energies of external orbitals of interacting subsystems:

$$\alpha = \frac{P'o/r_i'-P''o/r_i''}{(P'o/r_i'+P''o/r_i'')/2}100\%$$

$$\alpha = \frac{P'_S-P''_S}{P'_S+P''_S} 200\%$$

$$(20, 20a),$$

where P_S – structural parameter is found by the equation:

$$\frac{1}{P_s} = \frac{1}{N_1 P'_E} + \frac{1}{N_1 P''_E} + \dots$$

$$(20b),$$

here, N_1 and N_2 – number of homogeneous atoms in subsystems.

The nomogram of the dependence of structural interaction degree По всем полученным данным была построена номограмма зависимости степени структурного взаимодействия (ρ) on the coefficient α, unified for the wide range of structures was prepared based on all the data obtained. In Fig. 2, one can see such nomogram obtained using P_E-parameters calculated via the bond energy of electrons (w_i) for structural interactions of isomorphic type.

The mutual solubility of atom-components in many (over a thousand) simple and complex systems was evaluated using this technique. The calculation results are in compliance with theoretical and experimental data [8].

Isomorphism as a phenomenon is used to be considered as applicable to crystalline structures. But similar processes can obviously take place between molecular compounds, where their role and importance are not less than those of purely coulomb interactions.

In complex organic structures during the interactions, the main role can be played by separate "blocks" or fragments. Therefore, it is necessary to identify these fragments and evaluate their spatial-energy parameters. Based on the wave properties of P-parameter, the overall P-parameter of each fragment should be found by the principle of adding the reverse values of initial P-parameters of all atoms. The resultant P-parameter of the fragment block or all the structure is calculated by the rule of algebraic addition of P-parameters of the fragments constituting them.

The role of the fragments can be played by valent-active radicals, e.g., CH, CH_2, $(OH)^-$, NO, NO_2, $(SO_4)^{2-}$, etc. In complex structures, the given carbon atom usually has two or three side bonds. During the calculations by the principle of adding the reverse values of P-parameters, the priority belongs to those bonds for which the condition of interference minimum is better performed. Therefore, the fragments of the bond C-H (for CH, CH_2, CH_3 ...) are calculated first, then separately the fragments N-R, where R is the binding radical (for example – for the bond C-N).

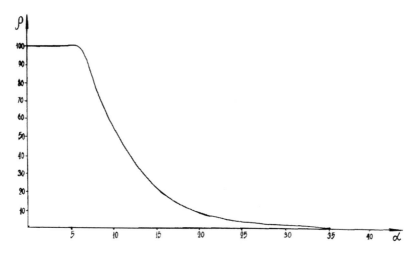

Figure 2. Dependence of the structural interaction degree (ρ) on the coefficient α.

Apparently, spatial-energy exchange interactions (SEI) based on equalizing electron densities of valent orbitals of atom-components have in nature the same universal value as purely electrostatic coulomb interactions, but they supplement each other. Isomorphism known from the time of E. Mitscherlich (1820) and D.I. Mendeleev (1856) is only the particular manifestation of this general natural phenomenon. The numerical side of the evaluation of isomorphic replacements of components both in complex and simple systems rationally fit in the frameworks of P-parameter methodology. More complicated is to evaluate the degree of structural SEI for molecular, including organic, structures. The technique for calculating P-parameters of molecules, structures and their fragments is successfully implemented. But such structures and their fragments are frequently not completely isomorphic with respect to each other. Nevertheless, there is SEI between them, the degree of which in this case can be evaluated only semi-quantitatively or qualitatively. By the degree of isomorphic similarity, all the systems can be divided into three types:

I Systems mainly isomorphic to each other – systems with approximately similar number of *dissimilar* atoms and summarily similar geometrical shapes of interacting orbitals.

II Systems with *the limited isomorphic similarity* – such systems, which:

1) either differ by the number of dissimilar atoms but have summarily similar geometrical shapes of interacting orbitals; or

2) have definite differences in geometrical shapes of orbitals but similar number of interacting dissimilar atoms.

III Systems not having isomorphic similarity – such systems, which considerably differ both by the number of dissimilar atoms and geometric shapes of their orbitals.

Then, taking into account the experimental data, all types of SEI can be approximately classified as follows:

Systems I

1. $\alpha < (0\text{-}6)\%$; $\rho = 100\%$. Complete isomorphism, there is complete isomorphic replacement of atom-components;

2. $6\% < \alpha < (25\text{-}30)\%$; $\rho = 98 - (0\text{-}3)$ %.

There is either wide or limited isomorphism according to nomogram 1.

3. $\alpha > (25\text{-}30)$ %; no SEI

Systems II

1. $\alpha < (0\text{-}6)\%$;

a) There is the reconstruction of chemical bonds, can be accompanied by the formation of a new compound;

b) Breakage of chemical bonds can be accompanied by separating a fragment from the initial structure but without attachments or replacements.

2. $6\% < \alpha < (25\text{-}30)\%$; limited internal reconstruction of chemical bonds without the formation of a new compound or replacements is possible.

3. $\alpha > (20\text{-}30)$ %; no SEI

Systems III

1. $\alpha < (0\text{-}6)\%$; a) Limited change in the type of chemical bonds of the given fragment, internal regrouping of atoms without the breakage from the main part of the molecule and without replacements;

b) Change in some dimensional characteristics of the bond is possible;

2. $6\% < \alpha < (25\text{-}30)\%$;

Very limited internal regrouping of atoms is possible;

3. $\alpha > (25\text{-}30)$ %; no SEI.

Nomogram (Fig. 2) is obtained for isomorphic interactions (systems of types I and II).

In all other cases, the calculated values α and ρ refer only to the given interaction type, the nomogram of which can be clarified by reference points of etalon systems. If we take into account the universality of spatial-energy interactions in nature, this evaluation can be significant for the analysis of structural rearrangements in complex biophysical-chemical processes.

Fermentative systems contribute a lot to the correlation of structural interaction degree. In this system, the ferment structure active parts (fragments, atoms, ions) have the value of P_E-parameter that is equal to P_E-parameter of the reaction final product. This means the ferment is structurally "tuned" via SEI to obtain the reaction final product, but it will not included into it due to the imperfect isomorphism of its structure (in accordance with III).

The most important characteristics of atomic-structural interactions (mutual solubility of components, energy of chemical bond, energetics of free radicals, etc.) were evaluated in many systems using this technique [8-15].

6. Types of Fundamental Interactions

According to modern theories, the main types of interactions of elementary particles, their properties and specifics are mainly explained by the availability of special complex "currents" – electromagnetic, proton, lepton, etc. Based on the foregoing model of spatial-energy parameter, the exchange structural interactions finally come to flowing and equalizing the electron densities of corresponding atomic-molecular components. The similar process is obviously appropriate for elementary particles as well. It can be assumed that in general case, interparticle exchange interactions come to the redistribution of their energy masses – M.

The elementary electrostatic charge with the electron as a carrier is the constant of electromagnetic interaction.

Therefore, for electromagnetic interaction, we will calculate the system proton-electron.

For strong internucleon interaction that comes to the exchange of π-mesons, let us consider the systems nuclides-π-mesons. Since the interactions can take place with all three mesons (π^-, π^0, π^+), we take the averaged mass in the calculations ($<M>=136,497$ MeV/s^2).

Rated systems for strong interaction:

P - (π^-, π^0, π^+);
(P-n) - (π^-, π^0, π^+);
(n-P-n) - (π^-, π^0, π).

Neutrino (electron, muonic) and its antiparticles were considered as the main representatives of weak interaction.

Dimensional characteristics of elementary particles (r) were evaluated in femtometer units (1 fm = 10^{-15} m) – by the data in [16].

At the same time, the classic radius: $r_e = e^2/m_e s^2$ was used for electron, where e – elementary charge, m_e – electron mass, s – light speed in vacuum.

The fundamental Heisenberg length ($6.690 \cdot 10^{-4}$ fm) was used as the dimensional characteristic of weak interaction for neutrino [16].

The gravitational interaction was evaluated via the proton P-parameter at the distance of gravitational radius ($1.242 \cdot 10^{-39}$ fm).

In the initial equation (3) for free atom, P_0-parameter is found by the principle of adding the reverse values q^2 and wr, where q – nucleus electric charge, w – bond energy of valent electron.

Modifying the equation (3), as applied to the interaction of free particles, we receive the addition of reverse values of parameters P=Mr for each particle by the equation:

$$1/P_0 = 1/(Mr)_1 + 1/(Mr)_2 + \ldots \tag{21},$$

where M – energy mass of the particle (MeV/s^2).

By the equation (21), using the initial data [16], P_0-parameters of coupled strong and electromagnetic interactions were calculated in the following systems:

1. nuclides-π-mesons – (P_n-parameters);
2. proton-electron – (P_e-parameter).

For weak and gravitational interactions, only the parameters $P_\upsilon = Mr$ and $P_\Gamma = Mr$ were calculated, as in accordance with the equation (21), the similar nuclide parameter with greater value does not influence the calculation results.

The relative intensity of interactions (Table 2) were found by the equations for the following interactions:

1) strong $\alpha_B = <P_n> / <P_n> = P_n / P_n = 1$ (22a)

2) electromagnetic $\alpha_B = P_e / <P_n> = 1/136.983$ (22b)

3) weak $\alpha_B = P_g / <P_n>$, $\alpha_B = 2.04 \cdot 10^{-10}$; $4.2 \cdot 10^{-6}$ (22c)

4) gravitational $\alpha_B = P_\Gamma / <P_n> = 5.9 * 10^{-39}$ (22d)

In the calculations of α_B, the value of P_n-parameter was multiplied by the value equaled $2\pi/3$, i.e., $<P> = (2\pi/3)P_n$. Number 3 for nuclides consisting of three different quarks is "a magic" number (see the next section for details). As it is known, number 2π has a special value in quantum mechanics and physics of elementary particles. In particular, only the value of 2π correlates theoretical and experimental data when evaluating the sections of nuclide interaction with each other [17].

As it is known [18], "very strong," "strong" and "moderately strong" nuclear interactions are distinguished. For all particles in the large group with relatively similar mass values of mass – unitary multiplets or supermultiplets – very strong interactions are the same [18]. In the frames of the given model, a very strong interaction between the particles corresponds to the maximum value of P-parameter P=Mr (coupled interaction of nuclides). Taking into consideration the equality of dimensional characteristics of proton and neutron, by the equation (21), we obtain the values of P_n-parameter equaled to 401.61; 401.88 and 402.16 (MeVfm/s2) for coupled interactions p-p, p-n and n-n, respectively, thus obtaining the average value $\alpha_B = 4.25$. It is a very strong interaction. For eight interacting nuclides $\alpha_B \approx 1.06$ – a strong interaction.

When the number of interacting nuclides increases, α_B decreases – moderately strong interaction. Since the nuclear forces act only between neighboring nucleons, the value α_B cannot be very small.

The expression of the most intensive coupled interaction of nuclides is indirectly confirmed by the fact that the life period of double nuclear system appears to be much longer than the characteristic nuclear time [19].

Thus, it is established that the intensity of fundamental interactions is evaluated via P_n-parameter calculated by the principle of adding the reverse values in the system nuclides-π-mesons. Therefore, it has the direct connection with Plank's constants:

$(2\pi/3)P_n \approx Er = 197.3$ MeVfm/s^2 (23)

$(2\pi/3)P_\Gamma \approx M_n\lambda_k =$ MeVfm/s^2 (23a)

where E and r – Plank's energy and Plank's radius calculated via the gravitational constant; M_n, λ_k – energy mass and nuclide Compton wave-length.

In the equation (21), the exchange interactions are evaluated via the initial P-parameters of particles equaled to the product of mass by the dimensional characteristic: P=Mr.

Since these P-parameters can refer to the particles characterizing fundamental interactions, their direct correlation defines the process intensity degree (α_B):

$$a_B = \frac{P_i}{P_n} = \frac{(Mr)_i}{(Mr)_n}$$

(24)

The calculations by the equation (24) using the known Plank's values and techniques are given in Table 3. As before, the energy and dimensional characteristics are taken from [16].

The results obtained are in accordance with theoretical and experimental data, e.g., [20, 21].

7. On Quark Screw Model

Let us proceed from the following theses and assumptions:

1) By their structural composition, macro- and micro-world resemble a complex matreshka. One part has some similarity with the other: solar system – atom – atom nucleus – quarks.
2) All parts of this "matreshka" are structural formations.
3) Main property of all systems – motion: translatory, rotary, oscillatory.
4) Description of these motions can be done in Euclid three-dimensional space with coordinates x,y,z.
5) Exchange energy interactions of elementary particles are carried out by the redistribution of their energy mass M (MeV/s^2).

Based on these theses, we suggest discussing the following screw model of the quark.

1) Quark structure is represented in certain case as a spherical one, but in general quark is a flattened (or elongated) ellipsoid of revolution. The revolution takes place around the axis (x), coinciding with the direction of angular speed vector, perpendicular to the direction of ellipsoid deformation.
2) Quark electric charge (q) is not fractional but integer, but it is redistributed in three-dimensional space with its virtual concentration in the directions of three coordinate axes: q/3.
3) Quark spherical or deformed structure has all three types of motion. Two of them —rotary and translatory—are in accordance with the screw model, which besides these two motions, also performs an oscillatory motion in one of three mutually perpendicular planes: xoy, xoz, yoz (Fig. 2).
4) Each of these oscillation planes corresponds to the symbol of quark color, e.g., xoy – red, xoz – blue, yoz – green.
5) Screw can be "right" or "left." This directedness of screw rotation defines the sign of quark electric charge. Let us assume that the left screw corresponds to positive and right – negative quark electric charge.
6) Total number of quarks is determined by the following scheme: for each axis (x,y,z) of translator motion, two screws (right and left) with three possible oscillation planes.

We have: 3·2·3=18. Besides, there are 18 antiquarks with opposite characteristics of screw motions. Total: 36 types of quarks.

7) These quark numbers can be considered as realized degrees of freedom of all three motions (3 translatory + 2 rotary + 3 oscillatory).

8) Translatory motion is preferable by its direction, coinciding with the direction of angular speed vector. Such elementary particles constitute our World. The reverse direction is less preferable – this is "Antiworld."

9) Motion along axis x in the direction of the angular speed vector, perpendicular to the direction of ellipsoid deformation, is apparently less energy consumable and corresponds to the quarks U and d, forming nuclides. Such assumption is in accordance with the values of energy masses of quarks in the composition of androns: 0.33; 0.33; 0.51; 1.8; 5; (?) in MeV/s^2 for d,u,s,c,b,t – types of quarks, respectively.

The quark screw model can be proved by other calculations and comparisons.

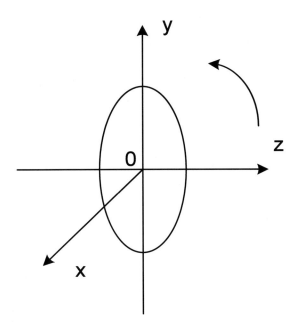

Figure 2. Structural scheme of quark in section yoz.

Calculation of Energy Mass of Free Nuclide (on the Example of Neutron)

Neutron has three quarks d_1-u-d_2 with electric charges -1, +2, -1, distributed in three spatial directions, respectively. Quark u cements the system electrostatically. Translatory motions of the screws d_1-u-d_2 proceed along axis x, but oscillatory ones – in three different mutually perpendicular planes (Pauli principle is realized).

Apparently, in the first half of oscillation period, u-quark oscillates in the phase with d_1-quark but in the opposite phase with d_2-quark. In the second half of the period, everything is vice versa. In general, such interactions define the geometrical equality of directed spatial-energy vectors, thus providing the so-called quark discoloration.

Table 2. Types of fundamental interactions

Interaction type		M, <M> (MeV/s²)	r (fm)	Elementary particles	M, <M> (MeV/s²)	r (fm)	P_n, P_e, P_ν, P_r (MeVfm/s²)	$\dfrac{2\pi}{3}P_n$ = <P_n>	$\alpha_в$, <$\alpha_в$> (calculat.) – by equat. (22)	$\alpha_в$ (experiment)
Electromagnetic	P	938.28	0.856	e^-	0.5110	2.8179	P_e=1.4374	-	1/136.983	1/137.04
Strong	P	938.28	0.856	π^-, π^0, π^+	136.497	0.78	P_n=94.0071	196.89	1	1
	P-n	938.92	0.856	π^-, π^0, π^+	136.497	0.78	P_n=94.015	196.90	1	1
	n-P-n	939.14	0.856	π^-, π^0, π^+	136.497	0.78	P_n=94.018	196.91	1	1
Weak				$\overline{\nu_e}$	<6·10⁻⁵	6.69·10⁻⁴	P_ν=4.014·10⁻⁸		<2.04·10⁻¹⁰	10⁻¹⁰ - 10⁻¹⁴
				$\overline{\nu_\mu}$	<1.2	6.69·10⁻⁴	P_ν=8.028·10⁻⁴		<4.2·10⁻⁶	10⁻⁵ - 10⁻⁶
Gravitational	P	938.28	1.242·10⁻³⁹				P_r=1.17·10⁻³⁶		5.9·10⁻³⁹	10⁻³⁸ - 10⁻³⁹

Table 3. Evaluation of the intensity of fundamental interactions using Plank's constants and parameter P=Mr

Interaction type	Particles, constants	M (MeV/s²)	r (fm)	Mr (MeVfm/s²)	$\alpha_в$=Mr/(Mr)$_p$ (calculation)	$\alpha_в$ (experiment)
Strong	Proton	938.28	λ=0.2103	197.3	1	1
	Plank's values	1.221·10²²	1.616·10⁻²⁰	197.3		
Electromagnetic	электрон	0.5110	2.8179	1.43995	1/137.02	1/137.036
Weak	$\overline{\nu_e}$	<6·10⁻⁵	6.69·10⁻⁴	<4.014·10⁻⁸	<2.03·10⁻¹⁰	10⁻¹⁰ - 10⁻¹⁴
	$\overline{\nu_\mu}$	<1.2	6.69·10⁻⁴	<8.028·10⁻⁴	<4.07·10⁻⁶	10⁻⁵ - 10⁻⁶
Gravitational	Proton	938.28	1.242·10⁻³⁹ Gravitational radius	1.165·10⁻³⁶	5.91·10⁻³⁹	10⁻³⁸ - 10⁻³⁹

The previously formulated rules of adding P-parameters spread to both types of P-parameters (P_0 and P_E). In this case, there is an addition of energy P_E-parameters, since the subsystems of interactions possess similar dimensional characteristics. As both interactions are realized inside the overall system, P_E-parameters are added algebraically and, more accurately in this case, geometrically by the following formula:

$$\frac{M}{2} = \sqrt{m_1^2 + m_2^2}$$

where, M – energy mass of free neutron, $m_1 = m_2 = 330$ MeV/s^2 masses of quarks u,d (in the composition of androns).

The calculation gives M=933.38 MeV/s^2. This is for strong interactions. Taking into account the role of quarks in electromagnetic interactions [21], we get the total energy mass of a free neutron: M=933.38 + 933.38/137 = 940.19 MeV/s^2. With the experimental value M = 939.57 MeV/s^2 the relative error in calculations is 0.06%.

Calculation of Bond Energy of Deuteron via the Masses of Free Quarks

The particle deuteron is formed during the interaction of a free proton and neutron. The bond energy is usually calculated as the difference of mass of free nucleons and mass of a free deuteron. Let us demonstrate the dependence of deuteron bond energy on the masses of free quarks. The quark masses are added algebraically in the system already formed: in proton $m_1 = 5+5+7=17$ MeV/s^2, in neutron $m_2=7+7+5=19$ MeV/s^2. As a dimensional characteristic of deuteron bond, we take the distance corresponding to the maximum value of nonrectangular potential pit of nucleon interaction. By the graphs experimentally obtained, we know that such distance approximately equals 1.65 fm. Exchange energy interactions of proton and neutron heterogeneous systems are evaluated based on the equation (21). Then we have:

$$1/ (M_C 1.65K) = 1/(17 \cdot 0.856) + 1/(19 \cdot 0.856),$$

where K=$2\pi/3$. Based on the calculations, we have M_C=2.228 MeV/s^2; this is practically coincide with reference data (M_C=2.225 MeV/s^2).

Modifying the basic theses of quark screw model, it can also be applied to other elementary particles (proton, electron, neutron, etc.). For instance, an electrically neutral particle neutron can be considered as a mini-atom, the analog of hydrogen atom.

Conclusion

1) The notion of spatial-energy parameter (P-parameter) is introduced based on the simultaneous accounting of important atomic characteristics and modified Lagrangian equation.

2) Wave properties of P-parameter are found; its wave equation formally similar to the equation of ψ-function is obtained.

3) Applying the methodology of P-parameter:
   a. most important characteristics of exchange energy interactions in different systems are calculated;
   b. intensities of fundamental interactions are calculated;
   c. initial theses of quark screw model are given.

References

[1] Batsanov S.S., Zvyagina R.A. Overlap integrals and challenge of effective charges – Novosibirsk: *Nauka,* 1966, – 386 p.

[1] Fischer C.F. Average-Energy of Configuration Hartree-Fock Results for the Atoms Helium to Radon. *Atomic Data,* 1972, № 4, p. 301–399.

[2] Waber J.T., Cromer D.T. Orbital Radii of Atoms and Ions. *J. Chem. Phys,* 1965, v. 42, No 12, p. 4116–4123.

[3] Clementi E., Raimondi D.L. Atomic Screening constants from S.C.F. Functions, 1. *J. Chem. Phys.,* 1963, v.38, No 11, p. 2686–2689.

[4] Clementi E., Raimondi D.L. Atomik Screening Constants from S.C.F. Functions, II. *J. Chem. Phys.,* 1967, v.47, No 4, p. 1300 – 1307.

[5] Gombash P. *Atom statistic theory and its application.* – M.: I.L., 1951, 398 p.

[6] Clementi E. Tables of atomic functions. *J.B.M. S. Res. Develop. Suppl.,* 1965, v. 9, No 2, p.76.

[7] Korablev G.A. Spatial Energy Principles of Complex Structures Formation. Netherlands, Leiden, Brill Academic Publishers and VSP, 2005, 426 p. (Monograph).

[8] Korablev G.A., Kodolov V.I., Lipanov A.M. Analog comparisons of Lagrangian and Hamiltonian functions with spatial-energy parameter. Chemical physics and mesoscopy, *URC RAS,* 2004, No1, v.6, p. 5-18.

[9] Korablev G.A., Zaikov G.E. Energy of chemical bond and spatial-energy principles of hybridization of atom orbitals. *J. of Applied Polymer Science. USA,* 2006, V.101, n.3, p.2101-2107.

[10] Korablev G.A., Zaikov G.E. P-Parameter as and Objective Characteristics of Electronegativity//*Reactions and Properties of Monomers Polymers,* Nova Science Publishers, Inc., New York, 2007, p.203-213.

[11] Korablev G.A., Zaikov G.E. Spatial-energy interactions of free radicals.// *Success in gerontology,* 2008, v.21, No4, p.535-563.

[12] Korablev G.A., Zaikov G.E. Formation of carbon nanostructures and spatial-energy criterion of stabilization. *Mechanics of composition materials and constructions,* 2009, RAS, v.15, No 1, p.106-118.

[13] Korablev G.A., Zaikov G.E. Energy of chemical bond and spatial-energy principles of hybridization of atom orbitals. *Chemical physics,* RAS, M.: 2006, v.25. No 7, p.24-28.

[14] Korablev G.A., Zaikov G.E. Calculations of activation energy of diffusion and self-diffusion. *Monomers, Oligomers, Polymers, Composites and Nanocomposites Research,* Nova Science Publishers, USA, 2008, pp. 441-448.

[15] Murodyan R.M. Physical and astrophysical constants and their dimensional and dimensionless combinations. *PEChAYa, M., Atomizdat,* 1977, v.8, iss.1, p.175-192.

[16] Barashenkov V.S. Sections of interactions of elementary particles, M.: *Nauka,* 1966, 532p.

[17] Yavorsky B.M., Detlav A.A. Reference-book in physics, M., *Nauka,* 1968, 940p.

[18] Volkov V.V. Exchange reactions with heavy ions. *PEChAYa, M., Atomizdat,* 1975, v.6, iss.4, p. 1040-1104.

[19] Bukhbinder I.L. Fundamental interactions. *Sorov educational journal,* №5, 1997, http://nuclphys.sinp.msu.ru/mirrors/fi.htm.

[20] Okun L.B. *Weak interactions.*http://www.booksite.ru/fulltext/1/001/008/103/116.htm.

In: Handbook of Research on Nanomaterials, Nanochemistry ...   ISBN: 978-1-61942-525-5
Editors: A. K. Haghi and G. E. Zaikov   © 2013 Nova Science Publishers, Inc.

Chapter XXIV

A Nanofiller Particles Aggregation in Elastomeric Nanocomposites: The Irreversible Aggregation Model

*Yu. G. Yanovskii[1], G. V. Kozlov[1],
S. Kubica[2] and G. E. Zaikov[3]*
[1] Institute of Applied Mechanics of Russian Academy of Sciences, Russia
[2] Institut Inzynierii Materialow Polimerowych I Barwnikow, Torun, Poland
[3] N.M. Emanuel Institute of Biochemical Physics of Russian Academy of Sciences,
Russia

Introduction

The aggregation of the initial nanofiller powder particles in more or less large particles aggregates always occurs in the course of technological process of making particulate-filled polymer composites in general [1] and elastomeric nanocomposites in particular [2].

The aggregation process tells on composites (nanocomposites) macroscopic properties [1, 3]. For nanocomposites, nanofiller aggregation process gains special significance, since its intensity can be the one that nanofiller particles aggregates size exceeds 100 nm – the value, which assumes (although and conditionally enough [4]) as an upper dimensional limit for nanoparticle.

In other words, the aggregation process can result in the situation when primordially supposed nanocomposite ceases to be the one.

Therefore, at present, several methods exist, which allowed suppressing nanoparticles aggregation process [2, 5].

Proceeding from this, in the present paper, theoretical treatment of disperse nanofiller aggregation process in butadiene-styrene rubber matrix within the frameworks of irreversible aggregation models was carried out.

Experimental

The elastomeric particulate-filled nanocomposite on the basis of butadiene-styrene rubber was an object of the study. Mineral shungite nanodimensional and microdimensional particles and also industrially produced technical carbon with mass contents of 37 mass % were used as a filler. The analysis of the received in milling process shungite particles were monitored with the aid of analytical disk centrifuge (CPS Instruments, Inc., USA), allowing determination with high precision the size and distribution by sizes within the range from 2 nm up to 50 mcm.

Nanostructure was studied on atomic-power microscopes Nano-DST (Pacific Nanotechnology, USA) and Easy Scan DFM (Nanosurf, Switzerland) by semi-contact method in the force modulation regime. Atomic-power microscopy results were processed with the aid of specialized software package SPIP (Scanning Probe Image Processor, Denmark). SPIP is a powerful programmes package for processing of images, obtained on SPM, AFM, STM, scanning electron microscopes, transmission electron microscopes, interferometers, confocal microscopes, profilometers, optical microscopes and so on. The given package possesses the whole functions number, which are necessary at images precise analysis, in the number of which the following are included:

1) the possibility of three-dimensional reflected objects obtaining, distortions automatized leveling, including Z-error mistakes removal for examination separate elements and so on;
2) quantitative analysis of particles or grains, more than 40 parameters can be calculated for each found particle or pore: area, perimeter, average diameter, the ratio of linear sizes of grain width to its height distance between grains, coordinates of grain center of mass a.a. can be presented in a diagram form or in a histogram form.

Results and Discussion

For theoretical treatment of nanofiller particles, aggregate growth processes and final sizes traditional irreversible aggregation models are inapplicable, since it is obvious that in nanocomposites aggregates, a large number of simultaneous growth takes place. Therefore, the model of multiple growth offered in paper [6] was used for nanofiller aggregation description.

In Fig. 1, the images of the studied nanocomposites, obtained in the force modulation regime, and corresponding to them nanoparticles aggregates fractal dimension d_f distributions are adduced. As it follows from the adduced values d_f (d_f=2.40-2.48), nanofiller particles aggregates in the studied nanocomposites are formed by a mechanism particle-cluster (P-Cl), i.e., they are Witten-Sander clusters [7]. The variant A was chosen, which according to mobile particles are added to the lattice, consisting of a large number of "seeds" with density of c_0 at simulation beginning [6]. Such model generates structures, which have fractal geometry on length short scales with value $d_f \approx 2.5$ (see Fig. 1) and homogeneous structure on length large scales. A relatively high particle concentration c is required in the model for uninterrupted network formation [6].

In case of "seeds" high concentration c_0 for the variant A, the following relationship was obtained [6]:

$$R_{max}^{d_f} = N = c/c_0, \tag{1}$$

where R_{max} is nanoparticles cluster (aggregate) greatest radius, N is nanoparticles number per one aggregate, c is nanoparticles concentration, c_0 is "seeds" number, which is equal to nanoparticles clusters (aggregates) number.

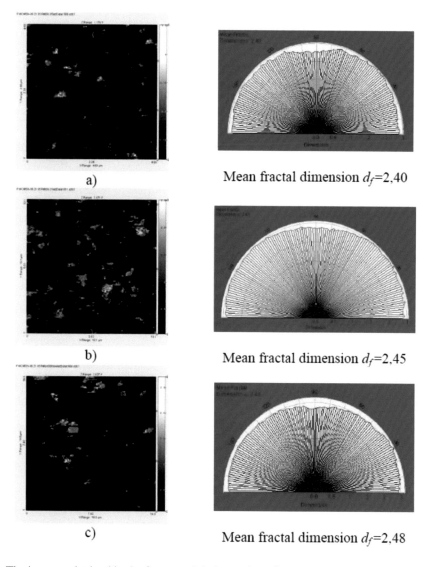

a)

Mean fractal dimension d_f=2,40

b)

Mean fractal dimension d_f=2,45

c)

Mean fractal dimension d_f=2,48

Figure 1. The images, obtained in the force modulation regime, for nanocomposites, filled with technical carbon (a), nanoshungite (b), microshungite (c) and corresponding to them fractal dimensions d_f.

The value N can be estimated according to the following equation [8]:

$$2R_{max} = \left(\frac{S_n N}{\pi \eta} \right)^{1/2},$$

(2)

where S_n is cross-sectional area of nanoparticles, from which aggregate consists, η is packing coefficient, equal to 0.74.

The experimentally obtained nanoparticles aggregate diameter $2R_{agr}$ was accepted as $2R_{max}$ (Table 1), and the value S_n was also calculated according to the experimental values of nanoparticles radius r_n (Table 1). In Table 1, the values N for the studied nanofillers, obtained according to the indicated method, were adduced. It is significant that the value N is a maximum one for nanoshungite despite larger values r_n in comparison with technical carbon.

Table 1. The parameters of irreversible aggregation model of nanofiller particles aggregates growth

Filler	R_{agr}, nm	r_n, nm	N	R_{max}^T, nm	R_{agr}^T, nm	R_c, nm
Technical carbon	34.6	10	35.4	34.7	34.7	33.9
Nanoshungite	83.6	20	51.8	45.0	90.0	71.0
Microshungite	117.1	100	4.1	15.8	158.0	255.0

Further, the equation (1) allows estimating the greatest radius R_{max}^T of nanoparticles aggregate within the frameworks of the aggregation model [6]. These values R_{max}^T are adduced in Table 1, from which their reduction in a sequence of technical carbon-nanoshungite-microshungite, which fully contradicts to the experimental data, i.e., to R_{agr} change (Table 1). However, we must not neglect the fact that the equation (1) was obtained within the frameworks of computer simulation, where the initial aggregating particles sizes are the same in all cases [6]. For real nanocomposites, the values r_n can be distinguished essentially (Table 1). It is expected that the value R_{agr} or R_{max}^T will be the higher, the larger is the radius of nanoparticles, forming aggregate, is, i.e., r_n. Then theoretical value of nanofiller particles cluster (aggregate) radius R_{agr}^T can be determined as follows:

$$R_{agr}^T = k_n r_n N^{1/d_f},$$

(3)

where k_n is proportionality coefficient, in the present work accepted empirically equal to 0.9.

The comparison of experimental R_{agr} and calculated according to the equation (3) R_{agr}^T values of the studied nanofillers particles aggregates radius shows their good correspondence (the average discrepancy of R_{agr} and R_{agr}^T makes up 11.4%). Therefore, the theoretical model [6] gives a good correspondence to the experiment only in case of consideration of aggregating particles real characteristics and, in the first place, their size.

Let us consider two more important aspects of nanofiller particles aggregation within the frameworks of the model [6]. Some features of the indicated process are defined by nanoparticles diffusion at nanocomposites processing. Specifically, length scale, connected with diffusible nanoparticle, is correlation length ξ of diffusion. By definition, the growth phenomena in sites, remote more than ξ, are statistically independent. Such definition allows connecting the value ξ with the mean distance between nanofiller particles aggregates L_n. The value ξ can be calculated according to the equation [6]:

$$\xi^2 \approx c^{-1} R_{agr}^{d_f - d + 2},$$
(4)

where c is nanoparticles concentration, d is dimension of Euclidean space, in which a fractal is considered (it is obvious that in our case $d=3$).

The value c should be accepted equal to nanofiller volume contents φ_n, which is calculated as follows [9]:

$$\varphi_n = \frac{W_n}{\rho_n},$$
(5)

where W_n is nanofiller mass contents, ρ_n is its density, determined according to the equation [3]:

$$\rho_n = 0.188(2r_n)^{1/3}.$$
(6)

The values r_n and R_{agr} were obtained experimentally (see histogram of Fig. 2). In Fig. 3, the relation between L_n and ξ is adduced, which, as it is expected, proves to be linear and passing through coordinates origin. This means that the distance between nanofiller particles aggregates is limited by mean displacement of statistical walks, by which nanoparticles are simulated. The relationship between L_n and ξ can be expressed analytically as follows:

$$L_n = 9.6\xi, \qquad \text{nm.}$$
(7)

The second important aspect of the model [6] in reference to nanofiller particles aggregation simulation is a finite nonzero initial particle concentration c or φ_n effect, which takes place in any real systems. This effect is realized at the condition $\xi \approx R_{agr}$, which occurs at the critical value $R_{agr}(R_c)$, determined according to the relationship [6]:

$$c \sim R_c^{d_f - d}.$$
(8)

The relationship (8) right side represents cluster (particles aggregate) mean density. This equation establishes that fractal growth continues only until cluster density reduces up to medium density in which it grows. The calculated according to the relationship (8) values R_c for the considered nanoparticles are adduced in Table 1, from which it follows that they give reasonable correspondence with this parameter experimental values R_{agr} (the average discrepancy of R_c and R_{agr} makes up 24%).

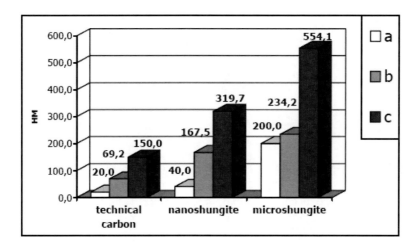

Figure 2. The initial particles diameter (a), their aggregates size in nanocomposite (b) and distance between nanoparticles aggregates (c) for nanocomposites, filled with technical carbon, nano- and microshungite.

Since the treatment [6] was obtained within the frameworks of a more general model of diffusion-limited aggregation, then its correspondence to the experimental data indicated unequivocally that aggregation processes in these systems were controlled by diffusion. Therefore, let us consider briefly nanofiller particles diffusion. Statistical walkers diffusion constant ζ can be determined with the aid of the relationship [6]:

$$\xi \approx \left(\zeta t\right)^{1/2},$$ (9)

where t is walk duration.

The equation (9) supposes (at t=const) ζ increase in a number technical carbon-nanoshungite-microshungite as 196-1069-3434 relative units, i.e., diffusion intensification at diffusible particles size growth. At the same time, diffusivity D for these particles can be described by the well-known Einstein's relationship [10]:

$$D = \frac{kT}{6\pi\eta r_n \alpha},$$ (10)

where k is Boltzmann constant, T is temperature, η is medium viscosity, α is numerical coefficient, which further is accepted equal to 1.

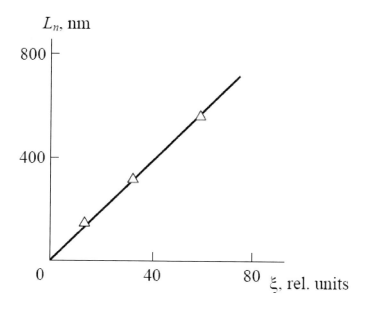

Figure 3. The relation between diffusion correlation length ξ and distance between nanoparticles aggregates L_n for studied nanocomposites.

In its turn, the value η can be estimated according to the equation [11]:

$$\frac{\eta}{\eta_0} = 1 + \frac{2.5\varphi_n}{1-\varphi_n},$$ (11)

where η_0 and η are initial polymer and its mixture with nanofiller viscosity, accordingly.

The calculation according to the equations (10) and (11) shows that within the above-indicated nanofillers, number the value D changes as 1.32-1.14-0.44 relative units, i.e., reduces in three times, which was expected. This apparent contradiction is due to the choice of the condition t=const (where t is nanocomposite production duration) in the equation (9). In real conditions, the value t is restricted by nanoparticle contact with growing aggregate and then instead of t the value t/c_0 should be used, where c_0 is seeds concentration, determined according to the equation (1). In this case, the value ζ for the indicated nanofillers changes as 0.288-0.118-0.086, i.e., it reduces in 3.3 times, which corresponds fully to the calculation according to the Einstein's relationship (the equation (10)). This means that nanoparticles diffusion in polymer matrix obeys classical laws of Newtonian rheology [10].

Conclusion

Disperse nanofiller particles aggregation in elastomeric matrix can be described theoretically within the frameworks of a modified model of irreversible aggregation particle-cluster. The obligatory consideration of nanofiller initial particles size is a feature of the indicated model application to real systems description. The indicated particles diffusion in polymer matrix obeys classical laws of Newtonian liquids hydrodynamics. The offered approach allows predicting nanoparticles aggregates final parameters as a function of the initial particles size, their contents and other factors number.

References

[1] Kozlov G.V., Yanovskii Yu.G., Zaikov G.E. *Structure and Properties of Particulate-Filled Polymer Composites: the Fractal Analysis.* New York, Nova Science Publishers, Inc., 2010, 282 p.

[2] Edwards D.C. *J. Mater. Sci.,* 1990, v. 25, № 12, p. 4175-4185.

[3] Mikitaev A.K., Kozlov G.V., Zaikov G.E. *Polymer Nanocomposites: the Variety of Structural Forms and Applications.* New York, Nova Science Publishers, Inc., 2008, 319 p.

[4] Buchachenko A.L. *Uspekhi Khimii,* 2003, v. 72, № 5, p. 419-437.

[5] Kozlov G.V., Yanovskii Yu.G., Burya A.I., Aphashagova Z.Kh. *Mekhanika Kompozitsionnykh Materialov i Konstruktsii,* 2007, v. 13, № 4, p. 479-492.

[6] Witten T.A., Meakin P. *Phys. Rev. B,* 1983, v. 28, № 10, p. 5632-5642.

[7] Witten T.A., Sander L.M. *Phys. Rev. B,* 1983, v. 27, № 9, p. 5686-5697.

[8] Bobryshev A.N., Kozomazov V.N., Babin L.O., Solomatov V.I. *Synergetics of Composite Materials.* Lipetsk, NPO ORIUS, 1994, 154 p.

[9] Sheng N., Boyce M.C., Parks D.M., Rutledge G.C., Abes J.I., Cohen R.E. *Polymer,* 2004, v. 45, № 2, p. 487-506.

[10] Happel J., Brenner G. The Hydrodynamics at Small Reynolds Numbers. Moscow, *Mir,* 1976, 418 p.

[11] Mills N.J. *J. Appl. Polymer Sci.,* 1971, v. 15, № 11, p. 2791-2805.

In: Handbook of Research on Nanomaterials, Nanochemistry ...   ISBN: 978-1-61942-525-5
Editors: A. K. Haghi and G. E. Zaikov   © 2013 Nova Science Publishers, Inc.

Chapter XXV

Calculations of Bond Energy in Cluster Aqueous Nanostructures

G. A. Korablev[1], N. V. Khokhriakov[1] and G. E. Zaikov[2]
[1] Izhevsk State Agricultural Academy, Russia
[2] Emmanuel Institute of Biochemical Physics, Russia

Introduction

Based on modified Lagrangian equation for relative movement of two interacting material points, the notion of spatial-energy parameter (P-parameter) is a complex characteristic of important atomic values responsible for interatomic interactions and directly connected with electron density in atom [1].

The value of relative difference of P–parameters of interacting atoms – components of α -coefficient of structural interactions was used as the main quantitative characteristic of structural interactions in condensed media:

$$\alpha = \frac{P_1 - P_2}{(P_1 + P_2)/2} 100\%$$ (1)

The nomogram of dependence of structural interaction degree upon the coefficient α (the same for a wide range of structures) was obtained applying the reliable experimental data. This approach allowed evaluating the degree and direction of structural interactions of phase-formation, isomorphism and solubility in multiple systems, including the molecular ones. In particular, the peculiarities of cluster-formation in the system $CaSO_4 - H_2O$ were studied [2].

To evaluate the direction and degree of phase-formation proccsses, the following equations were used [1]:

To calculate the initial values of P-parameters:

$$\frac{1}{q^2/r_i}+\frac{1}{W_in_i}=\frac{1}{P_э}\;;\;\frac{1}{P_0}=\frac{1}{q^2}+\frac{1}{(Wrn)_i}\quad P_э=P_0/r_i \qquad (2,3,4)$$

here: W_i - orbital energy of electrons [3]; r_i – orbital radius of i-orbital [4]; $q = Z^*/n^*$ - [5, 6]; n_i – number of electrons of the given orbital, Z^* and n^* - nucleus effective charge and effective main quantum number. The value P_0 will be called spatial-energy parameter (SEP), and P_E - effective P-parameter.

The calculation results by equations [2,3,4] for some elements are given in Table 1, where we can see that for hydrogen atom the values of P_E–parameters considerably differ at the distances of orbital (r_i) and covalent radii (R). The hybridization of valent orbitals of carbon atom were evaluated as an averaged value of P-parameters of $2S^2$ and $2P^2$-orbitals.

2) To calculate the value of P_S-parameter in binary and complex structures:

$$\frac{1}{P_S}=\frac{1}{N_1P_1}+\frac{1}{N_2P_2}+........ \qquad (5)$$

where N – number of homogeneous atoms in each subsystem.

The results of these calculations for some systems are given in Table 2.

3) To determine the bond energy (E) in binary and more complicated structures:

$$\frac{1}{E}\approx\frac{1}{P_E}=\frac{1}{P_1(N/K)_1}+\frac{1}{P_2(N/K)_2}+.......... \qquad (6)$$

Here (as applied to cluster systems): K_1 and K_2 – number of subsystems forming the cluster system; N_1 and N_2 – number of homogeneous clusters [7].

Thus, for $C_{60}(OH)_{10}$ $k_1 = 60$, $k_2 =10$.

It is assumed that the stable aqueous cluster (H_2O) can have the same static number of subsystems (k) as the number of subsystems in the system interacting with it [8]. For example, aqueous cluster of N $(H_2O)_{10}$ type interacts with fullerene $[C_6OH]_{10}$.

Apparently, cluster $[(C_2H_5OH)_6 - H_2O]_{10}$ can be formed similarly to cluster $[C_6OH]10$, which assumes the structural interaction of subsystems $(C_2H_5OH)_{60} - (H_2O)_{10}$. And the interaction of aqueous clusters can be considered as the interaction of subsystems $(H_2O)_{60} - N(H_2O)_{60}$.

Based on such notions and assumptions, the bond energy in corresponding systems was calculated by the equation (6); the results are given in Table 3.

The calculation data obtained by N.V. Khokhriakov based on his quantum-mechanical technique [9] are given here for comparison. Both techniques present comparable bond energy values (eV). Transfer multiplier: $(1\frac{kcal}{mol}=0.04336$ eV). Besides, the technique of P-parameter allows explaining why the energy value of cluster bonds of water molecule with fullerene $C_{60}(OH)_{10}$ two times exceeds the bond energy between the molecule in cluster water (Table 3).

Table 1. P-parameters of atoms calculated via the bond energy of electrons

Atom	Valent electrons	W (eV)	r_i (Å)	q^2 (eVÅ)	P_0 (eVÅ)	R (Å)	$P_E=P_0/R$ (eV)
H	1S¹	13.595	0.5295	14.394	4.7985	0.5295	9.0624
						0.28	17.137
C	2P¹	11.792	0.596	35.395	5.8680	0.77	7.6208
						0.69	8.5043
	2P²	11.792	0.596	35.395	10.061	0.77	13.066
	2S¹	19.201	0.620	37.240	9.0209	0.77	11.715
	2S²				14.524	0.77	18.862
	2S²+2P²				24.585	0.77	31.929
	$\frac{1}{2}\left(2S^2+2P^2\right)$						15.964
O	2P¹	17.195	0.4135	71.383	4.663	0.66	9.7979
	2P³	17.195	0.4135	71.383	11.858	0.66	17.967
						0.59	20.048
	2P⁴	17.195	0.4135	71.383	20.338	0.66	30.815

Table 2. Structural P-parameters

Radical, molecules	P_1 (eV)	P_2 (eV)	P_3 (eV)	P_4 (eV)	P_S (eV)	Orbitals of oxygen atom
OH	17.967	17.137			8.7712	2P²
OH	9.7979	9.0624			4.7080	2P¹
H₂O	2×17.138	17.967			11.788	2P²
	2×9.0624	17.967			9.0226	2P²
C₂H₅OH	2×15.964	2×9.0624	9.7979	9.0624	3.7622	2P¹

In accordance with the nomogram, the structure phase-formation can take place only with the relative difference of their P-parameters below 25%-30%, and the most stable structures are formed at $\alpha < (6\text{-}7)\%$.

Table 3. Calculations of bond energy – $E(eV)$

System	C_{60}	$(OH)_{10}$	$(H_2O)_{10}$		$P_E E$ (calculation)	
	P_1/κ_1	P_2/κ_2	P_3/κ_3	N_3	Equation (6)	Quantum-mechanical
$C_{60}(OH)_{10} -$ $- N(H_2O)_{10}$	15.964/60	8.7712/10	11.788/10	1	0.174	0.176
				2	0.188	0.209
				3	0.193	0.218
				4	0.196	0.212
				5	0.197	0.204
$(H_2O)_{60} -$ $- N(H_2O)_{60}$	P_1/κ_1	P_2/κ_2	N_2			
	9.0226/60	9.0226/60	1		0.0768	0.0863
			2		0.1020	0.1032
			3		0.1128	0.1101
			4		0.1203	0.1110
			5		0.1274	0.115
$(C_2H_5OH)_{60} -$ $- (H_2O)_{10}$	P_1/κ_1	P_2/κ_2				
	3.7622/60	9.0226/10			0.0586	0.0607
$(C_2H_5OH)_{10} -$ $- (H_2O)_{60}$	P_1/κ_1	P_2/κ_2				
	3.7622/10	9.0226/60			0.1074	$\approx 0{,}116$

Table 4 gives the values of coefficient α in systems H-C, H-OH and H-H_2O, which are within $0.44 - 7.09(\%)$.

But in the system H-C for carbon and hydrogen atoms, the interactions at the distances of covalent radii were considered, but for other systems – at the distances of orbital radius.

Thus, the interaction in the system H-C at the distances of covalent radius plays a role of fermentative action, which results in the transition of dimensional characteristics of water molecules from orbital radius to the covalent one, i.e., to the formation of the system C_{60} $(OH)_{10} - N(H_2O)_{10}$ with bond energy between the main components two times exceeding the one between water molecules themselves.

The broad possibilities of aqueous clusters in changing their spatial-energy characteristics apparently explain all other water properties with its different names: mineral, holly, live, spring, radioactive, etc.

Table 4. Spatial-energy interactions in the system H-R, where R= C, (OH), H₂O

System	$P_1(eV)$	$P_2(eV)$	$\alpha = \dfrac{\Delta P}{<P>} 100\%$	Type of spatial bond
H-C	17.137	15.964	7.09	Covalent
H-OH	9.0624	8.7712	3.27	Orbital
H-H_2O	9.0624	9.0226	0.44	Orbital

Conclusion

1) Structural interactions in the bond H-C at the distances of covalent radius play the role of fermentative action, which results in the transition of dimensional characteristics of water molecules from orbital radius to covalent one, i.e., to the system: $C_{60}(OH)_{10}$ - $N(H_2O)_{10}$.

2) Broad possibilities of aqueous clusters in the change of their spatial-energy characteristics apparently explain all other unique properties of water with different names: mineral, holly, live, spring, radioactive, etc.

References

[1] Korablev G.A. Spatial-Energy Principles of Complex Structures Formation, Leiden, the Netherlands, Brill Academic Publishers and VSP, 2005, 426 pages (Monograph).

[2] Korablev G.A., Yakovlev G.I., Kodolov V.I. Some peculiarities of cluster-formation in the system $CaSO_4$-H_2O. *Chemical physics and mesoscopy,* vol. 4, №2, 2002, p.188-196.

[3] Fischer C.F. Average-Energy of Configuration Hartree-Fock Results for the Atoms Helium to Radon.//*Atomic Data,*-1972, № 4, p. 301-399.

[4] Waber J.T., Cromer D.T. Orbital Radii of Atoms and Ions.//*J. Chem. Phys* -1965, -V 42, -№12, p. 4116-4123.

[5] Clementi E., Raimondi D.L. Atomic Screening constants from S.C.F. Functions, 1.//*J. Chem. Phys.* 1963, v.38, №11, p. 2686-2689.

[6] Clementi E., Raimondi D.L. Atomic Screening constants from S.C.F. Functions, 2.//*J. Chem. Phys.*-1967, v.47, №4, p. 1300-1307.

[7] Korablev G.A., Zaikov G.E. Energy of chemical bond and spatial-energy principles of hybridization of atom orbitals.//*J. Applied Polymer Science. USA,* 2006,V.101, n.3, p.2101-2107.

[8] Hodges M.P., Wales D.J. Glolal minima of protonated Water clusters. *Chemical Physics Letters,* 324, (2000), p.279-288.

[9] Khokhriakov N.V., Melchor Ferrer S. Electron properties of contacts in ideal carbon nanotubes. // *Chemical physics and mesoscopy,* 2002. Vol. 4. №2. P. 261-263.

In: Handbook of Research on Nanomaterials, Nanochemistry ...   ISBN: 978-1-61942-525-5
Editors: A. K. Haghi and G. E. Zaikov   © 2013 Nova Science Publishers, Inc.

Chapter XXVI

The Fractal Physics of Branched Polymers Synthesis: Polyhydroxyether

**G. V. Kozlov[1], D. A. Beeva[1], G. E. Zaikov[2],
A. K. Mikitaev[1] and S. Kubica[3]**

[1] Kabardino-Balkarian State University, Russia
[2] N.M. Emanuel Institute of Biochemical Physics
of Russian Academy of Sciences, Russia
[3] Institut Inzynierii Materialow Polimerowych I Barwnikow, Torun, Poland

Introduction

In the last 25 years, an interest of physicists in the theory of polymer synthesis has sharply increased (for example, papers [1-5], where the concept of a mean field was used). Simultaneously with the indicated papers, a number of publications concerning analytical study and computer simulation of reactions in different spaces [6-9], including fractal ones [10], has appeared. It has been clarified that the main factor defining chemical reactions course is a space connectivity degree, irrespective of its type. Also, a large amount of theoretical and applied researches on irreversible aggregation models of different kinds offered for the description of such processes as a flocculation, coagulation and polymerization are carried out [11]. These papers are intimately connected to the fractal analysis, intensively developing during the last years, as the aggregation within the frameworks of the indicated models forms fractal aggregates. Nevertheless, the application of these modern physical concepts for the description of polymers synthesis still has unitary character [12-15].

However, the application of the fractal analysis methods to synthesis process for today becomes a vital problem. Such necessity is due not to the convenience of the fractal analysis as mathematical approach, which supposes the existence of approaches alternate to it. The necessity of the indicated problem solution is defined only by physical reasons. The basic object during synthesis of polymers in solutions is the macromolecular coil, which represents a fractal object [16, 17]. As it is known [18], the description of fractal objects within the frameworks of an Euclidean geometry is incorrect, which predetermines the necessity of fractal analysis application. Besides, practically all kinetic curves at synthesis of polymers represent curves with a decreasing reaction rate, which is typical designation for fractal reactions [19, 20], i.e., reactions of fractal objects, or reactions in fractal space. Therefore, the purpose of the present review is fractal analysis methods application for synthesis kinetics description, and this process is final characteristics determination for branched polymers on the example of polyhydroxyther.

The Macromolecular Coil Structure Influence on Polyhydroxyether Synthesis

The studied polyhydroxyether (PHE) was synthesized by one-step method, namely, by a direct interaction of epichlorohydrin and 4,4'-dioxidiohenylpropane according to the scheme [21]:

In the work [21], the dependences of PHE synthesis main characteristics, namely, reduced viscosity η_{red} and conversion degree Q, on synthesis temperature T were studied. It was found out that T rising up to the definite limits influences favourably on the indicated process: a reaction rate rises, η_{red} and Q are increased. This effect can be observed in the narrow enough range of $T = 333\text{-}348$ K. At T lower than 333 K, PHE formation process decelerates sharply due to insufficient activity of epoxy groups at low temperatures. At $T > 353$ K, cross-linking processes proceed, which are due to the activity enhancement of secondary hydroxyls in polymer chain [21]. It is also supposed [21] that at the indicated temperatures of synthesis, PHE branched chains formation is possible.

As it is known [16], the macromolecular coil, which is the main structural unit at polymers synthesis in solution, represents a fractal, and its structure (coil elements distribution in space) can be described by the fractal dimension D. Proceeding from this, the authors [22] used the fractal analysis methods for the description of T effect on PHE synthesis course and its main characteristics.

The general fractal relationship for synthesis processes description can be written as follows [23]:

$$Q \sim t^{(3-D)/2}, \tag{1}$$

where Q is conversion degree, t is synthesis duration.

If the relationship (1) is expressed in a diagram form in double logarithmic coordinates, then from its slope in case of such plot linearity, the exponent in the indicated relationship and, hence, the value D, can be determined. The calculated by the indicated mode according to the curves $Q(t)$ for the initial part of these curves values D are adduced in Table 1. As one can see, reduction D at T growth is observed.

Within the frameworks of fractal analysis, a fractal (macromolecular coil) branching degree is characterized by spectral (fracton) dimension d_s, which is object connectivity degree characteristic [24]. For linear polymer d_s=1.0, for statistically branched one d_s=1.33 [24]. For macromolecular coil with arbitrary branching degree, the value d_s varies within the limits of 1.0-1.33. Between dimensions D and d_s, the following relationship exists, which takes into consideration the excluded volume effects [17]:

$$D = \frac{d_s(d+2)}{2+d_s},$$

(2)

where d is dimension of Euclidean space, in which a fractal is considered (it is obvious that in our case, d=3).

The equation (2) allows estimating the values d_s according to the known magnitudes D (Table 1). As it follows from the data, T increase results in d_s reduction, i.e., polymer chain branching degree decrease, and at T=348 K, PHE polymer chain is a linear one ($d_s \approx 1.0$).

A number of traditional estimations methods of polymer chain branching exists as well [25-27]. So, the branching factor g is defined as follows [27]:

$$g = \frac{R_\theta^2}{R_{l,\theta}^2},$$

(3)

where R_θ and $R_{l,\theta}$ are mean gyration radii of a branched polymer and its linear analog in θ-solvent at the same values of molecular weight MM and Kuhn segment size A, which characterizes chain thermodynamical rigidity.

Within the frameworks of fractal analysis, the relationship between coil gyration radius and molecular weight is given as follows [17]:

$$R_\theta \sim MM^{1/D_\theta},$$

(4)

$$R_{l,\theta} \sim MM^{1/D_{l,\theta}},$$

(5)

where D_θ and $D_{l,\theta}$ are macromolecular coil fractal dimensions of branched polymer and its linear analog in θ-solvent, accordingly.

The fractal equation can be obtained from the relationships (3)-(5) combination for g estimation [22]:

$$g = MM^{2\left[(D_{l,\theta}-D_\theta)/D_{l,\theta}D_\theta\right]}.$$

(6)

For linear macromolecule, the dimension $D_{l,\theta}$ is always equal to 2.0 [16]. For its branched analog, D_θ determination is more difficult, and this dimension value will depend on the branching degree. For statistically branched coil in paper [28], the following result was obtained:

$$D_\theta = \frac{4(d+1)}{7} \approx 2.286.$$ (7)

If to assume that D_θ changes proportionally to d_s and to use the boundary conditions $D_\theta = D_{l,\theta} = 2.0$ at $d_s = 1.0$ and $D_\theta = 2.286$ at $d_s = 1.33$, then according to the above-obtained d_s values (Table 1), the corresponding dimension D_θ can be calculated according to the formula [22]:

$$D_\theta = 2 + 0.858(d_s - 1).$$ (8)

The values g, calculated according to the equations (6) and (8), are cited in Table 1. As it was expected, the branching degree growth (g decrease) at T reduction is observed. The cited in Table 1 values g were calculated for $MM = 3 \times 10^4$.

In Fig. 1, the dependences $D(g)$ for PHE and also for polyphenylxalines (PPX) and bromide-containing aromatic copolyethersulfones (B-PES) were shown [29]. As one can see, for all adduced in Fig. 1 polymers D similar growth at g reduction is observed, which indicates on the observed effect community. The different D values for linear analogs at $g = 1.0$ are due to different solvents usage, i.e., to different level of interactions polymer-solvent [22].

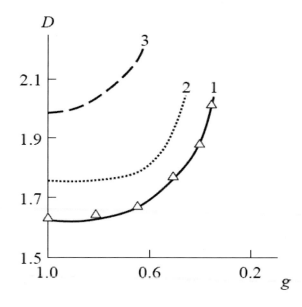

Figure 1. The dependences of macromolecular coil fractal dimension D on branching factor g for PHE (1), B-PES (2) and PPX (3) [22].

The equation (6) supposes chain branching degree increase (g reduction) at MM growth. In Fig. 2, the dependence $g(MM)$, calculated according to the equation (6), is adduced, which illustrates this law. The calculation g was fulfilled at $D_{l,\theta} = 2.0$ and $D_\theta = 2.249$ [22].

An interaction level between macromolecular coil elements and interactions polymer-solvent level can be characterized with the aid of the parameter ε, which is determined as follows [30]:

$$\left\langle \overline{h}^2 \right\rangle \sim MM^{1+\varepsilon},$$

(9)

where $\left\langle \overline{h}^2 \right\rangle$ is mean-square distance between macromolecule ends.

In paper [31], the following relationship between D and ε was obtained:

$$D = \frac{2}{\varepsilon+1}.$$

(10)

ε positive values characterize repulsion forces between coil elements, the negative ones – attraction forces. In Fig. 3, the dependence of parameter ε on the branching factor g for PHE is adduced. As it follows from the adduced plot, the dependence $\varepsilon(g)$ is linear and can be described by the following empirical relationship [22]:

$$\varepsilon = 0.45g - 0.15.$$

(11)

Therefore, from the equation (11), it follows that macromolecule branching degree increase (g decrease) results in attraction forces growth between coil elements and, as consequence, to coil compactization (D growth). Let us note that the entire variation range ε, corresponding to the same variation range $D=1-3$ [32], makes up 1/3-1.0 [31]. From the equation (11), the variation range ε for PHE can be obtained at the condition $g=0$-1 [25]: $\varepsilon=-0.15 - 0.30$. This corresponds to variation of $D=1.54$-2.35 according to the equation (10). Hence, PHE macromolecule can assume different structural states within the range of leaking coil-coil in θ-solvent [16].

g change results in the exponent a change in Kuhn-Mark-Houwink equation, describing relation between intrinsic viscosity [η] and MM [27]:

$$[\eta] = K \cdot MM^a,$$

(12)

where K and a are constants for the given polymer.

As it is known [14], the following relationship exists between D and a:

$$D = \frac{3}{1+a}.$$

(13)

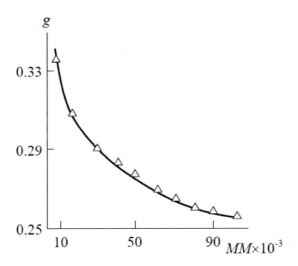

Figure 2. The dependence of branching factor g on molecular weight MM for PHE (at $D_{l,\theta}=2$ and $D_{\theta}=2.249$) [22].

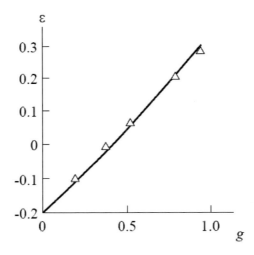

Figure 3. The dependence of parameter ε on branching parameter g for PHE [22].

The equations (10), (11) and (13) combination allows obtaining the dependence of a on g for PHE [22]:

$$a = 0.275 + 0.68g. \qquad (14)$$

Hence, a chain branching degree increase (g reduction) results in a decrease. This conclusion is confirmed experimentally in paper [25] on the example of two polyarylates D-1, received by equilibrium and interphase polycondensation. For the first from them (the branched one), the value a at the same condition is smaller than for the second (linear one). Besides, for θ-conditions in case of linear polyarylate $a=0.50$, which was expected [16], whereas for branched polyarylate at the same conditions $a=0.36$. According to the equation (13), this corresponds to $D_\theta \approx 2.206$, which corresponds well to the above-adduced Family estimation [28] (the equation (7)) and, according to the equation (8), corresponds to $d_s \approx 1.24$, i.e., to a branched polymer.

Using the equations (10) and (11) combination, the relationship between D and g can be obtained as follows [22]:

$$D = \frac{2}{0.85 + 0.45g} .$$

(15)

Calculated according to the equation (15), D values are also adduced in Table 1, from which their good correspondence to D values, estimated according to kinetic curves $Q(t)$ with the aid of the relationship (1), follows (the mean discrepancy of value D, received by two indicated methods, makes up $\sim 1.4\%$).

And in conclusion, let us consider the physical significance of PHE polymer chain branching degree decrease at synthesis temperature growth. The branching centers average number per one macromolecule m can be determined according to the equation [26]:

$$g = \left[\left(1 + \frac{m}{7} \right)^{1/2} + \frac{4m}{9\pi} \right]^{-1/2} .$$

(16)

As it is well-known [33], fractal objects are characterized by strong screening of internal regions by fractal surface. Therefore, accessible for reaction (in our case – for branching formation) sites are either on fractal (macromolecular coil) surface or near it. Such sites number N_u scales with coil gyration radius R_g as follows [33]:

$$N_u \sim R_g^{d_u} ,$$

(17)

where d_u is dimension of unscreened (accessible for reaction) surface, which is determined according to the equation [33]:

$$d_u = (D-1) + \frac{(d-D)}{d_w} ,$$

(18)

where d_w is dimension of random walk on fractal, which is estimated according to Aarony-Stauffer rule [34]:

$$d_w = D + 1 .$$

(19)

Besides, it is well known [35] that in case of chemical reactions of various kinds course, including reactions at polymers synthesis, the so-called steric factor p ($p \leq 1$) plays an essential role, which shows that not all collisions of reacting molecules occur with proper for chemical bond formation in these molecules orientation. The value p is connected with R_g as follows [36]:

$$p \sim \frac{1}{R_g}.$$

(20)

Therefore, it can be assumed that the number of accessible for branching formation sites of macromolecular coil m will be proportional to the product pN_u or [22]:

$$m \sim \frac{R_g^{d_u}}{R_g} \sim R_g^{d_u - 1}.$$

(21)

For this problem solution, it is necessary to determine R_g variation at T change. This can be made as follows. Using experimentally determined η_{red} values, $[\eta]$ can be calculated according to Shultze-Braschke empirical equation [37]:

$$[\eta] = \frac{\eta_{red}}{1 + K_\eta c_p \eta_{red}},$$

(22)

where K_η is coefficient, which is equal to 0.28 [37], c_p is polymer concentration.

Then the coefficient K in Kuhn-Mark-Houwink equation can be determined [25]:

$$K = \frac{21}{m_e} \left(\frac{1}{2500 \cdot m_e} \right)^a,$$

(23)

where m_e is mean weight of polymer elementary link (without substituents).

Further, the value MM is determined according to the equation (12) and polymerization degree N is calculated as follows:

$$N = \frac{MM}{m_0},$$

(24)

where m_0 is monomer link molecular weight (for PHE $m_0 = 284$).

And at last, the value R_g is calculated according to the following fractal relationship [31]:

$$R_g = 37.5 N^{1/D}, \text{Å}. \tag{25}$$

In Fig. 4, the dependence of m on $R_g^{d_u-1}$ is adduced, which proves to be linear. Since the value m increases at $R_g^{d_u-1} \sim pN_u$ growth, then this supposes the above-offered treatment identity [22].

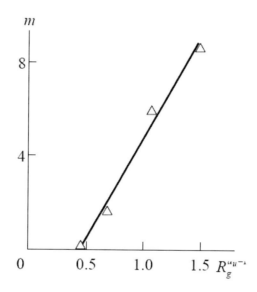

Figure 4. The dependence of branching number per one macromolecule m on parameter $R_g^{d_u-1}$ for PHE [22].

Hence, the above-stated results showed that fractal analysis notions allowed giving principally new treatment of phenomena, occurring at PHE synthesis at synthesis temperature variation. The notion of macromolecular coil structure, characterized by its fractal dimension D, forms the basis of this treatment. This structure coil elements interactions among each other and interactions polymer-solution, characterized by parameter ε in the relationship (9), are defined. In its turn, polymer chain branching degree, characterized by branching factor g, is an unequivocal function of D according to the equation (15). The equation (2) gives the physical sense of this correlation. Besides, the macromolecular coil structure defines kinetic curves course according to the relationship (1). It is important to note that fractal analysis methods allow the indicated effects quantitative treatment [22]. Proceeding from these general concepts, the authors [38] gave the description of PHE macromolecular coil structure influence on its synthesis rate at four temperatures in the above-indicated range of temperatures T.

As it follows from the data of Fig. 5, synthesis temperature T increase results in PHE synthesis reaction rate enhancement. Within the frameworks of fractal analysis, the reaction rate constant k_r can be determined according to the following relationship [39]:

$$t^{(D-1)/2} = \frac{c_1}{k_r(1-Q)},$$ (26)

where c_1 is constant.

Since in paper [38], the main studied parameters are given in relative units, then the value c_1 was accepted equal to one. The estimated at $t=3600$ s k_r values are adduced in Table 2.

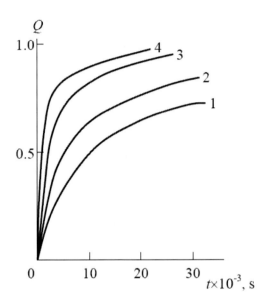

Figure 5. The kinetic curves of conversion degree – reaction duration (Q-t) for PHE at synthesis temperatures 333 (1), 338 (2), 343 (3) and 348 K (4) [38].

The calculation of polymerization degree N and macromolecular coil gyration radius R_g according to the equations (24) and (25), respectively, shows wide enough variation of the indicated parameters. In paper [40], the following fractal relationship for steric factor p estimation was obtained:

$$p = \frac{c_2}{t^{(D-1)/2}},$$ (27)

where c_2 is constant, which is also accepted equal to one according to the above-mentioned reasons. The values p for PHE at four magnitudes T are given in Table 2.

In Fig. 6, the relationship $k_r(p)$ for PHE is adduced, which has an expected character – p growth results in k_r increase. However, this dependence is not directly proportional. The authors [38] supposed that for obtaining a more general correlation reaction rate, macromolecular coil structure sites number on coil surface N_u, accessible for reaction (unscreened) should be taken into consideration. In other words, in such treatment, the dependence of k_r on complex characteristic pN_u should be plotted, which is shown in Fig. 7. From this Figure plot, it follows that the linear correlation $k_r(pN_u)$ is now obtained, i.e., k_r growth at T increase is defined by PHE macromolecular coil structure change.

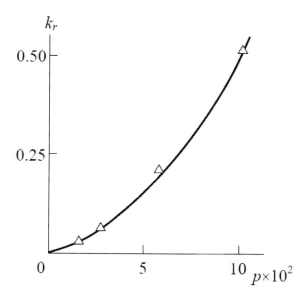

Figure 6. The relationship between reaction rate constant k_r and steric factor p for PHE [38].

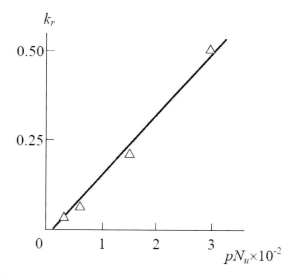

Figure 7. The dependence of reaction rate constant k_r on complex characteristic pN_u for PHE [38].

Let us note that $k_r=0$ is achieved at small but finite quantity pN_u. The calculation according to the equations (18), (26) and (27) shows that $D \approx 2.20$, i.e., fractal dimension of branched chain in θ-conditions, corresponds to this value pN_u [28]. After this dimension achievement, reaction rate decreases sharply.

In Fig. 8, the dependence of k_r on the branching factor g, calculated according to the equation (6) (see Table 1), is adduced in the form of $k_r(g^3)$. As one can see, this dependence is linear and passes through coordinate's origin. Hence, the chain branching degree increase, characterized by g reduction, defines PHE synthesis rate sharp decrease (a cubic dependence).

Hence, the above-stated results showed definite influence of macromolecular coil structure, characterized by its fractal dimension D, on PHE synthesis rate. D increase, i.e., coil compactization, decreases sharply reaction rate constant k_r the value in virtue of two factors influence: decrease of accessible for reaction sites number N_u and steric factor p reduction. As a matter of fact, D growth defines transition from PHE synthesis diffusive regime to kinetic one [41]. In its turn, polymer chain branching degree growth rises, D, and decreases sharply, k_r [38].

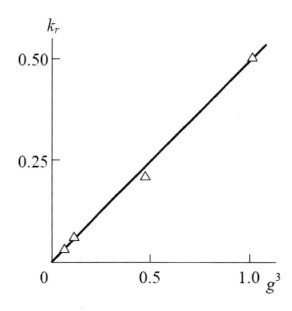

Figure 8. The dependence of reaction rate constant k_r on branching factor g for PHE [38].

In paper [21], strong dependence of the reduced viscosity η_{red} on reactionary medium components relation at PHE synthesis in mixture water-isopropanol was found. Besides, it was revealed that the process of PHE synthesis can be divided into three modes depending on the contents of isopropanol c_{is} in a mixture.

At the small contents of isopropanol $c_{is}<15$ vol. %, the synthesis process practically does not take place, and it turns into low-molecular oligomer. At $c_{is}=15$-60 vol. %, the linear PHE was received and its viscosity η (or molecular weight MM) grows at c_{is} increasing. And at last, at $c_{is}>60$ vol. % the cross-linked polymer is formed. In work [21], the indicated mode existence is explained only from the chemical point of view as follows.

PHE was synthesized by one-step method, namely, by a direct interaction of epichlorohydrin and bisphenol (see above). The first regime ($c_{is}\leq15$ vol. %) is explained by poor ability for mixing of epichlorohydrin with water, which complicates its access to bisphenolate, being found in water. Besides, in water, the formed products are dropped out as viscous resin, and the chain growth is either strongly slowed down or stops.

As it is known [42], the increase of reactionary ability of hydroxyl group will be promoted by OH-group introduction in hydrogen bond as the donor of protons owing to the displacement of electronic density under the scheme:

It allows assuming that at c_{is}=15-60 vol. %, there occurs an increase of bisphenols reactionary ability according to the above-mentioned mechanism, and this results in the increase of PHE production reaction rate and in higher values η_{red}.

And, at last, the production of the cross-linked PHE at c_{is}>60 vol. % is explained by the participation of isopropanol molecules in reaction, which was observed earlier [43].

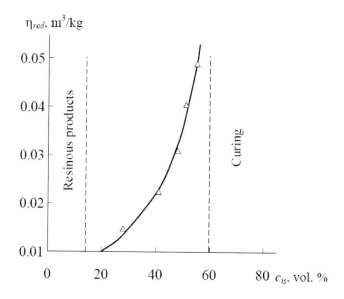

Figure 9. The dependence of mean reduced viscosity η_{red} on isopropanol volume contents c_{is} in reactionary mixture water-isopropanol at PHE synthesis. The vertical shaded lines indicate regimes boundary [44].

The above-cited analysis does not take into account the features of macromolecular coil structure, which is a basic element at synthesis of polymers in a solution [40]. As it was noted above, macromolecular coil is a fractal, and its structure (the distribution of its links in space) can be characterized by the fractal dimension D. In its turn, the value D is defined by two groups of interactions: by interactions of the coil links to each other and by the interactions polymer-solvent [28]. It is obvious that the reactionary medium change should result to the changes of interactions of the second group and, as consequence, to a variation D. Therefore, the authors [44] proposed the alternative explanation of three modes existence during PHE synthesis process with the fractal analysis representations [45, 46] participation.

In Fig. 9, the dependence of average value of the reduced viscosity η_{red} for PHE on the isopropanol contents c_{is} in reactionary medium water-isopropanol is adduced, and the boundaries of the above-mentioned synthesis three modes are also indicated (shaded vertical lines). From the data of Fig. 9, it follows that the first regime (synthesis practical absence) is completed at c_{is}=15 vol. %.

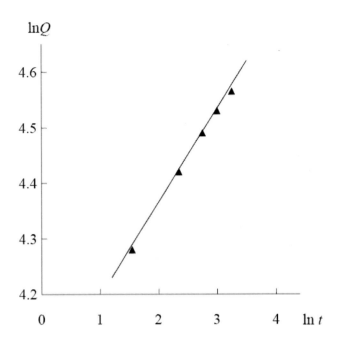

Figure 10. The dependence of conversion degree Q on reaction duration t in double logarithmic coordinates for PHE synthesis at T=348 K [44].

The relationship (1) allows obtaining synthesis cessation condition. At $D=d=3$ (d is the dimension of Euclidean space, in which a process is considered), Q=const. Besides, the initial conditions of synthesis will be $Q=0$ at $t=0$. Thus, at $D=d$, Q=const, and the reaction does not take place. In other words, PHE synthesis reaction will be realized in case of fractal reacting objects (macromolecular coils) only.

The value D can be determined according to the equation [47]:

$$D = 1.5 + 0.45\chi_1,\tag{28}$$

where χ_1 is Flory-Haggnis interaction parameter, which characterizes the level of interaction polymer-solvent.

In its turn, the value χ_1 is determined according to the following equation [27]:

$$\chi_1 = \frac{E_{ev}}{RT}\left(1 - \frac{\delta_p^2}{\delta_s^2}\right)^2 - \chi_{ent},\tag{29}$$

where E_{ev} is the solvent evaporation heat, R is universal gas constant, T is synthesis temperature, δ_p and δ_s are solubility parameters of polymer and solvent, accordingly, χ_{ent} is the entropic contribution to Flory-Haggins parameter, determined experimentally.

The value δ_s in the considered case is a solubility parameter of mixture water-isopropanol and its determination as a function of c_{is} carried out according to the technique [48]. Within the frameworks of the indicated technique, the value δ_s can be written as follows:

$$\delta_s^2 = \delta_f^2 + \delta_c^2,$$

(30)

where the solubility parameter component δ_f includes the energy of dispersive interactions and the energy of dipole bonds interaction; the component δ_c includes the energy of hydrogen bonds and the energy of interaction between atom with electrons deficiency of one molecule (acceptor) and atom with electrons abundance of other molecule (donor), which requires the certain orientation of these two molecules. For water δ_f=8.12 (cal/cm^3)$^{1/2}$ and δ_c=22.08 (cal/cm^3)$^{1/2}$, for isopropanol δ_f=7.70 (cal/cm^3)$^{1/2}$ and δ_c=5.25 (cal/cm^3)$^{1/2}$ [48].

The value δ_f^m (δ_c^m) for a solvents mixture can be determined according to the mixtures rule [48]:

$$\delta_f^m = \sum_{i=1}^{n} \phi_i \delta_{fi},$$

(31)

$$\delta_c^m = \sum_{i=1}^{n} \phi_i \delta_{ci},$$

(32)

where the index "m" designates mixture, the index i designates i-th mixture component, the complete number of which is equal to n, and ϕ_i is the component volumetric fraction.

The empirical constant χ_{ent} can be estimated according to the following considerations, but at first, the authors [44] showed such estimation necessity. The value D calculation for c_{is}=30 vol. % (δ_p=9.15 (cal/cm^3)$^{1/2}$, δ_s^m=18.8 (cal/cm^3)$^{1/2}$) gives the value D=3.15, which is physically impossible since $D<d$ [49]. This means that there exists significant entropic contribution to value χ_1 and, hence, D. In Fig. 5, a number of kinetic curves $Q(t)$ for PHE was adduced. Using kinetic curve $Q(t)$ at T=348 K plotting in double logarithmic coordinates, the exponent in the relationship (1) can be estimated and, hence, the value D, which is equal to ~ 2.60. Then according to the equation (28) at known E_{ev} (the values E_{ev}^m for mixtures were estimated according to the mixtures rule by analogy with the equations (31) and (32)), δ_s^m and δ_p can estimate the parameter χ_{ent}, which is equal to ~ 1.23. In further calculations, χ_{ent}=const is accepted.

In Fig. 11, the dependence $D(c_{is})$ calculated by the indicated method is adduced. At c_{is}=15 vol. % D=2.967, i.e., macromolecular coil structure, as indicated above, does not allow synthesis reaction proceeding. At $c_{is}<$15 vol. % $D\approx$3=const in virtue of the above-mentioned condition $D\leq d$.

As it is known [50], polymers cross-linking macroscopic process provides joining cluster formation (from one end of a reactionary bath up to another one). This is gelation process for cross-linking polymer, and the cluster dimension at this point is equal to ~ 2.5 [51, 52]. The gelation process proceeds in medium with cross-linked macromolecular coils (so-called microgels) plenty, which results in reactionary medium viscosity essential enhancement and D increasing. According to work [53], the fractal dimension value d_f for such "dense" solution is connected with value D in case of diluted solution by the following relationship:

$$D = \frac{(d+2)d_f}{2(1+d_f)}.$$

(33)

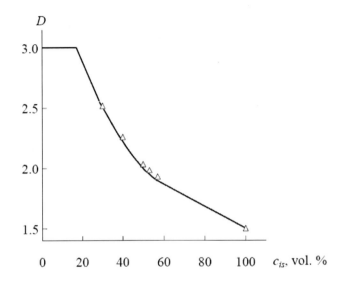

Figure 11. The dependence of macromolecular coil fractal dimension D on isopropanol volume contents c_{is} in reactionary mixture water-isopropanol for PHE [44].

For d_f=2.5, we shall receive D=1.786 according to the equation (33). At d_f=3, i.e., compact globule, for which any chemical reaction, including cross-linking process, is impossible, we shall receive D=1.875.

Hence, the cross-linking process must begin within the range of D=1.875-1.786. According to the equations (28)-(32) for this value D, we shall obtain $\delta_s^m \approx 14.8$ (cal/cm^3)$^{1/2}$, which corresponds to $c_{is} \approx 58$ vol. %. Such estimation corresponds completely to the experimental data, adduced in Fig. 9.

Hence, the above-stated results showed that macromolecular coil structure, characterized by its fractal dimension D, could be a critical factor in polymers synthesis regimes definition, in particular, PHE. Compact coil (D=d) does not allow reaction proceeding, and this condition defines the first regime-resin-like products formation. D decreasing defines polycondensation reaction proceeding possibility in diluted solutions, and for reaction proceeding on the gelation stage, one needs even smaller D value, defining d_f magnitude for "dense" solution. In the considered case, the value D is controlled by interactions polymer-solvent change.

Theoretical Analysis of Molecular Weight Change in Synthesis Process

As it has been noted above, the synthesis temperature increase within the range of T=333-348 K results in final characteristics enhancement of this process for PHE – conversion degree Q and reduced viscosity η_{red}, in the first approximation characterizing polymer molecular weight MM. In Fig. 12, the dependences of η_{red} on synthesis duration t are adduced for PHE at four different T. As it follows from these plots, at first η_{red} sharp increase at t growth is observed, and then values η_{red} achieve asymptotic branch. Thus, one can suppose that at some t values, η_{red} (or MM) magnitudes achieve their limit, depending on T.

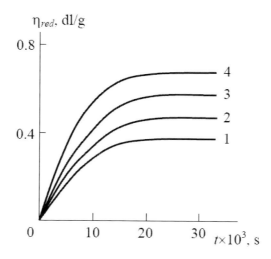

Figure 12. The dependences of reduced viscosity η_{red} on synthesis duration t for PHE at synthesis temperature T: 333 (1), 338 (2), 343 (3) and 348 K (4) [54].

This effect can be explained within the frameworks of irreversible aggregation models [36], in which a macromolecular coil in solution is considered as a fractal and its structure is characterized by dimension D [16]. The greatest attainable radius R_c of the coil can be estimated according to the following scaling relationship [36]:

$$R_c \sim c_0^{-1/(d-D)}, \tag{34}$$

where c_0 is reacting particles (monomers) initial concentration.

The authors [54] estimated applicability of the scaling relationship (34) for the limiting values MM determination and elucidated the macromolecular coil structure, characterized by dimension D, as far as T changes at PHE synthesis.

The range of D values (see Table 1) at PHE synthesis assumes that this process proceeds according to the mechanism cluster-cluster [36], i.e., a large macromolecular coil is formed by merging of smaller ones. This circumstance allows estimating R_c value according to the relationship (34). The value c_0 choice does not influence on the dependence $R_c(D)$ course, but it can change R_c growth rate at D reduction. Proceeding from these considerations, in paper

[54], the value $c_0=25$ was chosen. The experimental values of molecular weight MM^e for PHE can be estimated according to Kuhn-Mark-Houwink equation (the formula (12)), which after determination of constants K and a for PHE acquires the following form [54]:

$$[\eta] = 2.84 \times 10^{-4} \left(MM^{\,e} \right)^{0.714}. \tag{35}$$

Theoretical limiting value of molecular weight MM^T was determined according to the following scaling relationship [17]:

$$MM^T \sim R_c^D. \tag{36}$$

In Fig. 13, the comparison of the dependences MM^e and MM^T on synthesis temperature T for PHE is adduced. The constant coefficient in the relationship (36) was determined by method of experimental and theoretical values MM superposition. As one can see, a good correspondence of theory and experiment is obtained, which confirms application correctness of the relationship (34) for limiting values MM estimation at PHE synthesis.

Hence, the above-stated results showed that limiting values of molecular weight, attainable in PHE synthesis process at different T, could be described within the frameworks of irreversible aggregation cluster-cluster model by the usage of the relationship (34). MM indicated limiting value is controlled by macromolecular coil structure, characterized by its fractal dimension D.

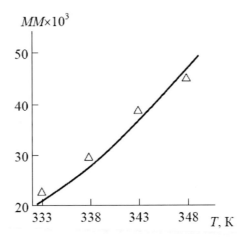

Figure 13. The comparison of theoretical (solid line) and experimental (points) dependences of limiting molecular weight MM on synthesis temperature T for PHE [54].

At present, it is known [55, 56] that polymer branching degree influences essentially on molecular weight change kinetics of polymer in its formation process. This process simulation for branched polymers by Monte-Karlo method shows correctness of the following scaling relationship [56]:

$$MM \sim t^{\gamma_t}, \tag{37}$$

where t is reaction duration.

For branched chains, growing in critical conditions, the following relationship was obtained [56]:

$$\gamma_t^{-1} = 1 - g',$$ (38)

where g' is the branching degree, determined as the exponent in the scaling relationship between branching centers number per one macromolecule m and MM [56]:

$$m \sim MM^{g'}.$$ (39)

Hence, the relationship (38) supposes the exponent γ_t growth at polymer branching degree g' increase. The authors [57] obtained the relationship between branching degree and molecular weight in case of real polymer synthesis on the example of PHE.

The experimental values MM for PHE were determined according to the equation (12), and the constants K and a – according to the formulas (13) and (23), respectively. The values K and a, obtained by the indicated mode, are adduced in Table 3. It is necessary to note that for the same polymer (PHE), different values of constants in Kuhn-Mark-Houwink were obtained, which is due to different structure of PHE macromolecular coils, received at different T.

Further, the dependences $MM(t)$ in double logarithmic coordinates, corresponding to the relationship (37), can be plotted for determination of the exponent γ_t (γ_t^e) experimental values in the indicated relationship. The plotted by the indicated mode dependences $MM(t)$ for four T are shown in Fig. 14, and the values γ_t^e are adduced in Table 3. As it follows from these data, the value γ_t^e grows at D increase (see Table 1), i.e., at polymer chain branching degree enhancement. This situation corresponds completely to papers [55, 56] conclusions. However, the equation (38) usage for theoretical value γ_t (γ_t^T) estimation in case of PHE is impossible. Since $\gamma_t^e < 1$ (see Table 3), then this means negative values g', which does not have physical significance. Therefore, for the estimation of PHE polymer chain branching degree, the authors [57] used another parameter, the branching factor g, which is determined according to the equation (3). Let us note the principal difference between parameters g' and g, which follows from the relationships (39) and (3) comparison. Polymer branching increase is characterized by g' increase within the range of 0-1 and g decrease within the same range. The values $g' = 0$ and $g = 1.0$ correspond to linear polymer [25, 27].

Proceeding from what is stated above and also from absolute values γ_t^e, the following form of the dependence γ_t^T (g) can be supposed [57]:

$$\gamma_t^T = 1 - g. \tag{40}$$

The comparison of γ_t^e and γ_t^T is adduced in Table 3, from which their satisfactory correspondence follows.

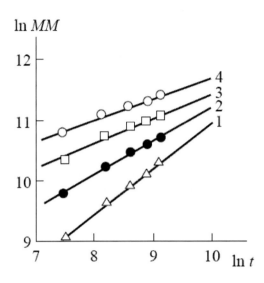

Figure 14. The dependences of molecular weight MM on reaction duration t in double logarithmic coordinates for PHE at synthesis temperatures T: 333 (1), 338 (2), 343 (3) and 348 K (4) [57].

In Fig. 15, comparison of the experimental (calculated according to Kuhn-Mark-Houwink equation) and theoretical (calculated according to the relationship (37) at $\gamma_t = \gamma_t^T$ and for linear PHE at $\gamma_t = \gamma_t^e$) dependences, $MM(t)$ is adduced. As one can see, the theory and experiment good correspondence was obtained.

Thus, the above-stated results showed correspondence of computer simulation of branched polymers formation and PHE synthesis in the aspect that the scaling relationship (37) describes correctly molecular weight change kinetics in both cases. However, there exist principal difference in the exponent γ_t determination in the indicated relationship, although in both cases, the value γ_t increases at polymer branching degree growth, whichever parameter it is characterized. The indicated discrepancy can be explained by different conditions of branched chains formation.

In paper [21], it has been shown that at PHE synthesis, reduced viscosity η_{red} (or polymer molecular weight MM) at reagents initial concentration c_0 growth occurs. Let us note that similar dependences were also observed at other polymers number synthesis [58]. However, there was no quantitative description of this effect.

The development of lately irreversible aggregation models, elaborated for the description of such processes as polymerization, flocculation, coagulation and so on, allows obtaining strict physical treatment of similar dependences. Nevertheless, the indicated models application to polymerization real processes are found still rarely enough [59], although it is obvious that such approach will allow obtaining more profound understanding of aggregation processes, including polymerization. Therefore, the authors [60] carried out the description of the dependence $\eta_{red}(c_0)$ or $MM(c_0)$ for PHE with the usage of irreversible aggregation models and fractal analysis methods.

In Fig. 16, the experimental dependence $\eta_{red}(c_0)$ for PHE (solid line) is adduced, from which monotone growth η_{red} from 0.23 up to 0.62 dl/g follows in the indicated range c_0. At $c_0 > 0.7$ mole/l polymer cross-linking process begins, which restricts an upper concentration limit of reaction proceeding [44].

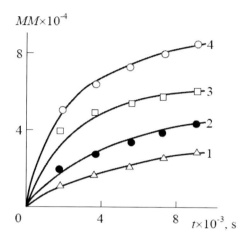

Figure 15. The comparison of experimental (lines) and theoretical (points) dependences of molecular weight MM on reaction duration t for PHE at synthesis temperatures T: 333 (1), 338 (2), 343 (3) and 348 K (4) [57].

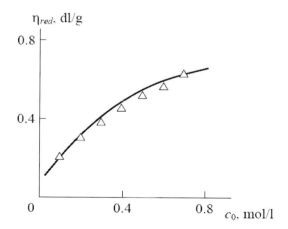

Figure 16. The comparison of experimental (solid curve) and theoretical (points) dependences of reduced viscosity η_{red} on initial reagents concentration c_0 for PHE [60].

For theoretical description of the dependence $\eta_{red}(c_0)$, a kinetic scaling for diffusion – limited aggregation at the condition $R_g \ll \xi$ (R_g is gyration radius, ξ is a scale of aggregation processes universality classes transition) was used [61]:

$$N \sim \left(c_0 t\right)^{D/(2+D-d)},$$

(41)

where N is particles number per aggregate (polymerization degree).

Since theoretical molecular weight MM^T can be written as m_0N, where m_0 is molecular weight of polymer repeating link (for PHE m_0=284), then proportionality coefficient in the relationship (41) can be obtained by the curves $MM^e(c_0)$ and $m_0N(c_0)$ superposition and estimated values MM^T. Then according to these values MM^T using the equations (35) and (22), the theoretical values η_{red}, corresponding to the model [61] are calculated. In Fig. 16, the comparison of the theoretical (points) and experimental (solid line) dependences $\eta_{red}(c_0)$ for PHE is adduced, from which the theory and experiment good correspondence follows.

Hence, the model of irreversible diffusion-limited aggregation allows correct quantitative description of reduced viscosity η_{red} (or molecular weight MM) change with variation of reagents initial concentration c_0. At the condition of value D, knowledge and polymer chemical constitution Kuhn-Mark-Houwink equation for it can be received theoretically, which allows avoiding laborious measurements. The relationship (41) predicts that value MM at the fixed t depends not only on c_0 but also on macromolecular coil structure [60].

In paper [21], it has been shown that characteristics of PHE, synthesized in reactionary medium water-isopropanol are dependant to a considerable extent on the indicated components ratio. The increase of isopropanol contents in mixture from 30 up to 55 vol. % results in reduced viscosity η_{red} growth from ~ 0.16 up to 0.50 dl/g. In paper [21], this effect was explained within the frameworks of Gammet model, which supposed that isopropyl spirit addition decreased interaction between molecules of water and bisphenol, resulting in enhancement of dioxicompaund nucleophyl reactionary ability at the expense of hydrogen bond formation [62].

The above-adduced analysis does not take into consideration features of macromolecular coil structure, which is the main element in polymers synthesis in solution. Therefore, the authors [63] explained the above-considered effect from the positions of irreversible aggregation models and fractal analysis.

From the relationship (41), it follows that value of polymerization degree N (and, hence, molecular weight MM) at the fixed c_0 and t is determined only by the fractal dimension D: the smaller D, the larger exponent in the relationship (41) and the higher N (MM). Knowing the value D (see Fig. 11) and PHE chemical constitution, Kuhn-Mark-Houwink equation can be received and the experimental values MM (MM^e) according to the experimentally determined η_{red} magnitudes can be calculated. Then the proportionality coefficient in the relationship (41) can be determined by N and MM^e superposition and thus obtained theoretical values MM (MM^T), predicted by the model of irreversible aggregation [61]. In Fig. 17, the comparison of the dependences $MM^e(c_{is})$ and $MM^T(c_{is})$ for PHE is adduced (in case of MM^e the error limits are adduced [21]), from which theory and experiment good correspondence follows. This indicates to PHE synthesis process description equivalency within the frameworks of irreversible aggregation models.

Since PHE, produced in different mixtures water-isopropanol, has different values D (i.e., different structure), then from the equations (13) and (23) it follows that for them, values a and K will be different, i.e., the relation between [η] and MM for the same polymer will be defined by different Kuhn-Mark-Houwink equations. Therefore, strictly speaking, polymer viscosity should be determined in the same solvent in which it was synthesized. This rule is confirmed by a well-known fact [25] that synthesized by different polycondensation modes polyarylates, having the same chemical constitution, have different values a (and, hence, D according to the equation (13)) and K and also distinguishing properties. In Fig. 17, the dependence $MM^e(c_{is})$, calculated according to the same Kuhn-Mark-Houwink, is adduced. As it follows from the data of Fig. 17, MM calculation, accounting for coil structure and without it, gives close enough results at large MM, but at small MM ($<10^4$), the discrepancy can be even quintuple one.

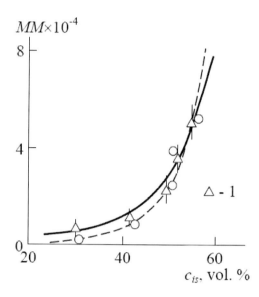

Figure 17. The comparison of experimental (1) and calculated according to the relationship (41) at using of different (2) and one (3) Kuhn-Mark-Houwink equations dependences of molecular weight MM on isopropanol contents c_{is} in reactionary mixture for PHE synthesis [63].

Hence, the above-stated results showed that PHE molecular weight changes at water-isopropanol ratio variation in reactionary mixture could be described quantitatively within the frameworks of irreversible aggregation models and fractal analysis. The indicated ratio change defines interactions polymer-solvent change, which results in macromolecular coil structure variations. An interaction polymer-solvent weakness, characterized by Flory-Haggins interaction parameter χ_1 reduction, results in D decreasing, which makes macromolecular coil more accessible for synthesis reaction proceeding.

The Branching Degree and Macromolecular Coil Structure

As it has been noted above, the branching degree of polymer chain can be characterized by several parameters. One of them is a number of branching centers per one macromolecule m. The branching degree g', determined from the scaling relationship (39), serves as another parameter. As a rule, the value $g' < 1$ [27], and this means that the number of branching centers is not proportional to the length of macromolecule or its polymerization degree, N.

From the chemical point of view, such effect is difficult to explain, since each monomer link in macromolecule has the same probability of branch formation, and then one can expect $m \sim N$.

However, in the real conditions of polymers synthesis, there are a number of causes that can, in principle, cause the ratio m/N to decrease.

One of such reasons can be the fact that the branching reactive centers, formed in the initial stages of synthesis, are proved to be "buried" inside a macromolecular coil and, consequently, are less accessible [55].

Such situation defines the necessity of the macromolecular coil structure allowance, which can be fulfilled with the aid of its fractal dimension D.

Therefore, the authors [64] carried out the description of the macromolecular coil structure influence on accessible for reaction branching centers number at polymer molecular weight change. This description is given within the frameworks of fractal analysis on the example of PHE.

Besides the characteristics indicated above, one more parameter, the branching factor g, can be used for estimation of polymer branching degree, which is determined according to the equation (3). Within the frameworks of fractal analysis, the formula (6) allows determining the value g, which supposes the dependence of g on molecular weight. The parameters g and m are connected by the relationship (16).

In Fig. 18, the dependence $m(MM)$, where the value m was calculated according to the equations (6) and (16), is adduced. As one can see, this dependence is a nonlinear one, i.e., the value m grows much weaker than MM. In Fig. 19, the same dependence is adduced in double logarithmic coordinates, which proves to be linear, and from its slope, the value g' ≈ 0.272 can be estimated.

The small value g' supposes strong influence of macromolecular coil structure on m value, and this effect can be estimated quantitatively as follows.

As it is known [33], one of the main features of the fractal object structure is strong screening of its internal regions by the surface.

Therefore, the accessible for chain branching reaction macromolecular coil sites are disposed either on its surface or near it. The number of such sites N_u is determined according to the scaling relationship (17).

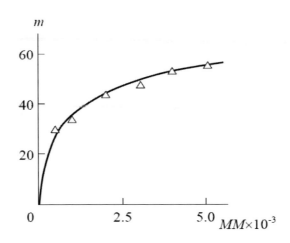

Figure 18. The dependence of branching center number per one macromolecule m on molecular weight MM for PHE synthesized at $T=333$ K [64].

The dependence $m(R_g^{d_u-1})$, corresponding to the relationship (21), is shown in Fig. 4. As one can see, it is linear and has an expected character: m increases with $R_g^{d_u-1}$ enhancement. The extrapolation of this plot to $m=0$ gives $R_g^{d_u-1} \approx 2.2$, which corresponds to the smallest size of PHE macromolecule for the beginning of the branching. This size is equal to ~ 9.9 Å. The volume of the repeating link of PHE V_0 can be estimated according to the equation [65]:

$$V_0 = \frac{m_0}{\rho N_A},\qquad(42)$$

where ρ is the polymer density (for PHE, $\rho \approx 1150$ kg/m^3 [6]), and N_A is Avogadro's number.

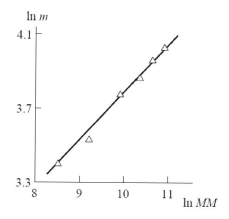

Figure 19. The dependence of branching center number per one macromolecule m on molecular weight MM in double logarithmic coordinates for PHE synthesized at $T=333$ K [64].

Then, believing that the cross-sectional area of PHE macromolecule is equal to 30.7 Å^2 [67] and using the volume $V_0 \approx 410$ Å^3, calculated according to the equation (42), the monomer link length for PHE, l_0, can be estimated to be equal to ~ 13.4 Å. The comparison of the smallest value R_g and l_0 indicates that PHE chain branching process begins already at the initial synthesis stage.

Let us note that the dependences $m(MM)$ with fractional exponent $g<1$ are typical for other polymers as well. So, in work [25], the constants of Kuhn-Mark-Houwink equation for θ-solvent in case of the branched polyarylate D-1 and its linear analog have been reported, which allows calculating intrinsic viscosities $[\eta]_0$ and $[\eta]_{l,\theta}$, accordingly, for arbitrary MM. Then, the value of g can be estimated according to the relationship [25]:

$$\frac{[\eta]_\theta}{[\eta]_{l,\theta}} = g^{2-a},$$ (43)

where a is an exponent in Kuhn-Mark-Houwink equation for a linear analog in θ-solvent ($a=0.5$ [25]).

Knowing the values g, m magnitudes can be calculated according to the equation (16). The estimations have shown that for polyarylate D-1 at MM increase from 5×10^4 up to 10×10^4 m growth from 2.0 up to 2.8 is observed, which corresponds to $g\approx0.66$ [64].

Taking into consideration that gyration radius, R_g, scales to MM according to the equations (4) and (5), from the relationship (21) one obtains [64]:

$$m \sim MM^{(d_u-1)/D}.$$ (44)

The comparison of the relationships (39) and (44) allows receiving the following equation [64]:

$$g = c_{ch}\left(\frac{d_u-1}{D}\right).$$ (45)

The proportionality coefficient, c_{ch}, in the equation (45) has a clear physical significance: it defines the greatest density of "chemical" branching centers per one macromolecule. Parameter $(d_u-1)/D$ defines this density decrease by macromolecular coil structural features. For PHE, the value $c_{ch}\approx1.41$. The values of g, $(d_u-1)/D$ and $c_{ch}[(d_u-1)/D]$ for PHE at three synthesis temperatures T are listed in Table 4, from which a good correspondence of the first and the third parameters from the indicated ones follows.

For branched polyarylate D-1, D value can be determined according to the equation (13). Further, according to the equations (18) and (19), by using $a=0.36$ [25] and $D\approx2.20$, d_u can be determined and c_{ch} can be calculated for D-1 according to the equation (45). In this case, $c_{ch}\approx3.22$, i.e., the greatest "chemical" branching centers density for D-1 is much higher than for PHE.

And in conclusion, the integral dependence of m on chemical and physical factors can be written [64]:

$$m \sim MM^{c_{ch}[(d_u-1)/D]}.$$ (46)

The dependence corresponding to the relationship (46) is shown in Fig. 20. As one can see, now this correlation is linear and passes through the coordinate's origin. This allows asserting that all factors controlling m value are taken into consideration [64].

Hence, the fractal analysis methods are efficient for clear structural identification of both chemical and physical factors, controlling a chain branching degree. The number of effective branching centers per one macromolecule m is controlled by four factors: polymer molecular weight MM, maximum "chemical" density of reactive centers c_{ch}, dimension of unscreened surface d_u of macromolecular coil and its fractal dimension D. The equation (45) allows determining the critical value $D(D_{cr})$, below which $g=0$ (i.e., branching does not occur): $D_{cr}=1.10$ [64].

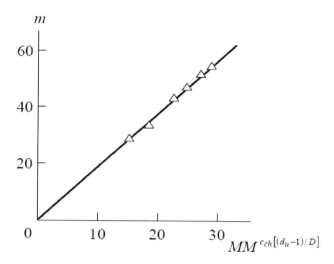

Figure 20. The dependence of branching center number per one macromolecule m on complex parameter $MM^{c_{ch}[(d_u-1)/D]}$ for PHE synthesized at $T=333$ K [64].

Table 1. The characteristics of branched PHE [22]

The synthesis temperature, K	D, the equation (1)	D, the equation (15)	d_s	g
333	1.98	1.93	1.29	0.394
338	1.89	1.88	1.24	0.478
343	1.69	1.67	1.03	0.774
348	1.56	1.54	1.0	1.0

Table 2. The characteristics of macromolecular coil of PHE, produced at different synthesis temperatures [38]

T, K	k_r, relative units	N	R_g, Å	p
333	0.027	54.4	282	0.0181
338	0.057	97.4	427	0.0262
343	0.205	162.6	764	0.0593
348	0.505	227.4	1215	0.1010

Table 3. The scaling parameters of PHE, synthesized at different temperatures [57]

T, K	a	K	γ_t^e	γ_t^T
333	0.515	1.4×10^{-3}	0.75	0.606
338	0.587	6.0×10^{-4}	0.54	0.522

| 343 | 0.775 | 7.0×10^{-5} | 0.40 | 0.226 |
| 348 | 0.923 | 1.3×10^{-5} | 0.33 | 0 |

Table 4. The experimental and theoretical characteristics of PHE chain branching at different synthesis temperatures [64]

T, K	g	$\dfrac{d_u - 1}{D}$	$c_{ch}(\dfrac{d_u - 1}{D})$
333	0.272	0.193	0.270
338	0.229	0.145	0.203
343	0.130	0.105	0.147

References

[1]   Belyi A.A., Ovchinnikov A.A. *Doklady AN SSSR*, 1986, v. 288, № 1, p. 151-155.

[2]   Burlatskii S.F. *Doklady AN SSSR*, 1986, v. 288, № 1, p. 155-159.

[3]   Aleksandrov I.V., Pazhitnov A.V. *Khimicheskaya Fizika*, 1987, v. 6, № 9, p. 1243-1247.

[4]   Burlatskii S.F., Ovchinnikov A.A., Pronin K.A. *Zhurnal Eksperimentalnoi i Teoreticheskoi Fiziki*, 1987, v. 92, № 2, p. 625-637.

[5]   Burlatskii S.F., Oshanin G.S., Likhachev V.N. *Khimicheskaya Fizika*, 1988, v. 7, № 7, p. 970-978.

[6]   Grassberger P., Procaccia I. *J. Chem. Phys.*, 1982, v. 77, № 12, p. 6281-6284.

[7]   Havlin S., Weiss G.H., Kiefer J.E., Dishon M. *J. Phys. A*, 1984, v. 17, № 4, p. L347-L350.

[8]   Redner S., Kang K. *J. Phys. A*, 1984, v. 17, № 5, p. L451-L455.

[9]   Kang K., Redner S. *Phys. Rev. Lett.*, 1984, v. 52, № 12, p. 955-958.

[10]   Meakin P., Stanley H.E. *J. Phys. A*, 1984, v. 17, № 2, p. L173-L177.

[11]   Shogenov V.N., Kozlov G.V. The Fractal Clusters in Physics-Chemistry of Polymers. *Nal'chik, Polygraphservice and T*, 2002, 268 p.

[12]   Kaufman J.H., Baker C.K., Nazzal A.I., Flickner M., Melray O.R., Kapitulnik A. *Phys. Rev. Lett.*, 1986, v. 56, № 18, p. 1932-1935.

[13]   Chu B., Wu C., Wu D.-Q., Phillips J.C. *Macromolecules*, 1987, v. 20, № 10, p. 2642-2644.

[14]   Karmanov A.P., Monakov Yu.B. *Vysokomolek. Soed. B*, 1995, v. 37, № 2, p. 328-331.

[15]   Kozlov G.V., Shustov G.B., Zaikov G.E. *J. Appl. Polymer Sci.*, 2009, v. 111, № 7, p. 3026-3030.

[16]   Baranov V.G., Frenkel S.Ya., Brestkin Yu.V. *Doklady AN SSSR*, 1986, v. 290, № 2, p. 369-372.

[17]   Vilgis T.A. *Phys. Rev. A*, 1987, v. 36, № 3, p. 1506-1508.

[18] Rammal R., Toulouse G. *J. Phys. Lett.* (Paris), 1983, v. 44, № 1, p. L13-L22.

[19] Klymko P.W., Kopelman R. J. *Phys. Chem.,* 1983, v. 87, № 23, p. 4565-4567.

[20] Kopelman R., Klymko P.W., Newhouse J.S., Anacker L.W. *Phys. Rev. B,* 1984, v. 29, № 6, p. 3747-3748.

[21] Nagaeva D.A. Dussert. ... kand khim nauk, Moscow, *MkhTI,* 1989, 161 p.

[22] Kozlov G.V., Beeva D.A., Mikitaev A.K. *Novoe v Polimerakh i Polimernykh Compositakh,* 2011, № , p.

[23] Novikov V.U., Kozlov G.V. *Uspekhi Khimii,* 2000, v. 69, № 4, p. 378-399.

[24] Alexander S., Orbach R. *J. Phys. Lett.* (Paris), 1982, v. 43, № 17, p. L625-L631.

[25] Askadskii A.A. The *Physics-Chemistry of Polyarylates.* Moscow, Khimiya, 1968, 216 p.

[26] Korshak V.V., Pavlova S.-S.A., Timofeeva G.I., Kroyan S.A., Krongauz E.S., Travnikova A.P., Raubah H., Shultz G., Gnauk R. *Vysokomolek. Soed.* A, 1984, v. 24, № 9, p. 1868-1876.

[27] Budtov V.P. Physical Chemistry of Polymer Solutions. Sankt-Peterburg, *Khimiya,* 1992, 384 p.

[28] Family F. J. *Stat. Phys.,* 1984, v. 36, № 5/6, p. 881-896.

[29] Kozlov G.V., Mikitaev A.K., Zaikov G.E. *Polymer Research J.,* 2008, v. 2, № 4, p. 381-388.

[30] Pavlov G.M., Korneeva E.V. *Biofizika,* 1995, v. 40, № 6, p. 1227-1233.

[31] Kozlov G.V., Afaunov V.V., Temiraev K.B. *Manuscript deposited to VINITI RAN,* Moscow, January 8, 1998, № 9-B98.

[32] Kozlov G.V., Afaunova Z.I., Zaikov G.E. *Polymer International,* 2005, v. 54, № 4, p. 1275-1279.

[33] Meakin P., Coniglio A., Stanley E.H., Witten T.A. *Phys. Rev. A,* 1986, v. 34, № 4, p. 3325-3340.

[34] Sahimi M., McKarnin M., Nordahl T., Tirrell M. *Phys. Rev. A,* 1985, v. 32, № 1, p. 590-595.

[35] Barns F.S. *Biofizika,* 1996, v. 41, № 4, p. 790-802.

[36] Kokorevich A.G., Gravitis Ya.A., Ozol-Kalnin V.G. *Khimiya Drevesiny,* 1989, № 1, p. 3-24.

[37] Brown D., Sherdron G., Kern V. Practical Hand-book by Synthesis and Study of Polymers Properties. Ed. Zubkov V.P. Moscow, *Khimiya,* 1976, 256 p.

[38] Kozlov G.V., Shustov G.B., Mikitaev A.K. Elektronnyi Zhyrnal "Issledovano v Rossii," 010, p. 78-80, 2009, http://zhurnal.ape.relarn.ru/articles/2009/010.pdf.

[39] Kozlov G.V., Shustov G.B., Zaikov G.E. *J. Appl. Polymer Sci.,* 2004, v. 93, № 5, p. 2343-2347.

[40] Kozlov G.V., Shustov G.B. In book: The Achievements in Polymers Physics-Chemistry Field. Ed. Zaikov G.E. a.a. Moscow, *Khimiya,* 2004, p. 341-411.

[41] Jullien R., Kolb M. *J. Phys. A,* 1984, v. 17, № 12, p. L639-L643.

[42] Sorokin M.F., Chebotareva M.A. *Plast. Massy,* 1985, № 5, p. 8-10.

[43] Shvets V.F., Lebedev N.N. *Proceedings MKhTI,* 1963, № 42, p. 72-77.

[44] Kozlov G.V., Zaikov G.E. J. *Balkan Tribologic. Assoc.,* 2003, v. 9, № 2, p. 196-202.

[45] Kozlov G.V., Zaikov G.E. *Fractal Analysis and Synergetics of Catalysis in Nanosystems.* New York, Nova Biomedical Books, 2008, 163 p.

[46] Naphadzokova L.Kh., Kozlov G.V. *Fractal Analysis and Synergetics of Catalysis in Nanosystems.* Moscow, Publishers of Academy of Natural Sciences, 2009, 230 p.

[47] Kozlov G.V., Shustov G.B., Dolbin I.V. *Proceedings I Internat. Sci. Conf. "Modern Problems of Organic Chemistry, Ecology and Biotechnology."* Luga, 2001, p. 17-18.

[48] Wiehe I.A. *Ind. Engng. Chem. Res.,* 1995, v. 32, № 2, p. 661-673.

[49] Mandelbrot B.B. The Fractal Geometry of Nature. San-Francisco, W.H. *Freeman and Company,* 1982, 459 p.

[50] Hess W., Vilgis T.A., Winter H.H. *Macromolecules,* 1988, v. 21, № 8, p. 2536-2542.

[51] Botet R., Jullien R., Kolb M. *Phys. Rev. A,* 1984, v. 30, № 4, p. 2150-2152.

[52] Kobayashi M., Yoshioka T., Imai M., Itoh Y. *Macromolecules,* 1995, v. 28, № 22, p. 7376-7385.

[53] Muthukumar M. *J. Chem. Phys.,* 1985, v. 83, № 6, p. 3161-3168.

[54] Kozlov G.V., Grineva L.G., Mikitaev A.K. *Mater. of VI Internat. Sci.-Pract. Conf. "New Polymer Composite Materials."* Nal'chik, KBSU, 2010, p. 200-204.

[55] Alexandrowich Z. *Phys. Rev. Lett.,* 1985, v. 54, № 13, p. 1420-1423.

[56] Alexandrowich Z. In book: *Fractals in Physics. Ed. Pietronero L., Tosatti E.* Amsterdam, Oxford, New York, Tokyo, North-Holland, 1986, p. 172-178.

[57] Kozlov G.V., Shustov G.B., Mikitaev A.K. *Mater. of V Internat. Sci.-Pract. Conf. "New Polymer Composite Materials."* Nal'chik, KBSU, 2009, p. 117-123.

[58] Korshak V.V., Vinogradova S.V. A Nonequilibrium Polycondensation. Moscow, *Nauka,* 1972, 696 p.

[59] Magomedov G.M., Kozlov G.V., Zaikov G.E. *Structure and Properties of Cross-linked Polymers. Shawbury, A Smithers Group Company,* 2011, 492 p.

[60] Kozlov G.V., Mikitaev A.K. *Mater. of VI Internat. Sci.-Pract. Conf. "New Polymer Composite Materials."* Nal'chik, KBSU, 2010, p. 205-210.

[61] Henschel H.G.E., Deutch J.M., Meakin P. *J. Chem. Phys.,* 1984, v. 81, № 5, p. 2496-2502.

[62] Gammet L. *The Principles of Physical Organic Chemistry.* Moscow, Mir, 1972, 326 p.

[63] Kozlov G.V., Mikitaev A.K. *Mater. of VI Internat. Sci.-Pract. Conf. "New Polymer Composite Materials."* Nal'chik, KBSU, 2010, p. 192-199.

[64] Kozlov G.V., Burya A.I., Shustov G.B. *Chemical Industry and Chemical Engineering Quarterly,* 2008, v. 14, № 3, p. 181-184.

[65] Kozlov G.V., Sanditov D.S. Anharmonic Effects and Physical-Mechanical Properties of Polymers. Novosibirsk, *Nauka,* 1994, 261 p.

[66] Beeva D.A., Mikitaev A.K., Zaikov G.E., Beev A.A. In book: *Molecular and High Molecular Chemistry.* Ed. Monakov Yu.B., Zaikov G.E. New York, Nova Science Publishers, Inc., 2006, p. 49-54.

[67] Aharoni S.M. *Macromolecules,* 1985, v. 18, № 12, p. 2624-2630.

In: Handbook of Research on Nanomaterials, Nanochemistry ...   ISBN: 978-1-61942-525-5
Editors: A. K. Haghi and G. E. Zaikov   © 2013 Nova Science Publishers, Inc.

Chapter XXVII

The Experimental Estimation of Nanofiller "Chains" Structure in Elastomeric Nanocomposites

G. V. Kozlov[1], Yu. G. Yanovskii[1],
S. Kubica[2] and G. E. Zaikov[3]

[1] Institute of Applied Mechanics of Russian Academy of Sciences, Russia
[2] Institut Inzynierii Materialow Polimerowych I Barwnikow, Torun, Poland
[3] N.M. Emanuel Institute of Biochemical Physics of Russian Academy of Sciences,
Moscow, Russia

Introduction

It is well known [1, 2] that in particulate-filled elastomeric nanocomposites (rubbers), nanofiller particles form linear spatial structures ("chains"). At the same time, in polymer composites filled with disperse microparticles (microcomposites), particles (aggregates of particles) of filler form a fractal network, which defines polymer matrix structure (analog of fractal lattice in computer simulation) [3]. This results in different mechanisms of polymer matrix structure formation in micro- and nanocomposites. If in the first filler particles (aggregates of particles), fractal network availability results in "disturbance" of polymer matrix structure, which is expressed in the increase of its fractal dimension d_f [3], then in case of polymer nanocomposites at nanofiller, contents change the value d_f is not changed and equal to matrix polymer structure fractal dimension [4]. As it has to been expected, composites indicated classes structure formation mechanism change defines their properties change, in particular, reinforcement degree.

At present, there are several methods of filler structure (distribution) determination in polymer matrix, both experimental [5, 6] and theoretical [3]. All the indicated methods describe this distribution by fractal dimension D_n of filler particles network. However, correct determination of any object fractal (Hausdorff) dimension includes three obligatory conditions. The first from them is the above-indicated determination of fractal dimension numerical magnitude, which should not be equal to object topological dimension. As it is known [7], any real (physical) fractal possesses fractal properties within a certain scales range [8]. And at last, the third condition is the correct choice of measurement scales range itself. As it has been shown in papers [9, 10], the minimum range should exceed at any rate one self-similarity iteration.

The present paper's purpose is dimension D_n estimation, both experimentally and theoretically, and checking two above-indicated conditions' fulfillment, i.e., obtaining of nanofiller particles (aggregates of particles) network ("chains") fractality strict proof in elastomeric nanocomposites on the example of particulate-filled butadiene-styrene rubber.

Experimental

The elastomeric particulate-filled nanocomposite on the basis of butadiene-styrene rubber (BSR) was an object of the study. The technical carbon of mark № 220 (TC) of industrial production, nano- and microshungite (the mean filler particles size makes up 20, 40 and 200 nm, accordingly) were used as a filler. All fillers' content makes up 37 mass %. Nano- and microdimensional disperse shungite particles were obtained from industrially extractive material by processing according to the original technology. A size and polydispersity of the received in milling process shungite particles were monitored with the aid of analytical disk centrifuge (CPS Instruments, Inc., USA), allowing to determine with high precision the size and distribution by sizes within the range from 2 nm up to 50 mcm. Nanostructure was studied on atomic-power microscopes Nano-DST (Pacific Nanotechnology, USA) and Easy Scan DFM (Nanosurf, Switzerland) by semi-contact method in the force modulation regime. Atomic-power microscopy results were processed with the aid of specialized software package SPIP (Scanning Probe Image Processor, Denmark). SPIP is a powerful program package for processing of images, obtained on SPM, AFM, STM, scanning electron microscopes, transmission electron microscopes, interferometers, confocal microscopes, profilometers, optical microscopes and so on. The given package possesses the whole functions numbers, which are necessary at images precise analysis, in the number of which the following are included:

1) the possibility of three-dimensional reflecting objects obtaining, distortions automatized leveling, including Z-error mistakes removal for examination separate elements and so on;
2) quantitative analysis of particles or grains; more than 40 parameters can be calculated for each found particle or pore: area, perimeter, average diameter, the ratio of linear sizes of grain width to its height distance between grains, coordinates of grain center of mass a.a. can be presented in a diagram form or in a histogram form.

Results and Discussion

The first method of dimension D_n experimental determination uses the following fractal relationship [11, 12]:

$$D_n = \frac{\ln N}{\ln \rho}, \tag{1}$$

where N is a number of particles with size ρ.

Particles sizes were established on the basis of atomic-power microscopy data (see Fig. 1). For each from the three studied nanocomposites, no less than 200 particles were measured, the sizes of which were united into ten groups, and mean values N and ρ were obtained. The dependences $N(\rho)$ in double logarithmic coordinates were plotted, which proved to be linear, and the values D_n were calculated according to their slope (see Fig. 2). It is obvious that at such approach, fractal dimension D_n is determined in two-dimensional Euclidean space, whereas real nanocomposite should be considered in three-dimensional Euclidean space. The following relationship can be used for D_n re-calculation for the case of three-dimensional space [13]:

$$D3 = \frac{d + D2 \pm \left[(d - D2)^2 - 2\right]^{1/2}}{2}, \tag{2}$$

where $D3$ and $D2$ are corresponding fractal dimensions in three- and two-dimensional Euclidean spaces, $d=3$.

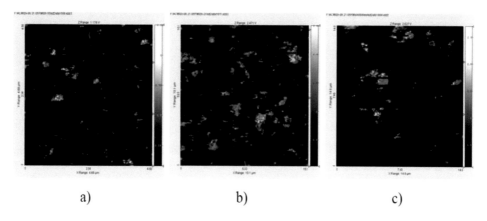

a)   b)   c)

Figure 1. The electron micrographs of nanocomposites BSR/TC (a), BSR/nanoshungite (b) and BSR/microshungite (c), obtained by atomic-power microscopy in the force modulation regime.

The calculated according to the indicated method dimensions D_n are adduced in Table 1. As it follows from the data of this Table, the values D_n for the studied nanocomposites are varied within the range of 1.10-1.36, i.e., they characterize more or less branched linear formations ("chains") of nanofiller particles (aggregates of particles) in elastomeric nanocomposite structure. Let us remind that for particulate-filled composites polyhydroxiether/graphite, the value D_n changes within the range of ~ 2.30-2.80 [5], i.e., for these materials, filler particles network is a bulk object but not a linear one [7].

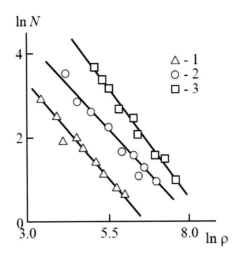

Figure 2. The dependence of nanofiller particles number N on their size ρ for nanocomposites BSR/TC (1), BSR/nanoshungite (2) and BSR/microshungite (3).

Table 1. The dimensions of nanofiller particles
(aggregates of particles) structure in elastomeric nanocomposites

The nanocomposite	D_n, the equations (1)	D_n, the equations (3)	d_0	d_{surf}	φ_n	D_n, the equations (7)
BSR/TC	1.19	1.17	2.86	2.64	0.48	1.11
BSR/nanoshungite	1.10	1.10	2.81	2.56	0.36	0.78
BSR/microshungite	1.36	1.39	2.41	2.39	0.32	1.47

Another method of D_n experimental determination uses the so-called "quadrates method" [14]. Its essence consists in the following. On the enlarged nanocomposite microphotograph (see Fig. 1), a net of quadrates with quadrate side size α_i, changing from 4.5 up to 24 mm with constant ratio $\alpha_{i+1}/\alpha_i = 1.5$, is applied and then quadrates number N_i, into which nanofiller particles hit (fully or partly), is calculated. Five arbitrary net positions concerning microphotograph were chosen for each measurement. If nanofiller particles network is fractal, then the following relationship should be fulfilled [14]:

$$N_i \sim S_i^{-D_n/2},\tag{3}$$

where S_i is quadrate area, which is equal to α_i^2.

In Fig. 3, the dependences of N_i on S_i in double logarithmic coordinates for the three studied nanocomposites, corresponding to the relationship (3), is adduced. As one can see, these dependences are linear, which allows determining the value D_n from their slope. The determined according to the relationship (3) values D_n are also adduced in Table 1, from which a good correspondence of dimensions D_n, obtained by the two above-described methods follows (their average discrepancy makes up 2.1% after these dimensions re-calculation for three dimensional space according to the equation (2)).

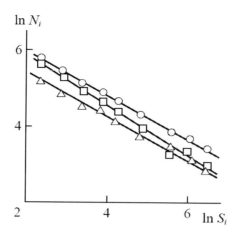

Figure 3. The dependences of covering quadrates number N_i on their area S_i, corresponding to the relationship (3), in double logarithmic coordinates for nanocomposites on the basis of BSR. The designations are the same as in Fig. 2.

As it has been shown in paper [15], at the relationship (3), the usage for self-similar fractal objects the condition should be fulfilled:

$$N_i - N_{i-1} \sim S_i^{-D_n}.$$ (4)

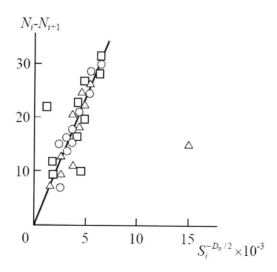

Figure 4. The dependences of (N_i-N_{i+1}) on the value $S_i^{-D_n/2}$, corresponding to the relationship (4), for nanocomposites on the basis of BSR. The designations are the same as in Fig. 2.

In Fig. 4, the dependence corresponding to the relationship (4) for the three studied elastomeric nanocomposites is adduced. As one can see, this dependence is linear, passes through coordinates origin, which according to the relationship (4) is confirmed by nanofiller particles (aggregates of particles) "chains" self-similarity within the selected αi range. It is obvious that this self-similarity will be a statistical one [15]. Let us note that the points, corresponding to $\alpha_i=16$ mm for nanocomposites BSR/TC and BSR/microshungite, do not correspond to a common straight line. Accounting for electron microphotographs of Fig. 1 enlargement, this gives the self-similarity range for nanofiller "chains" of 464-1472 nm. For nanocomposite BSR/nanoshungite, which has no points deviating from a straight line of Fig. 4, α_i range makes up 311-1510 nm, which corresponds well enough to the above-indicated self-similarity range.

In papers [9, 10], it has been shown that measurement scales S_i minimum range should contain at least one self-similarity iteration. In this case, the condition for ratio of maximum S_{max} and minimum S_{min} areas of covering quadrates should be fulfilled [10]:

$$\frac{S_{max}}{S_{min}} > 2^{2/D_n}. \tag{5}$$

Hence, accounting for the above-defined restriction, let us obtain $S_{max}/S_{min}=121/20.25=5.975$, which is larger than values $2^{2/D_n}$ for the studied nanocomposites, which are equal to 2.71-3.52. This means that measurement scales range is chosen correctly.

The self-similarity iterations number μ can be estimated from the inequality [10]:

$$\left(\frac{S_{max}}{S_{min}}\right)^{D_n/2} > 2^{\mu}. \tag{6}$$

Using the above-indicated values of the included in the inequality (6) parameters, $\mu=1.42-1.75$ is obtained for the studied nanocomposites, i.e., in our experiment conditions, self-similarity iterations number is larger than unity, which again is confirmed by the value D_n estimation correctness [6].

And let us consider, in conclusion, the physical grounds of smaller values D_n for elastomeric nanocomposites in comparison with polymer microcomposites, i.e., the causes of nanofiller particles (aggregates of particles) "chains" formation in the first. The value D_n can be determined theoretically according to the equation [3]:

$$\varphi_{if} = \frac{D_n + 2.55d_0 - 7.10}{4.18}, \tag{7}$$

where φ_{if} is interfacial regions relative fraction, d_0 is nanofiller initial particles surface dimension.

The dimension d_0 estimation can be carried out with the aid of the relationship [4]:

$$S_u = 410\left(\frac{D_p}{2}\right)^{d_0-d}, \tag{8}$$

where S_u is nanofiller initial particles specific surface in m^2/g, D_p is their diameter in nm, d is dimension of Euclidean space, in which a fractal is considered (it is obvious, in our case $d=3$).

The value S_u can be calculated according to the equation [16]:

$$S_u = \frac{6}{\rho_n D_p},$$ (9)

where ρ_n is nanofiller density, which is determined according to the empirical formula [4]:

$$\rho_n = 0.188(D_p)^{1/3}.$$ (10)

The results of value d_0 theoretical estimation are adduced in Table 1. The value φ_{if} can be calculated according to the equation [4]:

$$\varphi_{if} = \varphi_n(d_{surf} - 2),$$ (11)

where φ_n is nanofiller volume fraction, d_{surf} is fractal dimension of nanoparticles aggregate surface.

The value φ_n is determined according to the equation [4]:

$$\varphi_n = \frac{W_n}{\rho_n},$$ (12)

where W_n is nanofiller mass fraction, and dimension d_{surf} is calculated according to the equations (8)-(10) at diameter D_p replacement on nanoparticles aggregate diameter D_{agr}, which is determined experimentally (see Fig. 5).

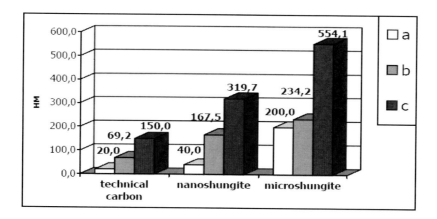

Figure 5. The initial particles diameter (a), their aggregates size in nanocomposite (b) and distance between nanoparticles aggregates (c) for nanocomposites on the basis of BSR, filled with technical carbon, nano- and microshungite.

The results of dimension D_n theoretical calculation according to the equations (7)-(12) are adduced in Table 1, from which theory and experiment good correspondence follows. The equation (7) indicates unequivocally the cause of filler in nano- and microcomposites different behaviour.

The high (close to 3, see Table 1) values d_0 for nanoparticles and relatively small ($d_0=2.17$ for graphite) values d_0 for microparticles at comparable values φ_{if} for composites of the indicated classes [3, 4].

Conclusion

Therefore, the present paper's results have shown that nanofiller particles (aggregates of particles) "chains" in elastomeric nanocomposites are physical fractal within self-similarity (and, hence, fractality [12]) range of ~ 500-1450 nm. In this range, their dimension D_n can be estimated according to the equations (1), (3) and (7).

The cited examples demonstrate the necessity of the measurement scales range correct choice. As it has been noted earlier [17], linearity of the plots, corresponding to the equations (1) and (3), and D_n nonintegral value do not guarantee object self-similarity (and, hence, fractality). The nanofiller particles (aggregates of particles) structure low dimensions are due to the initial nanofiller particles surface high fractal dimension.

References

[1]   Lipatov Yu.S. The Physical Chemistry of Filled Polymers. Moscow, *Khimiya,* 1977, 304 p.

[2]   Bartenev G.M., Zelenev Yu.V. The Physics and Mechanics of Polymers. Moscow, *Vysshaya Shkola,* 1983, 391 p.

[3]   Kozlov G.V., Yanovskii Yu.G., Zaikov G.E. *Structure and Properties of Particulate-Filled Polymer Composites: the Fractal Analysis.* New York, Nova Science Publishers, Inc., 2010, 282 p.

[4]   Mikitaev A.K., Kozlov G.V., Zaikov G.E. *Polymer Nanocomposites: the Variety of Structural Forms and Applications.* New York, Nova Science Publishers, Inc., 2008, 319 p.

[5]   Kozlov G.V., Mikitaev A.K. *Mekhanika Kompozitsionnykh Materialov i Konstruktsii,* 1996, v. 2, № 3-4, p. 144-157.

[6]   Kozlov G.V., Yanovskii Yu.G., Mikitaev A.K. *Mekhanika Kompozitnykh Materialov,* 1998, v. 34, № 4, p. 539-544.

[7]   Balankin A.S. *Synergetics of Deformable Body.* Moscow, Publishers of Ministry Defence SSSR, 1991, 404 p.

[8]   Hornbogen E. *Intern. Mater. Rev.,* 1989, v. 34, № 6, p. 277-296.

[9]   Pfeifer P. *Appl. Surf. Sci.,* 1984, v. 18, № 1, p. 146-164.

[10]   Avnir D., Farin D., Pfeifer P. *J. Colloid Interface Sci.,* 1985, v. 103, № 1, p. 112-123.

[11]   Ishikawa K. *J. Mater. Sci. Lett.,* 1990, v. 9, № 4, p. 400-402.

[12] Ivanova V.S., Balankin A.S., Bunin I.Zh., Oksogoev A.A. Synergetics and Fractals in Material Science. Moscow, *Nauka*, 1994, 383 p.

[13] Vstovskii G.V., Kolmakov L.G., Terent'ev V.F. *Metally*, 1993, № 4, p. 164-178.

[14] Hansen J.P., Skjeitorp A.T. *Phys. Rev. B,* 1988, v. 38, № 4, p. 2635-2638.

[15] Pfeifer P., Avnir D., Farin D. *J. Stat. Phys.,* 1984, v. 36, № 5/6, p. 699-716.

[16] Bobryshev A.N., Kozomazov V.N., Babin L.O., Solomatov V.I. Synergetics of Composite Materials. Lipetsk, *NPO ORIUS,* 1994, 154 p.

[17] Farin D., Peleg S., Yavin D., Avnir D. *Langmuir,* 1985, v. 1, № 4, p. 399-407.

Index

D

I

J

K

L

Q

R

S

T

U

V

W